126
Advances in Biochemical Engineering/Biotechnology

Series Editor: T. Scheper

Editorial Board:
S. Belkin · I. Endo · S.-O. Enfors · W.-S. Hu
B. Mattiasson · J. Nielsen · G. Stephanopoulos · G. T. Tsao
R. Ulber · A.-P. Zeng · J.-J. Zhong · W. Zhou

Advances in Biochemical Engineering/Biotechnology

Series Editor: T. Scheper

Recently Published and Forthcoming Volumes

Tissue Engineering III: Cell–Surface Interactions for Tissue Culture
Volume Editors: Kasper, C., Witte, F., Pörtner, R.
Vol. 126, 2012

Biofunctionalization of Polymers and their Applications
Volume Editors: Nyanhongo G.S., Steiner W., Gübitz, G.M.
Vol. 125, 2011

High Resolution Microbial Single Cell Analytics
Volume Editors: Müller S., Bley, T.
Vol. 124, 2011

Bioreactor Systems for Tissue Engineering II
Strategies for the Expansion and Directed Differentiation of Stem Cells
Volume Editors: Kasper C., van Griensven M., Pörtner, R.
Vol. 123, 2010

Biotechnology in China II
Chemicals, Energy and Environment
Volume Editors: Tsao, G.T., Ouyang, P., Chen, J.
Vol. 122, 2010

Biosystems Engineering II
Linking Cellular Networks and Bioprocesses
Volume Editors: Wittmann, C., Krull, R.
Vol. 121, 2010

Biosystems Engineering I
Creating Superior Biocatalysts
Volume Editors: Wittmann, C., Krull, R.
Vol. 120, 2010

Nano/Micro Biotechnology
Volume Editors: Endo, I., Nagamune, T.
Vol. 119, 2010

Whole Cell Sensing Systems II
Volume Editors: Belkin, S., Gu, M.B.
Vol. 118, 2010

Whole Cell Sensing Systems I
Volume Editors: Belkin, S., Gu, M.B.
Vol. 117, 2010

Optical Sensor Systems in Biotechnology
Volume Editor: Rao, G.
Vol. 116, 2009

Disposable Bioreactors
Volume Editor: Eibl, R., Eibl, D.
Vol. 115, 2009

Engineering of Stem Cells
Volume Editor: Martin, U.
Vol. 114, 2009

Biotechnology in China I
From Bioreaction to Bioseparation and Bioremediation
Volume Editors: Zhong, J.J., Bai, F.-W., Zhang, W.
Vol. 113, 2009

Bioreactor Systems for Tissue Engineering
Volume Editors: Kasper, C., van Griensven, M., Poertner, R.
Vol. 112, 2008

Food Biotechnology
Volume Editors: Stahl, U., Donalies, U. E. B., Nevoigt, E.
Vol. 111, 2008

Protein-Protein Interaction
Volume Editors: Seitz, H., Werther, M.
Vol. 110, 2008

Biosensing for the 21st Century
Volume Editors: Renneberg, R., Lisdat, F.
Vol. 109, 2007

Biofuels
Volume Editor: Olsson, L.
Vol. 108, 2007

Tissue Engineering III: Cell–Surface Interactions for Tissue Culture

Volume Editors:
Cornelia Kasper · Frank Witte · Ralf Pörtner

With contributions by

I. Bartsch · A. R. Boccaccini · A. Bruinink · S. Cohen
E. M. Czekanska · D. Das · L.-C. Gerhardt · B. Giere
C. Goepfert · C. I. Günter · M. Harder · J. S. Hayes · R. Janssen
C. Kasper · F. K. Kasper · K. Kim · S. Kress · R. Luginbuehl
H.-G. Machens · S. Michaelis · A. G. Mikos · M. M. Morlock
M. Mour · A. Neumann · R. Pörtner · T. Re'em · R. G. Richards
R. Robelek · A. F. Schilling · M. Schulze · E. Tobiasch
J. Wegener · B. Weyand · K. Wiegandt · M. Wilhelmi · J. Will
E. Willbold · T. Winkler · F. Witte · D. M. Yoon · Z. Zhang

Editors
Prof. Dr. Cornelia Kasper
Department of Biotechnology
University of Natural Resources
 and Life Sciences
Muthgasse 18
1190 Vienna
Austria

Prof. Dr. Ralf Pörtner
Technical University of Hamburg-Harburg
Institute of Bioprocess and Biosystems
 Engineering
Denickestrasse 15
21073 Hamburg
Germany

Prof. Dr. Frank Witte
Hannover Medical School
 CrossBIT—Center for Biocompatibility
 and Implant-Immunology
Feodor-Lynen-Str. 31
30625 Hannover
Germany

ISSN 0724-6145
ISBN 978-3-642-28281-2
DOI 10.1007/978-3-642-28282-9
Springer Heidelberg New York Dordrecht London

e-ISSN 1616-8542
e-ISBN 978-3-642-28282-9

Library of Congress Control Number: 2012934459

© Springer-Verlag Berlin Heidelberg 2012
This work is subject to copyright. All rights are reserved by the Publisher, whether the whole or part of the material is concerned, specifically the rights of translation, reprinting, reuse of illustrations, recitation, broadcasting, reproduction on microfilms or in any other physical way, and transmission or information storage and retrieval, electronic adaptation, computer software, or by similar or dissimilar methodology now known or hereafter developed. Exempted from this legal reservation are brief excerpts in connection with reviews or scholarly analysis or material supplied specifically for the purpose of being entered and executed on a computer system, for exclusive use by the purchaser of the work. Duplication of this publication or parts thereof is permitted only under the provisions of the Copyright Law of the Publisher's location, in its current version, and permission for use must always be obtained from Springer. Permissions for use may be obtained through RightsLink at the Copyright Clearance Center. Violations are liable to prosecution under the respective Copyright Law.
The use of general descriptive names, registered names, trademarks, service marks, etc. in this publication does not imply, even in the absence of a specific statement, that such names are exempt from the relevant protective laws and regulations and therefore free for general use.
While the advice and information in this book are believed to be true and accurate at the date of publication, neither the authors nor the editors nor the publisher can accept any legal responsibility for any errors or omissions that may be made. The publisher makes no warranty, express or implied, with respect to the material contained herein.

Printed on acid-free paper

Springer is part of Springer Science+Business Media (www.springer.com)

Series Editor

Prof. Dr. T. Scheper

Institute of Technical Chemistry
University of Hannover
Callinstraße 5
30167 Hannover, Germany
scheper@iftc.uni-hannover.de

Volume Editors

Prof. Dr. Cornelia Kasper

Department of Biotechnology
University of Natural Resources
 and Life Sciences
Muthgasse 18
1190 Vienna
Austria

Prof. Dr. Frank Witte

Hannover Medical School
 CrossBIT—Center for Biocompatibility
 and Implant-Immunology
Feodor-Lynen-Str. 31
30625 Hannover
Germany

Prof. Dr. Ralf Pörtner

Technical University of Hamburg-Harburg
Institute of Bioprocess and Biosystems
 Engineering
Denickestrasse 15
21073 Hamburg
Germany

Editorial Board

Prof. Dr. S. Belkin

Interfaculty Biotechnology Program
Institute of Life Sciences
The Hebrew University of Jerusalem
Jerusalem 91904, Israel
shimshon@vms.huji.ac.il

Prof. Dr. I. Endo

Saitama Industrial Technology Center
3-12-18, Kamiaoki Kawaguchi-shi
Saitama, 333-0844, Japan
a1102091@pref.saitama.lg.jp

Prof. Dr. W.-S. Hu

Chemical Engineering
and Materials Science
University of Minnesota
421 Washington Avenue SE
Minneapolis, MN 55455-0132, USA
wshu@cems.umn.edu

Prof. Dr. B. Mattiasson

Department of Biotechnology
Chemical Center, Lund University
P.O. Box 124, 221 00 Lund, Sweden
bo.mattiasson@biotek.lu.se

Prof. Dr. S.-O. Enfors

Department of Biochemistry
and Biotechnology
Royal Institute of Technology
Teknikringen 34,
100 44 Stockholm, Sweden
enfors@biotech.kth.se

Prof. Dr. G. Stephanopoulos

Department of Chemical Engineering
Massachusetts Institute of Technology
Cambridge, MA 02139-4307, USA
gregstep@mit.edu

Prof. Dr. G. T. Tsao

Professor Emeritus
Purdue University
West Lafayette, IN 47907, USA
tsaogt@ecn.purdue.edu
tsaogt2@yahoo.com

Prof. Dr. Roland Ulber

FB Maschinenbau und Verfahrenstechnik
Technische Universität Kaiserslautern
Gottlieb-Daimler-Straße
67663 Kaiserslautern, Germany
ulber@mv.uni-kl.de

Prof. Dr. A.-P. Zeng

Technische Universität Hamburg-Harburg
Institut für Bioprozess- und Biosystemtechnik
Denickestrasse 1
21073 Hamburg, Germany
aze@tu-harburg.de

Prof. Dr. J. Nielsen

Chalmers University of Technology
Department of Chemical and Biological
Engineering
Systems Biology
Kemivägen 10
41296 Göteborg
Sweden
nielsen@chalmers.se

Prof. Dr. J.-J. Zhong

Bio-Building #3-311
College of Life Science & Biotechnology
Key Laboratory of Microbial Metabolism,
Ministry of Education
Shanghai Jiao Tong University
800 Dong-Chuan Road
Minhang, Shanghai 200240, China
jjzhong@sjtu.edu.cn

Dr. W. Zhou

Sr. Director, BioProcess Engineering
Technology Development
Genzyme Corporation
45 New York Avenue
Framingham, MA 01701-9322, USA
Weichang.Zhou@genzyme.com

Honorary Editor

Prof. Dr. K. Schügerl
University of Hannover
Germany

Founding Editors

Prof. Dr. Armin Fiechter[†]
Zurich, Switzerland

Prof. Dr. Tarun Ghose
New Delhi, India

Advances in Biochemical Engineering/Biotechnology

Advances in Biochemical Engineering/Biotechnology is included in Springer's ebook package *Chemistry and Materials Science*. If a library does not opt for the whole package the book series may be bought on a subscription basis. Also, all back volumes are available electronically.

For all customers who have a standing order to the print version of *Advances in Biochemical Engineering/Biotechnology*, we offer free access to the electronic volumes of the Series published in the current year via SpringerLink.

If you do not have access, you can still view the table of contents of each volume and the abstract of each article by going to the SpringerLink homepage, clicking on "Chemistry and Materials Science," under Subject Collection, then "Book Series," under Content Type and finally by selecting *Advances in Biochemical Bioengineering/Biotechnology*

You will find information about the
- Editorial Board
- Aims and Scope
- Instructions for Authors
- Sample Contribution

at springer.com using the search function by typing in *Advances in Biochemical Engineering/Biotechnology*.

Color figures are published in full color in the electronic version on SpringerLink.

Aims and Scope

Advances in *Biochemical Engineering/Biotechnology* reviews actual trends in modern biotechnology.

Its aim is to cover all aspects of this interdisciplinary technology where knowledge, methods and expertise are required for chemistry, biochemistry, microbiology, genetics, chemical engineering and computer science.

Special volumes are dedicated to selected topics which focus on new biotechnological products and new processes for their synthesis and purification. They give the state-of-the-art of a topic in a comprehensive way thus being a valuable source for the next 3–5 years. It also discusses new discoveries and applications.

In general, special volumes are edited by well-known guest editors. The series editor and publisher will however always be pleased to receive suggestions and supplementary information. Manuscripts are accepted in English.

In references *Advances in Biochemical Engineering/Biotechnology* is abbreviated as *Adv. Biochem. Engin./Biotechnol.* and is cited as a journal.

Special volumes are edited by well-known guest editors who invite reputed authors for the review articles in their volumes.

Impact Factor in 2010: 2.139; Section "Biotechnology and Applied Microbiology": Rank 70 of 160

Attention all Users of the "Springer Handbook of Enzymes"

Information on this handbook can be found on the internet at springeronline.com.

A complete list of all enzyme entries either as an alphabetical Name Index or as the EC-Number Index is available at the above-mentioned URL. You can download and print them free of charge.

A complete list of all synonyms (more than 57,000 entries) used for the enzymes is available in print form (ISBN 978-3-642-14015-0) and electronic form (ISBN 978-3-642-14016-7).

Save 15%

We recommend a standing order for the series to ensure you automatically receive all volumes and all supplements and save 15% on the list price.

Preface

In the first place the editors of this special volume would like to thank all authors for their excellent contributions to this volume addressing the exciting field of cell–surface interactions, with special focus on tissue generation. We would also like to thank Prof. Dr. Thomas Scheper, Dr. Marion Hertel and Karin Bartsch for providing the opportunity to compose this volume and Springer for organizational and technical support.

The growth of three dimensional tissues is a rapidly expanding field in modern medicinal biotechnology. Many different aspects play a role in the formation of 3D tissue structures and the source of the used cells is especially important. To prevent tissue rejection or immune response, nowadays, preferentially autologous cells are used. In particular, stem cells from different sources are gaining exceptional importance, as they can be differentiated into different tissues by using special medium compositions and supplements. In the field of biomaterials, numerous scaffold materials already exist but also new composites are being developed based on polymeric, natural or xenogenic sources. A very important issue in tissue engineering is the formation of tissues under well defined, controlled and reproducible conditions. Therefore, a substantial number of new bioreactors have been developed. Two volumes previously published in this series addressed "Bioreactor Systems for Tissue Engineering" (Vol. 112) and "Strategies for the Expansion and Directed Differentiation of Stem Cells" (Vol. 123). Here we focus on the interaction of cells and materials. The knowledge and expertise of the authors covers disciplines like material sciences, engineering, biotechnology and clinical sciences. Recent advances in material development, evaluation and design of biocompatibility, analytical tools for effects of cell–surface interactions, as well as cutting edge applications of new materials (stem cell differentiation, cardiac, cartilage and bone tissue engineering) are also discussed.

We hope that this state-of-the-art volume is helpful for your research. Please enjoy reading it, as much as we enjoyed preparing it.

Spring 2012

Cornelia Kasper
Frank Witte
Ralf Pörtner

Contents

The Cell–Surface Interaction 1
J. S. Hayes, E. M. Czekanska and R. G. Richards

**Studying Cell–Surface Interactions In Vitro: A Survey
of Experimental Approaches and Techniques** 33
Stefanie Michaelis, Rudolf Robelek and Joachim Wegener

**Harnessing Cell–Biomaterial Interactions for Osteochondral
Tissue Regeneration** 67
Kyobum Kim, Diana M. Yoon, Antonios G. Mikos and F. Kurtis Kasper

Interaction of Cells with Decellularized Biological Materials 105
Mathias Wilhelmi, Bettina Giere and Michael Harder

**Evaluation of Biocompatibility Using In Vitro Methods:
Interpretation and Limitations** 117
Arie Bruinink and Reto Luginbuehl

**Artificial Scaffolds and Mesenchymal Stem Cells
for Hard Tissues** 153
Margit Schulze and Edda Tobiasch

Bioactive Glass-Based Scaffolds for Bone Tissue Engineering 195
Julia Will, Lutz-Christian Gerhardt and Aldo R. Boccaccini

Microenvironment Design for Stem Cell Fate Determination 227
Tali Re'em and Smadar Cohen

Stem Cell Differentiation Depending on Different Surfaces 263
Sonja Kress, Anne Neumann, Birgit Weyand and Cornelia Kasper

Designing the Biocompatibility of Biohybrids 285
Frank Witte, Ivonne Bartsch and Elmar Willbold

Interaction of Cartilage and Ceramic Matrix 297
K. Wiegandt, C. Goepfert, R. Pörtner and R. Janssen

Bioresorption and Degradation of Biomaterials 317
Debarun Das, Ziyang Zhang, Thomas Winkler, Meenakshi Mour,
Christina I. Günter, Michael M. Morlock, Hans-Günther Machens
and Arndt F. Schilling

Index 335

The Cell–Surface Interaction

J. S. Hayes, E. M. Czekanska and R. G. Richards

Abstract The realm of surface-dependent cell and tissue responses is the foundation of orthopaedic-device-related research. However, to design materials that elicit specific responses from tissues is a complex proposition mainly because the vast majority of the biological principles controlling the interaction of cells with implants remain largely ambiguous. Nevertheless, many surface properties, such as chemistry and topography, can be manipulated in an effort to selectively control the cell–material interaction. On the basis of this information there has been much research in this area, including studies focusing on the structure and composition of the implant interface, optimization of biological and chemical coatings and elucidation of the mechanisms involved in the subsequent cell–material interactions. Although a wealth of information has emerged, it also advocates the complexity and dynamism of the cell–material interaction. Therefore, this chapter aims to provide the reader with an introduction to the basic concepts of the cell–material interaction and to provide an insight into the factors involved in determining the cell and tissue response to specific surface features, with specific emphasis on surface microtopography.

Keywords Mechanotransduction · Microroughness · Surface topography · Tissue–implant interface

Abbreviations

APC	Anodic–plasma–chemical
BMP	Bone morphogenetic protein
CaP	Calcium phosphate
CoCrMo	Cobalt–chromium–molybdenum
ECM	Extracellular matrix
EPSS	Electropolished stainless steel
ERK	Extracellular-signal-regulated kinase
FAK	Focal adhesion kinase

J. S. Hayes (✉)
Regenerative Medicine Institute, National Centre for Biomedical Engineering Science, National University of Ireland, Galway, Republic of Ireland
e-mail: jessica.hayes@nuigalway.ie

E. M. Czekanska · R. G. Richards
AO Research Institute, Davos, Switzerland

GFOGER	Glycine–phenylalanine–hydroxyproline–glycine–glutamine–arginine
MAPK	Mitogen-activated protein kinase
mRNA	Messenger RNA
MSC	Mesenchymal stem cell
RGD	Arginine–glycine–aspartic acid
rhBMP-2	Recombinant human bone morphogenetic protein 2
TAN	Titanium–6% aluminium–7% niobium
TAV	Titanium–4% aluminium–6% vanadium
TGFβ	Transforming growth factor β

Contents

1	General Introduction...	2
2	Surface Conditioning upon Implantation of a Device	3
	2.1 Initial Interactions upon Implantation...	3
	2.2 The Role of Integrins	4
3	Cell Meets Surface: Factors Involved in the Surface-Dependent Response.....................	6
	3.1 Surface Chemistry	6
	3.2 Surface Topography..	11
	3.3 Roughness Spectrum	15
	3.4 Mechanotransduction	19
4	Tissue–Implant Interface	22
	4.1 Interface Structure and Composition	22
5	Summary and Conclusions	25
References..		25

1 General Introduction

Literally thousands of studies exist that have looked to define and control the cell/tissue–implant interface. The large majority of these stem from the observation that microrough surfaces are conducive for the naturally occurring phenomenon of osseointegration (the integration of an implant to bone) and, therefore, focus on ways to manipulate and control this interaction for determining specific tissue responses and the magnitude of these responses. The basis of this field of research is that microrough surfaces will enhance osseointegration; however, this is an oversimplistic statement. Although it is true that osseointegration can be increased with microrough surfaces, little consensus exists as to what constitutes a 'rough' surface. Furthermore, the mechanisms involved in this response are only now starting to become apparent. Essentially, the cell–material interaction is a complex relationship with many questions still left unanswered. Several concepts and events, including protein adsorption, cell adhesion and signal mechanotransduction, are substrate-dependent. This can make understanding the cell–material interaction more challenging as the system under study is dynamic from both the cell and the material side. Furthermore, extrapolation of information collated in vitro to results observed in in vivo studies is not as straightforward as the statement that 'micro-rough surfaces enhance osseointegration' may lead one to believe. The aim of this chapter, therefore, is to introduce the reader to the basic concepts of the cell–material interaction and to provide an insight into the factors involved in determining the cell and tissue response to specific surface features, with specific emphasis on surface microtopography and the osteoblast/bone response.

2 Surface Conditioning upon Implantation of a Device

The nanoseconds subsequent to a device being implanted determine the fate of the implant. This instantaneous reaction is a result of tissue biomolecules interfacing with surface properties such as hydrophobicity, charge, chemistry and topography. All of these properties help determine which proteins adsorb to the surface and the types of intermolecular forces that ensue.

2.1 Initial Interactions upon Implantation

The primary biological reaction to an implanted device is the formation of a water layer via hydroxyl groups of converged dissociated water molecules, within which naturally occurring ions such as calcium (Ca^{2+}) and sodium (Na^+) become incorporated [1, 2]. The formation of a surface water layer with hydrated ions is specific and dynamic depending on different surface chemical properties. Therefore, surfaces with varying topography and chemical composition will ultimately produce layers of different biological compositions.

Upon contact with blood, the implant becomes covered in a protein-enriched film, which adheres to the surface via weak temporary bonds or stronger permanent covalent bonds [1]. Blood has more than 2,000 proteins. The proteins that come from the blood provide a provisional matrix for the cells to adhere to. Cells never interact directly with the actual implant surface. The surface itself initially determines which proteins absorb to it and also determines the orientation of their attachment. Blood proteins on the surface adsorb and desorb according to electrostatic and hydrophobic interactions with the surface, and their concentration, size and stability to ensure the formation of thermodynamically stable properties. Albumin (66 kDa) is the most concentrated protein in blood; therefore, it generally dominates the initial surface interactions. Fibrinogen (340 kDa), in lower concentration within blood, is much slower to arrive at the implant surface owing to its larger size. Upon arrival at the surface, however, it usually dominates the surface protein coating because of its higher affinity for the surface, exchanging with the smaller and more weakly bound albumin [2, 3].

The desirable effect upon implantation of a device is the formation of a haematoma. This dynamic structure contains factors important in recruiting cells essential for inflammatory-mediated response, bone repair and angiogenesis [4–7]. To achieve this outcome, platelets undergo the process of degranulation upon adhering to a surface. This involves the release of intracellular contents such as potent platelet activators, which in turn recruit additional platelets to the wound site [8, 9]. Macrophages and other inflammatory cells such as granulocytes, lymphocytes and monocytes infiltrate the haematoma and function to prevent infection and to secrete cytokines and growth factors such as fibroblast growth factor 2, vascular endothelial factor, macrophage colony stimulating factor, interleukins, bone morphogenetic proteins (BMPs) and tumour necrosis factor α (see [5] for a review). Furthermore, activated platelets secrete a myriad of growth factors, such as platelet-derived growth factor and transforming growth factor β (TGFβ). The latter factors possess chemotactic activity, thus serving as migratory signals for repair cells such as osteoblasts, fibroblasts, monocytes, neutrophils and leukocytes [10, 11].

2.2 The Role of Integrins

The interactions of surface-bound proteins with cells are mediated via integrins. These cell membrane glycoproteins recognize and bind a variety of cell-surface-associated and extracellular matrix (ECM)-associated ligands. Integrin receptors are heterodimers composed of α and β subunits that occur in distinct combinations which then bind specific ligands. A subset of the α subunits have an additional structural domain located towards the N-terminal. This is known as the α-A domain (also termed the α-I domain). Integrins carrying this domain either bind to collagens (e.g. integrins $\alpha_1\beta_1$, and $\alpha_2\beta_1$), or act as cell–cell adhesion molecules (integrins of the β_2 family). Integrins that do not have this inserted domain on the α subunit do have an A-domain in their ligand binding site, which is found on the β subunit.

In both cases, the A-domains carry up to three divalent cation binding sites. One is permanently occupied in physiological concentrations of divalent cations, and carries either a calcium or a magnesium ion, the principal divalent cations in blood at median concentrations of 1.4 mM (calcium) and 0.8 mM (magnesium). The other two sites become occupied by cations when ligands bind—at least for those ligands involving an acidic amino acid in their interaction sites. The interaction of integrins with ligands is based on the ability of the receptors to recognize the arginine–glycine–aspartic acid (RGD) tripeptide sequence [12].

Osteoblasts express various integrins that bind numerous ECM ligands. Specifically, integrins $\alpha_2\beta_1$ and $\alpha_5\beta_1$ are crucial in osteoblast function. The α_2 subunit is one of the type I collagen receptors. It also plays an important role in osteoblast differentiation by activating Osf2 (Cbfa1) transcription factor and induction of osteoblast-specific osteocalcin gene expression [13]. The α_5 integrin subunit is responsible for the interactions with fibronectin. Furthermore, the $\alpha_5\beta_1$ integrin has been reported to be necessary for osteoblast proliferation and differentiation. It is responsible for bone-like nodule formation in vitro by osteoprogenitor cells grown on various synthetic biomaterials. In a study by Petrie et al. [14], coatings consisting of defined multimer constructs with monomer, dimer, tetramer and pentamer recombinant fragments of fibronectin were prepared. The authors aimed to assess how nanoscale ligand clustering affects integrin binding, stem cell responses, tissue healing and biomaterial integration. Clinical-grade titanium was grafted with polymer brushes that presented monomers, dimers, trimers or pentamers of the $\alpha_5\beta_1$ integrin-specific fibronectin III (7–10) domain. Their results indicate that coatings with trimer and pentamer modifications enhanced integrin-mediated adhesion in vitro, osteogenic signalling, and differentiation in human mesenchymal stem cells (MSCs) [14]. Keselowsky et al. [15] showed that the expression of the α_5 integrin subunits regulates alkaline phosphatase activity, but this function is not affected by surface microtopography. However, the microrough surfaces increase the expression of the β_1 integrin subunit, which is essential in osteoblastic differentiation through the protein kinase C pathway [16]. More recent data obtained using molecular beacons targeted to the β_1 integrin subunit messenger RNA (mRNA) in order to visualize surface-dependent changes in its expression in individual MG63 cells showed that effects of the substrate on β_1 mRNA previously observed in confluent cultures were also evident in preconfluent cultures, supporting the hypothesis that β_1 integrin is important in proliferation as well as differentiation of osteoblasts [17].

Many in vitro studies have focused on identifying integrins on osteoblast cell membranes to determine their role in the mechanism of initial attachment of cells to the implant surface [17–19]. Gronowicz and McCarthy [20] have revealed that the initial attachment to metal materials involves integrins. SaOs-2 cells were incubated on titanium–4% aluminium–6% vanadium (TAV), polystyrene, glass and cobalt–chromium–molybdenum (CoCrMo) with antibodies to the fibronectin receptor $\alpha_5\beta_1$ and the vitronectin receptor ($\alpha_v\beta_3/\alpha_v\beta_5$). On samples with fibronectin receptor antibody on TAV and CoCrMo the attachment was reduced by 63 and 49%, respectively. In contrast, no significant effect was seen on samples with the antibody to the vitronectin receptor. Further analysis of changes in integrin expression within 24 h on samples without antibodies revealed that the α_5 integrin subunit has the highest expression level after 24 h of adhesion on TAV compared with polystyrene-, glass-, CoCrMo- and fibronectin-covered surfaces. The α_2 and α_v subunits were also detected, but their expression was lower on TAV than on polystyrene, whereas the production of α_1 was inhibited only in cells cultured on polystyrene, but not in cells cultured on the other surfaces.

Further studies on primary human osteoblasts confirmed these findings. Sinha and Tuan [21] reported differences in integrin expression on polished and rough (R_a not given) TAV and CoCrMo after 12 h in culture. For cells on the polished TAV surface, α_2, α_3, α_4, α_6, α_v, β_1 and β_3 were detected. Results for rough TAV showed that α_3 and α_6 were not expressed. Additionally, in both cases α_5 was not detected. Interestingly, $\alpha_5\beta_1$ integrin, the receptor for fibronectin and regulator of differentiation, is expressed mainly on the polystyrene surface, whereas on titanium surfaces $\alpha_2\beta_1$ integrin, binding to collagen and laminin, is primarily expressed [22]. Moreover, Olivares-Navarette et al. [23] demonstrated that $\alpha_2\beta_1$ integrin regulates differentiation of cells cultured on titanium implants in long-term culture. For the attachment of cells to the proteinaceous coating, heterodimers $\alpha_v\beta_1$, receptor for fibronectin and vitronectin, $\alpha_6\beta_1$, interacting with laminin, and multifunctional receptors $\alpha_3\beta_1$ and $\alpha_v\beta_3$ are essential [12, 22]. Additionally, Schneider and Burridge [24] indicated that fibronectin enhances formation of focal contacts and stress fibres. Furthermore, they localized β_1 integrin subunits within focal contacts on surfaces precoated with fibronectin, whereas β_3 was evident on surfaces precoated with vitronectin. However, these contradictions may result from the different cell models used in these studies.

3 Cell Meets Surface: Factors Involved in the Surface-Dependent Response

In a presidential address to the American Biomaterials Society, Buddy Ratner [25] meaningfully noted that current biomaterials have been developed as a result of trial-and-error optimization rather than specific design. However, as acknowledged by Brunette [26], to design materials that elicit specific responses from tissues is a complex proposition. The main reason for this is that the vast majority of the biological principles controlling the interaction of cells with implants remain largely ambiguous. For instance, the early 1980s saw the introduction of the inhibition of epithelial down growth onto implants via contact inhibition [27]. Even in a well-studied phenomenon such as contact guidance, elucidating the controlling mechanisms of cell response remains challenging.

3.1 Surface Chemistry

It is no great stretch of the imagination to see why the chemical composition of an implant surface has attracted interest. The excellent biopassivity, corrosion resistance and repassivation ability of metal implant materials are a direct consequence of the chemical stability and integrity of the oxide film. Further, importance has been assigned to the oxide layer since essentially it is this that interacts with proteins and cells upon implantation and persists at the interface for the life of the fixation [28]. In fact, the sensitivity of cells to the chemical composition of a device is to the extent that even different grades of titanium are detected at a cell level [29]. Some studies suggest that by increasing the thickness of the oxide layer, one can increase bony ingrowth proportionally [30], whereas others acknowledge the beneficial effects of the oxide layer but do not report major differences in fibroblast cell response based on oxide thickness alterations [31]. Clearly, surface chemistry is an important factor in terms of the cell–material interaction. Here we briefly discuss the oxide layer of commercial metal implants and explore the effect of chemistry.

3.1.1 The Oxide Layer

Metal implants such as titanium and stainless steel are termed 'biocompatible' on the basis of the presence of a surface oxide layer. It is this oxide layer that allows the implants to have a high degree of corrosion resistance and separates the delicate biological environment from the highly reactive and incompatible bulk material of an implant. In implant-quality electropolished stainless steel (EPSS), the passive film consists mainly of iron, nickel and chromium in addition to smaller quantities of elements such as molybdenum and manganese. The distinguishing passive film of EPSS is formed through the reaction of chromium within the steel surface with oxygen. The naturally occurring oxide layer of EPSS is generally in the region of a few nanometres (2–3-nm) thick [32].

In contrast, titanium and its alloys form much thicker (5–6-nm) naturally occurring oxide layers, the composition of which is dominated by titanium, oxygen and carbon, with the most stable stoichiometry of the oxide layer being TiO_2. For dual-phase alloys such as titanium–6% aluminium–7% niobium (TAN), the oxide layer consists of Al_2O_3 and Nb_2O_5, with aluminium being enriched within the alpha phase of the oxide and niobium being enriched within the beta phase [32]. Anodizing is a commercially available process used for increasing the oxide thickness of clinical implants. With this method it is possible to increase the oxide thickness by approximately 2–3 nm per volt to produce an oxide layer of approximately 200 nm.

When the oxide film is mechanically abraded, this allows the release of metal ions from the highly reactive and incompatible bulk material. The release of potentially toxic metal ions persists until the oxide layer is regenerated, which for EPSS devices tested in 0.9% saline solution takes approximately 35 min, compared with approximately 8 min for titanium and titanium alloys [33, 34]. Recent in vitro and in vivo studies have produced a convincing case identifying many of the components of implant materials, such as chromium, cobalt, iron and nickel, as the main culprits of toxicity. In particular, it has been shown that the potential negative effects of released ionic and particulate implant components can impact a variety of systems, such as excretory, reproduction, vascular, immune and integumentary and nervous systems (see [35] for a review).

3.1.2 Protein Adsorption

The importance of surface chemistry is demonstrated with greatest efficacy when considering protein adsorption. Cells bind to titanium and its alloys through a series of adhesive molecules such as vitronectin and fibronectin [36]; thus, alterations in surface chemistry will effectively influence protein adsorption and conformation, and ultimately initial cell attachment. Yang et al. [37] found that increased cell attachment was directly proportional to the amount of preadsorbed protein; however, this mechanism was also protein-type dependent. In this study, both fibronectin and albumin were assessed and it was reported that the concentration of fibronectin adsorbed onto the titanium surfaces was higher than the concentration albumin adsorbed. Considering the hydrophilic nature of commercially pure titanium, and the fact that albumin is known to display improved binding to hydrophobic surfaces, this finding seems coherent. Time also appeared to be a factor as the positive effect of preadsorbed fibronectin was observed to be highest after 15 min; however, after 180 min this effect on cell attachment was negated [37]. This outcome seems logical since the rapid adsorption of proteins onto devices is considered to be one of the first events to occur upon implantation [38], an effect that is suggested to then diminish as the effect of topography comes to the fore. However, this system works in synergy rather than exclusively. More recently, Rapuano and McDonald [39] showed that negatively charged surface oxide functional groups in TAV can modulate fibronectin integrin receptor activity by altering the adsorbed protein's conformation.

Interestingly Howlett et al. [40] found that vitronectin was essential for osteoblast cell attachment onto titanium, stainless steel, alumina, and poly(ethlyene terephthalate). However, this outcome was fibronectin-independent, a result which contradicts others identifying fibronectin as the principal component [41]. One must keep in mind, however, that this effect may also be cell-type-dependent and/or species-dependent (Howlett et al. [40] used cells derived from human bone, whereas Horbett and Schway [41] used a mouse cell line). Furthermore, it was suggested by Howlett et al. [40] that perhaps fibronectin (in this model) plays a role in cell adhesion rather than initial attachment and that vitronectin is a more effective competitor compared with other serum proteins for surface binding. Nevertheless, the 90-min time point included by the authors may have clouded the outcome slightly. For instance, Meyer et al. [38] showed that protein and lipid adsorption was already detectable after 5 min implantation, the earliest time point studied.

3.1.3 Surface Chemical Modifications

To induce cell adhesion and spreading, the surface chemistry of a material can be altered. This can be achieved by different treatments involving chemical and biochemical surface modifications. Biological coatings, such as immobilized ECM proteins of implant surfaces, give more promising results than chemical modifications with calcium phosphate (CaP), for instance. Although CaP modifications do promote integration of an implant with surrounding bone, this osseointegration is relatively slow and results in poor mechanical anchorage when the device is inserted into osteoporotic bone [42, 43]. Furthermore, coating techniques with commercial hydroxyapatite or CaP are demanding and problematic to obtain the maximal biological response [44]. Additionally, in long-term implantation they can delaminate [45]. Thus, large variability in the quality of different hydroxyapatite coatings from different companies or even from different batches causes concerns about the long-term reliability of these coated implants [46]. Recently, Schlegel et al. [47] used anodic–plasma–chemical (APC) treatment to improve the adhesive strength of the CaP layer. APC treatment is an anodization technique that allows porous oxide layer formation with incorporation of CaP directly into the oxide. Results from Schlegel et al. showed no significant difference in bone remodelling and removal torque between APC-treated implants and surfaces coated in a standard manner. However, the histological results indicated some delamination of standard coated CaP and hydroxyapatite surfaces. Thus, the APC treatment results in higher strength of bonding to the implant surface and allows the drawbacks of standard CaP coatings to be overcome.

More recently, O'Hare et al. [48] investigated a novel surface modification for incorporating biomolecules such as hydroxyapatite within the oxide layer of a metal substrate. The CoBlast method [49] is an advanced version of microblasting in which both an abrasive and a dopant are applied to the substrate surface simultaneously without the need for any form of presurface treatment. O'Hare et al. [48] showed that CoBlasting resulted in a stable surface that was observed to support enhanced osteoblast attachment and viability in vitro compared with hydroxyapatite alone or metal substrate controls. Implantation of the CoBlast surface in a rabbit femoral model confirmed that the surface promoted in vivo formation of early-stage lamellar bone growth after 28 days. Techniques such as this may provide a way for chemical modifications to become more reliable and reproducible, which is advantageous in bone-compromised patients in need of rapid osseointegration, although testing in a larger-animal model would be required.

3.1.4 Surface Biological Modifications

Currently available biochemical surface modifications include immobilization of ECM proteins such as collagen or peptide sequences modulating bone cell adhesion; immobilization of DNA for structural reinforcement; deposition of cell signalling agents (bone growth factors) to trigger new bone formation; and enzyme-modified titanium surfaces for enhanced bone mineralization [46]. Presently, coating implant surfaces with the RGD sequence is the most common peptide-based strategy. The RGD sequence has been identified as a cell attachment motif present on several plasma and ECM proteins, including collagen type I, fibronectin, vitronectin, bone sialoprotein and osteopontin. These proteins interact with integrins, including the predominant osteoblast integrin $\alpha_5\beta_1$ [50].

Many recent studies have demonstrated a significant impact of RGD sequence–integrin receptor interactions on osteoblast adhesion, migration, gene and protein expression and mineralization [51]. Results obtained by Zreiqat et al. [51] showed that modifying TAV surfaces with RGD peptide upregulated bone protein levels of osteocalcin, type I collagen and bone sialoprotein at 7 days, compared with control surfaces, such as coated a control peptide RGE and a control amino acid surface bound with cysteine. Additionally, alkaline phosphatase production on RGD-modified TAV was significantly higher at day 14 on the RGD surface compared with the control surfaces. These results indicate that the implant surfaces modified with RGD peptide regulate and promote bone formation and calcification in vitro.

Rammelt et al. [52] compared different organic coatings to assess their influence on bone remodelling and healing. Briefly, in their study titanium implants coated with collagen type I, RGD peptides and chondroitin sulphate were implanted into the tibia of rat. The histological and immunohistochemical evaluation of the effect of RGD on bone remodelling at the implant surface revealed that the addition of RGD proteins enhances bone healing and direct bone contact with the implant surface after 4 weeks. Furthermore, the RGD sequence directly activates macrophages, osteoblasts and osteoclasts, which results in faster bone remodelling activity around titanium implants. This leads to earlier transformation of the newly formed woven bone into lamellar bone around day 14.

Titanium surfaces may also be coated with the more selective collagen-mimetic peptide glycine–phenylalanine–hydroxyproline–glycine–glutamine–arginine (GFOGER). It was shown that this sequence selectively promotes binding of $\alpha_2\beta_1$ integrin [53], a crucial receptor for osteoblast differentiation that interacts with type I collagen to activate the Runx2 transcription factor to regulate osteoblast differentiation [13]. Reyes et al. [43] showed significantly higher osteoblast-specific gene expression, such as expression of Runx2 transcription factor, osteocalcin and bone sialoprotein, in rat bone marrow stromal cells for samples treated with GFOGER peptide compared with uncoated samples. Additionally, enhanced implant–osteoblast contact in vitro was observed together with increased alkaline phosphatase activity and calcium content on coated titanium and confirmed the positive effect on osteoblastic differentiation. Also, the GFOGER peptide coating influenced and enhanced osseointegration of titanium implants in vivo. However, although these results are promising; they were obtained from a rat model system. Thus, further studies focusing on the evaluation of the response of human osteoblasts to GFOGER-coated implants would give more clinically relevant results.

BMPs are a group of growth factors belonging to the TGFβ superfamily that have prominent roles in a variety of bone-related processes, one of which includes osteoblast differentiation. Moreover, several BMPs, in particular BMP-2, have been shown to be involved in the substrate-dependent cell reponse [54]. BMP-2-coated TiO_2 nanotubes of various diameters have also been shown to induce significantly increased levels of osteoblast differentiation in MSCs in vitro [55]. However, understandably, biological modification requires an appreciation of the type, delivery and concentration of the biomolecule to be used as several studies have reported conflicting results when evaluating implant surfaces modified by

BMP coatings. Liu et al. [56] investigated the effects of BMP-2 and its mode of delivery on the osteoconductivity of dental implants (uncoated titanium surface or a CaP-coated surface) in the maxillae of miniature pigs. The results indicated that the bone volume within the peri-implant space was highest on coated and uncoated implants bearing no BMP-2, whereas the lowest bone volume was observed on coated implants bearing only adsorbed BMP-2. Therefore, it appears the method of biomolecule delivery is of utmost importance for increasing osseointegration.

Wikesjö et al. [57] studied the ability of recombinant human BMP-2 (rhBMP-2) coated onto a titanium porous oxide implant surface to stimulate local bone formation in a dog model. Their method involved creating 5 mm critical size supra-alveolar, peri-implant defects into which implants coated with rhBMP-2 at 0.75, 1.5, or 3.0 mg/mL or an uncoated control were implanted. The histological results showed newly formed bone for implants coated with 0.75 or 1.5 mg/mL rhBMP-2; however, implants coated with 3.0 mg/mL rhBMP-2 were noted to produce immature trabecular bone formation and impaired peri-implant bone remodelling, resulting in implant displacement. Therefore, under these experimental conditions, clinically relevant local bone formation was induced with rhBMP-2 modification of the surfaces, but higher concentrations resulted in inadequate bone remodelling and subsequent osseointegration.

3.2 Surface Topography

Although a wide range of surface chemistry modification techniques have been introduced recently, the 3D morphology of an implant is thought to be a major factor in determining implant performance and success [58]. In this section we explore the cell–material interaction on a microscale in vitro and in vivo.

3.2.1 Surface Topography In Vitro

The cell–material interaction is a finely balanced relationship that can be influenced by subtle changes in microtopography. Altering surface microtopography can have ramifications for a wide variety of cellular responses in vitro, such as cytocompatibility, and if they are negatively affected, it could be deleterious for implantation. Previous studies have highlighted the influence of surface microtopography on a wide range of factors that would directly contribute to the cytocompatibility of a device. For instance, Meredith et al. [59] reported that commercially available standard microrough TAN selectively inhibits fibroblast proliferation. Subsequently, it was elucidated that the niobium-rich particles associated with the beta phase of the TAN surface produced significantly fewer focal adhesion sites compared with EPSS and standard titanium. However, when this surface was polished smoother, without changing the surface chemistry, the negative effects on proliferation and focal adhesion structure were negated [59].

In contrast, standard TAN does not negatively influence osteoblast attachment and proliferation in this manner (Fig. 1), even in its standard microrough form [60].

The effect of altering surface microtopography has also been shown to be beneficial for selective cell adhesion [61, 62]. Differences in surface microtopography have also been implicated in controlling proliferation, with several studies indicating that microrough surfaces have reduced proliferative capacity compared with smooth surfaces [63–66]. However, with this reduced proliferative capacity emerges as a more differentiated osteoblast phenotype on the microrough surfaces as indicated by alkaline phosphatase activity [63, 65, 66]. Moreover, it is suggested that the process of matrix mineralization is dependent on surface microroughness [65]. This roughness-dependent response is also seen for ECM components with surfaces of varying roughness displaying varied synthesis of collagen type I, vitronectin and fibronectin [40, 66].

Both cytokine and growth factors involved in modulating fracture healing response have been shown to be differentially influenced by surface microtopography. Boyan et al. [67] have extensively shown that local factors such as TGFβ_1 and prostaglandin E$_2$ display a surface-roughness-dependent response. Kieswetter et al. [68] also identified the relationship between surface microroughness and TGFβ_1 levels, reporting a 3–5 times higher activity on coarse sandblasted and titanium-plasma-sprayed surfaces, respectively, compared with tissue culture plastic. In a similar trend to prostaglandin E$_2$ production, TGFβ_1 is also found at low levels on smooth surfaces, whereas a marked increase is reported for microrough substrates. This growth factor has been shown to be pivotal to bone formation for many reasons, some of which include its ability to stimulate MSC proliferation, matrix production and the downregulation of osteoclast activity.

Osseointegration at the bone–implant interface requires key regulatory pathways which influence osteoblastogenesis, promotion of osteoblastic differentiation and maturation. Some studies claim to identify 'roughness response genes' via microarrays and although some of the data may be valuable, given the differences between relatively similar studies, it is often difficult to consolidate the findings [69, 70]. Other studies prefer to focus on specific genes or transcription factors known to be fundamental for osteoblast differentiation. Most of the time, this includes real-time PCR, which is a powerful and sensitive method for detecting changes at an mRNA level. It should be noted, however, that confirmation at the protein level is also an important consideration as it is known, for instance, that changes in mRNA levels may not be efficiently translated to similar changes in protein level. Many of these studies have highlighted specific bone-related markers involved in osteoblast differentiation and mineralization that are regulated in a substrate-dependent manner [71–74]. Furthermore, osteospecific MSC fate determination also appears to be substrate-dependent [75, 76].

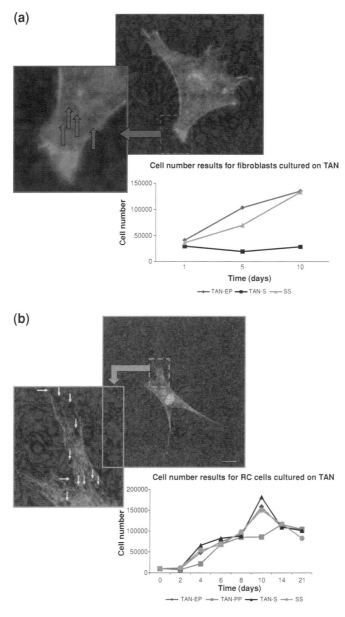

Fig. 1 Differences in cytocompatibility of surfaces. **a** Fibroblasts cultured on electropolished and standard titanium–6% aluminium–7% niobium (TAN) for 1, 5 and 10 days. The beta-phase particles of the standard TAN surface are shown to disrupt the cytoskeleton (*blue arrows*) and this was associated with a decrease in cell number. However, polishing of the surface negates this effect and cells can proliferate [59]. **b** Rat calvaria osteoblasts cultured on standard, electropolished and paste-polished TAN. Although similar disruption is observed to the cytoskeleton as on standard TAN (*white arrows*), this was not observed to influence osteoblast proliferation [71]. *TAN-EP* electropolished TAN, *TAN-PP* paste-polished TAN, *TAN-S* standard microrough TAN, *SS* Stainless steel

3.2.2 Surface Topography In Vivo

Recent short- and long-term in vivo studies using implant devices with varying surface topographies have shown that the mechanisms involved in this 'niche' are possibly more ill-defined than in the in vitro situation and, in some cases, conflicting results are reported. For instance, Hayakawa et al. [77] observed a difference in bone response to grit-blasted and smooth substrates in rabbit cortical bone. However, this difference was not reported for the same samples when they were investigated in a trabecular bone model. Conversely, Pearce et al. [78] showed that standard microrough commercially pure titanium and TAN had higher torque removal in both a trabecular and a cortical sheep bone model compared with polished samples of the same materials over periods of 6, 12 and 18 weeks. Interestingly, this study also made evident a distinct difference in peak torque removal for both bone types.

There are literally hundreds of studies that have focused on enhancing osseointegration by controlling surface microroughness in vivo; therefore, here we would like to focus on other applications. For instance, what if strong, rapid bone bonding or soft tissue adhesion is an undesirable outcome of implantation? Such cases would include fracture fixation in the hand or shoulder, where tendons and connective tissues are required to glide over an implant, the face, where strong tissue adhesion to the implant may cause irritation and disfiguration, and in paediatric patients, where device implantation is transient and will ultimately require removal. In instances such as these, the occurrence of direct bone bonding would be a hindrance for desirable implant function.

Several studies have found that microtopographical manipulation of an implant surface can provide a degree of resolution for these issues. Under normal circumstances, direct bonding of bone to an implant is the desired outcome and the occurrence of a fibrous tissue interface is often viewed as an unwanted, negative outcome. However, in situations such as fracture fixation of the hand, tissue is required to glide freely over an implant. The current state of the art describes how titanium and its alloys have more of a tendency for intratendon inflammation compared with EPSS. This occurrence can cause painful tendon-implant adhesion and damage possibly causing limited palmar flexion and even tendon rupture. Although EPSS is produced for clinics with an innate mirrorlike smooth surface, there remains a preference for the use of titanium and its alloys because of reduced artefact production in MRI, superior resistance to corrosion and subsequent metal sensitivity reactions and superior biocompatibility. With regard to fracture fixation in paediatric and trauma patients, a similar problem occurs as current commercial metal implants naturally induce rapid bony overgrowth. This makes implant removal extremely difficult and fraught with complications.

The AO Foundation group [59, 71, 78–81] in particular has produced evidence that by reducing the microroughness of current clinic metal implants (titanium and its alloys), one can achieve both implant removal and prevention of tissue adhesion to hand devices. In both instances the studies used the commercial process of polishing to reduce the microroughness of the implants. Electropolishing is a

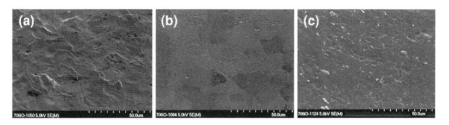

Fig. 2 Scanning electron micrographs of titanium and its polished counterparts. **a** Standard commercially pure titanium as used in clinics ($R_a \approx 1$ μm). **b** Electropolished titanium—implants are submerged in an electrolyte solution and a charge is applied. Material is removed from the surface at a rate that is dependent on the electrical conductivity of the metal. **c** Paste polished titanium—a mechanically abrasive method that physically removes material from the surface which is based on the relative hardness of the metal. Polishing results in an R_a of approximately 0.2–0.3 μm

method that involves submerging the implants in an electrolyte solution to which a charge is applied. Material is removed from the surface at a rate that is dependent on the electrical conductivity of the metal or its alloying elements. In contrast, paste polishing is a mechanically abrasive method that physically removes surface features. The removal of material is based on the relative hardness of the metal and it alloying elements. Essentially, a 'hard' material will provide resistance to the physical abrasion, which helps produce a homogenous smoothened surface (Fig. 2). Both these techniques were employed to reduce the surface microroughness of titanium and two of its alloys, TAN and titanium–15% molybdenum. It was found that compared with microrough control surfaces, polishing of titanium and its alloys significantly increased the occurrence of soft tissue capsule formation in hand fracture fixation devices in an in vivo rabbit model in the tibia [79]. Furthermore, polishing significantly reduces the force required for removal of conventional and locked screws as well as intramedullary nails from sheep tibial bone after short-term (6, 12 and 18 weeks) and long-term (6, 12 and 18 months) implantation [78, 80, 81]. Histologically, it was observed that polished implants supported fibro-osseointegration or the occurrence of a very thin fibrous layer (sometimes only one to three cells thick) between the bone and implant without loss of implant stability. These results also challenge, therefore, the general notion that direct bone bonding is required for stable fixation (Figs. 3, 4).

3.3 Roughness Spectrum

Cell–material interaction studies have clearly defined that a microrough surface is inductive for osteoblast differentiation, and that this phenomenon is echoed by microrough implants in vivo. However, a clear definition describing what constitutes a 'rough' or 'smooth' surface (millimetres, micrometres, nanometres) is distinctly lacking. One of the main reasons for this omission is that different

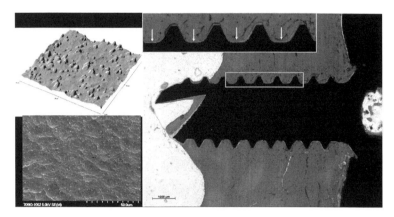

Fig. 3 Standard surface of microrough TAN as used in clinics. This surface has a characteristic 3D morphology with R_a approximately 1 μm. Consequently, as depicted in the histological section, this surface normally supports direct osseointegration (*white arrows*) [35]

groups use different scales of microroughness based on their own experimental experience. For instance, Boyan and Schwartz [82] state that if the average roughness of a surface is greater than the size of an individual osteoblast, then essentially this surface may been seen as smooth since the distance between peaks is too great to be detected. The interpretation of Boyan and Schwartz of a roughness spectrum suggests that an average roughness less than 2 μm will support a fibroblast-like morphology, whereas an average roughness more than 2 μm but with a peak-to-peak distance exceeding 10 μm (the suggested size of an average osteoblast) will also be perceived by a cell as smooth and will consequently induce a fibroblast-like morphology. In contrast, if the average roughness is greater than 2 μm but the peak-to-peak distance is less than 10 μm, then the osteoblast cells are unable to spread, and as a result they adopt a more typical osteoblast cuboidal morphology (Fig. 5). This observation is fundamentally the same point that was made much earlier by Brunette [83], who reported that if the peak-to-peak distance is less than the length of the cell body, then osteoblasts assume their characteristic cuboidal morphology, but if it is greater, a well-spread fibroblast-like morphology is assumed. Although the theory itself makes sense, one point that may go against the spectrum set by Boyan and Schwartz is the fact that osteoblasts in practice (depending on their origin, i.e. species, primary vs. cell line) can range in size; therefore, the defined spectrum laid out may only be effective for specific osteoblasts, i.e. MG-63 as used by Boyan and and Schwartz.

Richards [3] identified a spectrum of roughness between 0.2–2 μm which is believed to provoke the optimal differences in cell behaviour for smooth versus rough samples. Below 200 nm (even as low as 10 nm) cells react in vitro with varying degrees of phenotypic change, but these changes have limited in vivo support. This may be because the proteins attaching to the surface upon

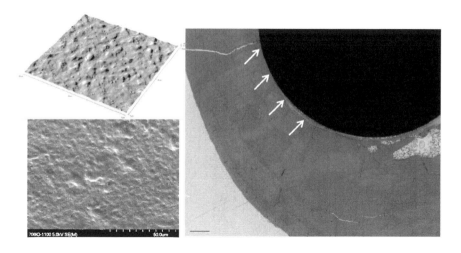

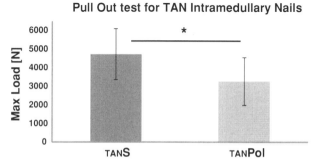

Fig. 4 Atomic force and scanning electron microscopy images of the polished TAN surface show how this technique is successful at reducing the characteristic surface roughness of the material. This modification produces a fibro–osseous interface (*white arrows*) which significantly impacts the ease with which the implants can be removed (*$p = 0.05$)

implantation completely mask such fine topographies and, therefore, their effective cue to relevant cells is masked or becomes insignificant. Numerous studies have shown major changes in cell behaviour above the 2-μm discontinuity size, but Richards postulated that the effect appears to become more and more marginal upon cells as the size increases. Despite this, most studies still include samples with average roughness of approximately 5 μm and designate their 'smooth' surface to be less than 0.6 μm [67, 84]. Clearly, this requires a degree of standardization for studies to be comparable. So theoretically, there is a consensus that a roughness spectrum exists; however, there is clear overlap and outright disagreement regarding the boundaries of the effective limits of the spectrum.

Although these issues can make discerning valuable information regarding interesting surface properties and cellular reactions difficult, the issue is further clouded with the introduction of nanotopography. Several groups have

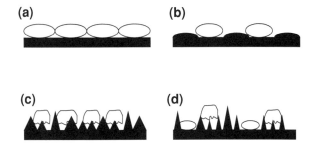

Fig. 5 How a cell may respond to the roughness spectrum. Cells must be able to perceive the surface roughness for them to respond in a surface-dependent manner. **a** When a surface is smooth, osteoblast cells will adopt a fibroblast-like morphology, becoming very flat and well spread. **b** On a wavy surface where the distance between peaks is more than the average cell size, cells will perceive the surface as smooth and will behave similarly to the behaviour shown in **a**. **c** If, however, a surface has frequent surface irregularities, producing a microrough surface, the cells are unable to spread and adopt typical osteoblast morphology. **d** On surfaces with mixed topographies, cell behaviour will reflect the average of rough and smooth microtopographies

convincingly shown that cells (MSCs, osteoblasts) can detect, interact and respond to nanotopographical features in vitro (see [85] for a review). In fact, this sensitivity has been described as far as the 15-nm range [86]. However, can features of this size be detected in vivo? And if so, how much actual influence do they have on determining the cell/tissue–material interaction? Recent studies tentatively indicate that cell and tissue interaction can be determined by nanotopography. For instance, Bjursten et al. [87] have recently shown that titanium dioxide nanotubes significantly enhance bone bonding, as measured by torque removal and percentage of bone contact, in an in vivo rabbit tibial model compared with grit-blasted titanium. However, it is difficult from the data presented to differentiate if the effect of microtopography was fully negated since the surface morphology of the nanotube surface also appeared to have microscale morphology. Furthermore, surface roughness measurements were made using scanning electron microscopy alone and did not include any validated quantitative methods.

Meirelles et al. [88] have also shown that nano-titania and nano-hydroxyapatite surfaces support bone on-growth in a rabbit model. However, it is worth noting that the 'nano' surfaces had S_a (mean arithmetic height measurement) of 121 nm for titanium and 170 nm for hydroxyapatite surfaces compared with 225 nm for the polished control. Although Meirelles et al. removed microstructures via grinding, again the surface morphology did appear to have a level of microroughness, which is supported by the fact that the height measurements were reflective of a microtopography rather than a true nanotopography, i.e. tens rather than hundreds of nanometres. Several methods such as photolithography exist for producing nanometric surfaces, so perhaps in time more convincing evidence will emerge that supports the theory that nanotopographical surface features can determine tissue–implant interaction.

3.4 Mechanotransduction

Focal contacts are protein complexes that constitute the primary attachment which is essential for long-term adhesion. Apart from integrins, focal adhesions consist of specific proteins such as talin, α-actinin and vinculin filaments that interact with the cytoskeleton on the cytoplasmic side. Through focal adhesions, cells react to extrinsic chemical and mechanical signals from the cell–cell contact or cell-ECM components. Signal propagation is achieved via direct and indirect mechanotransduction, both of which are explored here in basic terms. The reader will be directed to more in-depth sources throughout.

3.4.1 Direct Mechanotransduction

Direct mechanotransduction utilizes conformational changes in focal adhesions and cytoskeletal conformation to pass information about the ECM topography to the nucleus as mechanical signals [89]. The phenomenon of how the cell relays mechanical signals from the environment may be explained by two theories—cellular tensegrity [90] and percolation [91]. The theory of cellular tensegrity (tensional integrity) of Ingber [90, 92] was adapted from civil engineering principles that define tensegrity systems that stabilize their shape by continuous tension and not by continuous compression. According to Ingber's theory, a cell is a prestressed tensegrity structure where microtubules act as load bearers and microfilaments are under tension. In addition, intermediate filaments serve as a tensile mode that interconnects and stiffens the entire cytoskeleton and nuclear lattice through tension. The tensional prestress, generated by actomyosin interactions in cortical and contractile stress fibres anchored to the focal adhesions, is a major determinant of cell and nuclear stability [90]. This model assumes that mechanical signals can be transferred across the cell membrane by ECM receptors and transduced into a chemical response at the site of the bound receptor.

The theory of percolation of Forgacs [91] involves an interconnected network system composed of cytoskeleton units, akin to a spider's web, for transducing mechanical signals. This network spans the distance from the membrane to the nucleus, where it connects with the nucleus laminin. The physical properties of cytoplasm determine the speed, whereas the interconnected network allows redundancy, which means the signal can arrive through several channels. This model provides speed and redundancy in signal transduction from the membrane to the nucleus. Fundamentally, in this model a threshold of components is required for a critical concentration to be reached and if this threshold is achieved, propagation of the signal via the cytoskeleton to the nucleus will be accomplished.

Several fundamental differences exist between the theories. Firstly, tensegrity does not allow for functional redundancy, which is required for reliable signalling to be achieved. Secondly, tensegrity structures adhere to strict rules of stability, i.e. they contain the absolute minimum of structural components required for stability. Finally, signal propagation through prestressed components is likely to produce a dampened signal as prestress will hamper the deflection incurred by each component [91].

Some insight into the theories behind mechanotransduction of the cytoskeleton provides a basis for understanding how changes in focal adhesion distribution and cell shape induced by a material can alter cellular function. Studies have shown that nuclei react to changes in cell morphology caused by the topography of the ECM and applied stress signals. Specifically, it has been shown that in response to tension the intermediate filaments of the cytoskeleton reorient, causing a distortion in the nucleus which results in nucleoli shifting along the appropriate axis [93]. Dahl et al. [94] showed that the nuclear lamina network is an elastic structure; however, it appears to have a compression limit. This feature suggests that the lamina functions as a molecular 'shock absorber'. Their experiments, using dextran to swell the nucleus and micropipettes to compress it, concluded that the DNA within the nucleus is afforded a degree of protection by the nuclear envelope, but that this retains a degree of flexibility for adequate mechanotransduction.

Recently, Dalby et al. [89] have put forth a modified version of self-induced mechanotransduction. Specifically, they suggest that by altering nuclear morphology and consequently chromosomal positioning with changes in topography, this will directly influence the probability of gene transcription. Investigations into nuclear and laminin morphology changes as a reaction to various nanotopographies revealed differences in genome regulation and gene expression in support of this theory. Fibroblasts cultured on nanocolumns (centre-to-centre spacing 184 nm) and nanopits (spacing 300 nm) react to both materials with reduced spreading, which affects cytoskeletal organization, resulting in relaxation of the nucleus size. Moreover, the interphase chromosome positioning by centromere analysis of chromosomes 3, 11 and 16 showed a reduction in the centromere pair distance, with a significant difference for chromosome 3 for cells cultured on surfaces with nanocolumns and nanopits and for chromosome 11 for cells cultured on surfaces with nanopits. Additionally, the reduction in cell spreading increases the number of gene downregulations [95]. On the basis of these results it appears that mechanotransduction as a reaction to topography is more likely to combine features of both the percolation model and the tensegrity model, with the likelihood of the involvement of additional factors.

3.4.2 Indirect Mechanotransduction

Signal propagation is also achieved through indirect mechanotransduction. Activation of the extracellular-signal-regulated kinase (ERK)/mitogen-activated protein kinase (MAPK) pathway is the main method by which indirect mechanotransduction is achieved. This pathway has fundamental roles in relaying extracellular information to the nucleus [96], cellular differentiation and cell cycle regulation. Further to this, the ERK/MAPK pathway has demonstrated a key function in the response of osteoblast cells to a variety of signals, including

ECM–integrin binding [97] and mechanical loading [98]. Additionally, integrin-mediated activation of the ERK/MAPK pathway results in phosphorylation and stimulation of the osteoblast differentiation master control gene *RUNX2* [99]. Focal adhesion kinase (FAK) is a non-receptor kinase that is linked to the β integrin subunit and is principally involved in integrin-dependent signalling. FAK is recruited to the focal adhesion site through integrin clustering. FAK influences transcriptional events through adhesion-dependent phosphorylation of downstream signalling molecules which results in the binding and activation of tyrosine kinase, which subsequently activates MAPK signalling pathways. MAPK pathways are essentially a chain of proteins involved in signal propagation from focal adhesion sites to the nucleus. Binding $\alpha_5\beta_1$ to fibronectin activates ERK1/2, a subclass of MAPK, which functions as a mediator of cellular differentiation [100]. It is likely, however, that other MAPKs may also be involved. For example, FAK is required for signalling through Jun NH_2-terminal kinase for cell cycle regulation. Consequently, integrins can differentially modulate cell proliferation and phenotypic expression through distinct pathways.

Surface topography has been identified as an influential factor in ERK/MAPK signalling changes [101], essentially through modulation of integrin clustering and adhesion formation. Hamamura et al. [102] recently showed that geometrical alterations in ECM environments can alter the phosphorylation pattern of p130Cas, FAK, ERK1/2 and p38 MAPK for osteoblast cells cultured on 3D collagen matrices. Specifically, they showed this using a whole-genome array that revealed that cells grown in the 3D collagen matrix partly suppress genes associated with cell adhesion and cell cycling. Furthermore, Western blot analysis revealed that the expression of phosphorylated p130Cas, FAK and ERK1/2 was decreased in cells grown in a 3D collagen matrix. Conversely, the phosphorylation of p38 MAPK was at an elevated level in the 3D matrix and its upregulation was linked to an increase in mRNA levels of dentin matrix protein 1 and bone sialoprotein.

Prior to this, Kokobu et al. [103] demonstrated that for human gingival fibroblasts ERK 1/2 is translocated to the nucleus in cells in manner dependent on surface topography and culture time. It appears that surface topography also differentially influences Src involvement in the ERK pathway [104]. Schwartz et al. [105] showed that ERK/MAPK activation is required for maintenance of control levels of alkaline phosphatase; however, they did not find this outcome to be reliant on surface microroughness.

Other studies indicate that FAK and ERK phosphorylation can also be affected by nanotopography. Salaszynk et al. [106] suggested that FAK activity is necessary for osteoblast differentiation of MSCs. In support of this, Biggs et al. [107] showed that osteoblast differentiation and function is correlated to focal adhesion growth and FAK-mediated activation of the ERK/MAPK pathway in MSCs.

Interestingly, it also appears that the ERK/MAPK pathway is involved in the molecular response to aseptic loosening of implants. One study suggests that the MAPK signalling pathway controls NF-κB-mediated transcriptional activation in response to wear debris particles [108]. Beidelschies et al. [109] revealed that activation of ERK1/2/Egr-1 and NF-κB pathways is responsible for the ability of adherent endotoxin to potentiate cytokine production, osteoblast differentiation and bone loss induced by wear particles.

4 Tissue–Implant Interface

4.1 Interface Structure and Composition

The formation of bone at the implant interface was first defined by Osborne and Newesley [110] in their description of distance and contact osteogenesis. Distance osteogenesis describes the phenomenon whereby bone surfaces adjacent to the implant provide a population of osteogenic cells, which through the process of appositional bone growth encroach upon the implant. Ultimately, therefore, the implant becomes surrounded by bone rather than bone forming de novo on the implant surface. It is postulated that the outcome of distance osteogenesis is that the surface will never actually have direct bone bonding since it will always be obscured by ECM and prevailing cells. In contrast to distance osteogenesis, contact osteogenesis describes the phenomenon whereby bone forms de novo on the implant surface. Essentially, therefore, the implant must become populated by osteogenic cells prior to matrix production can be initiated. Although both theories describe distinct methods for bone to become juxtaposed to an implant surface, it is likely that both methods are actually involved in implant osseointegration.

Since the peri-implant site becomes primarily encased in blood, the migration of the cells will be via the fibrin network that is produced during clot formation, and it is these cells that will ultimately afford the basis for the osteogenic cell population required for differentiation. What is interesting is that since fibrin is a by-product released into the implantation site to promote healing, one could therefore assume that this protein would adhere to virtually all surfaces, and thus osteoconduction could theoretically occur for any biomaterial; however, this is not the case. As is the case with dermal wound healing, cell migration is associated with subsequent wound contraction; thus, migration of cells on a provisional fibrin matrix results in its subsequent withdrawal (Fig. 6), preventing further migration, and ultimately osteogenesis (see [111] for a review). It seems increasingly apparent, therefore, that implant design is important in providing the correct degree of anchorage for a transitory scaffold for the purpose of cell migration. Therefore, the ability of a biomaterial surface to retain fibrin attachment during this retraction phase is crucial in determining if migrating cells will reach the device. It is suggested that the complexity of a microrough surface provides a 3D topography so that fibrin remains sufficiently attached to the implant to withstand retraction, allowing cell migration.

Once an osteogenic cell population is present at the surface, the next critical step is the initial formation of a mineralized matrix. The method of de novo bone formation at the implant interface is described by four principal phases and is

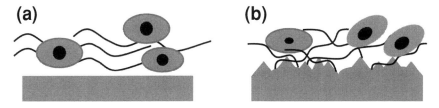

Fig. 6 The ability of a biomaterial to retain fibrin attachment during the retraction phase of wound healing is crucial in determining if migrating cells will reach the device. Owing to the lack of anchorage, smooth surfaces do not retain fibrin matrix during wound healing contraction. In contrast, microrough surfaces withstand the forces of wound healing contraction, thereby retaining the fibrin matrix, which supports continuous cell migration and direct tissue contact. (Modified from 111])

supported by both in vitro [112, 113] and in vivo [114] experiments. Initially, differentiating osteogenic cells secrete a non-collagenous organic matrix containing both osteopontin and bone sialoprotein, which serve as CaP nucleation sites where crystals can grow in size. Concomitant with this growth at the boundary is the introduction of collagen fibre assembly. Collagen is thus deposited onto this layer, which subsequently mineralizes (Fig. 7), but is separated from the substratum by a collagen-free calcified tissue layer (approximately 0.5 μm thick). It is suggested that the presence of a heterogeneous population at the bone–implant boundary and an afibrillar interfacial zone is comparable to cement lines and the lamina limitans [115].

As previously outlined, the biopassivity of an implant is partly attributable to the oxide layer. It is this layer, often only a few nanometres thick, which is crucial to the application of metal devices in vivo. The major differences between the oxide layers of EPSS and titanium and its alloys involve the actual thickness of the layer and the chemistry, both of which result in evoking distinct biological responses. This distinction is highlighted most poignantly by an early investigation by Albrektsson and Hansson [116] which studied in much detail the ultrastructural differences evoked by EPSS and titanium screws in rabbit bone. Specifically, they reported a continuous one to two cell thick layer separating the EPSS device from bone, whereas titanium had direct anchorage to the bone. These observations are also supported by recent studies that have observed fibro-osseointegration at the tissue–EPSS interface versus osseointegration at the titanium interface [78, 80, 81]. Albrektsson and Hansson [116] also described the presence of inflammatory cells adjacent to EPSS devices, and a proteoglycan coat void of collagen filaments was observed within the interface. However, in contrast, titanium had a proteoglycan layer at the interface with collagen bundles in close proximity. The importance of this direct apposition of the proteoglycan layer on titanium is believed to directly result in the accelerated osseointegration properties of this material due to the enhanced degradation of the hyaluronan network which is formed as a result of the wound healing response [117].

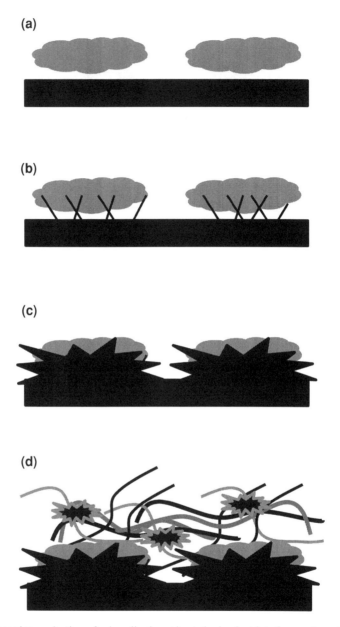

Fig. 7 Stepwise production of mineralized matrix at the implant interface. **a** Secretion of the non-collagenous proteins osteopontin and bone sialoprotein. **b** Calcium phosphate nucleation at calcium binding sites. **c** Crystal growth and propagation. **d** Collagen production and mineralization with matrix separated from the substratum by a collagen-free calcified tissue layer (approximately 0.5 μm). (Modified from [111])

5 Summary and Conclusions

The cell–material interaction is a complex interplay of a variety of biological processes that are differentially influenced by changes in surface properties, in particular surface microtopography. Alterations on a microscale and even a nanoscale to a surface have been shown to directly impact on initial adhesion to nuclear signal transduction and ultimately influence cell phenotype. Although many details regarding the exact mechanisms involved in this substrate-dependent control are still under investigation, a wealth of information has emerged that has been useful in refining an empirical approach towards biomaterial development for determining specific cell and tissue responses. Nevertheless, with the development of new, more sensitive analytical techniques and refinement of fabrication processes, additional valuable information is yet to emerge. As this happens, the realm of biologic-biomaterial interaction research will undoubtedly welcome a influx of novel approaches that will potentially expand the application of this technology beyond orthopaedic medicine.

References

1. Kasemo B, Lausmaa J (1994) Material–tissue interfaces: the role of surface properties and processes. Environ Health Perspect 102(5):41–45
2. Kasemo B, Gold J (1998) Implant surfaces and interface processes. Adv Dent Res 13:8–20
3. Richards RG (2008) The relevance of implant surfaces in hand fracture fixation. In: Osteosynthesis in the hand: current concepts. FESSH instructional course 2008. Karger, Basel, pp 20–30
4. Probst A, Speigel HA (1997) Cellular mechanisms of bone repair. J Invest Surg 10:77–86
5. Schindeler A, McDonald MM, Bokko P, Little DG (2008) Bone remodeling during fracture repair: the cellular picture. Semin Cell Dev Biol. doi:10.1016/j.semcdb.2008.07.004
6. Deuel TF, Senior RM, Huang JS, Griffin GL (1982) Chemotaxis of monocytes and neutrophils to platelet-derived growth factor. J Clin Invest 69:1046–1049
7. Postlethwaite AE, Keski-Oja J, Moses HL, Kang AH (1987) Stimulation of the chemotactic migration of human fibroblasts by transforming growth factor beta. J Exp Med 165:251–256
8. Togrul E, Bayram H, Gulsen M, Kalaci A, Ozbarlas S (2005) Fractures of the femoral neck in children: long-term follow up in 62 hip fractures. Injury 36:123–130
9. Schmalzreid TP, Grogan TJ, Neumeier PA, Dorey FJ (1991) Metal removal in a pediatric population: benign procedure or necessary evil? J Pediatr Orthop 11:72–76
10. Kahle WK (1994) The case against routine metal removal. J Pediatr Orthop 14:229–237
11. Hamilton P, Doig S, Williamson O (2004) Technical difficulty of metal removal after LISS plating. Injury 35:626–628
12. Ruoslahti E (1991) Integrins. J Clin Invest 87(1):1–5
13. Xiao G, Wang D, Benson MD, Karsenty G, Franceschi RT (1998) Role of the alpha2-integrin in osteoblast-specific gene expression and activation of the Osf2 transcription factor. J Biol Chem 273(49):32988–32994
14. Petrie TA, Raynor JE, Dumbauld DW, Lee TT, Jagtap S, Templeman KL, Collard DM, García AJ (2010) Multivalent integrin-specific ligands enhance tissue healing and biomaterial integration. Sci Transl Med 2(45):45–60

15. Keselowsky BG, Wang L, Schwartz Z, Garcia AJ, Boyan BD (2007) Integrin alpha(5) controls osteoblastic proliferation and differentiation responses to titanium substrates presenting different roughness characteristics in a roughness independent manner. J Biomed Mater Res A 80(3):700–710
16. Wang L, Zhao G, Olivares-Navarrete R, Bell BF, Wieland M, Cochran DL, Schwartz Z, Boyan BD (2006) Integrin beta1 silencing in osteoblasts alters substrate-dependent responses to 1,25-dihydroxy vitamin D3. Biomaterials 27(20):3716–3725
17. Lennon FE, Hermann CD, Olivares-Navarrete R, Rhee WJ, Schwartz Z, Bao G, Boyan BD (2010) Use of molecular beacons to image effects of titanium surface microstructure on beta1 integrin expression in live osteoblast-like cells. Biomaterials 31(30):7640–7647
18. Pegueroles M, Aguirre A, Engel E, Pavon G, Gil FJ, Planell JA, Migonney V, Aparicio C (2011) Effect of blasting treatment and Fn coating on MG63 adhesion and differentiation on titanium: a gene expression study using real-time RT-PCR. J Mater Sci Mater Med 22(3):617–627
19. Galli D, Benedetti L, Bongio M, Maliardi V, Silvani G, Ceccarelli G, Ronzoni F, Conte S, Benazzo F, Graziano A, Papaccio G, Sampaolesi M, Cusella De Angelis MG (2011) In vitro osteoblastic differentiation of human mesenchymal stem cells and human dental pulp stem cells on poly-L-lysine-treated titanium-6-aluminium-4-vanadium. J Biomed Mater Res A 97A(2)118–126. doi:10.1002/jbm.a.32996
20. Gronowicz G, McCarthy MB (1996) Response of human osteoblasts to implant materials: integrin-mediated adhesion. J Orthop Res 14:878–887
21. Sinha RK, Tuan RS (1996) Regulation of human osteoblast integrin expression by orthopaedic implant materials. Bone 18(5):451–457
22. Siebers MC, Brugge PJ, Walboomers XF, Jansen JA (2005) Integrins as linker proteins between osteoblasts and bone replacing materials. A critical review. Biomaterials 26(2):137–146
23. Olivares-Navarette R, Raz P, Zhao G, Chen J, Wieland M, Cochran DL, Chaudhri RA, Ornoy A, Boyan BD, Schwartz Z (2008) Integrin alpha2beta1 plays a critical role in osteoblast response to micron-scale surface structure and surface energy of titanium substrates. Proc Natl Acad Sci USA 105(41):15767–15772
24. Schneider G, Burridge K (1994) Formation of focal adhesions by osteoblasts adhering to different substrata. Exp Cell Res 214(1):264–269
25. Ratner BD (1993) New ideas in biomaterials science—a path to engineered biomaterials. J Biomed Mater Res (7):837–850
26. Brunette DM (2001) Principles of cell behaviour on titanium surfaces and their application to implanted devices. In: Brunette DM, Tengvall P, Textor M, Thomsen P (eds) Titanium in medicine. Springer, Heidelberg, pp 486–512
27. Brunette DM, Kenner GS, Gould TR (1983) Grooved titanium surfaces orient growth and migration of cells from human gingival explants. J Dent Res 62:1045–1048
28. Eisenbarth E, Velten D, Schenk-Meuser K, Linez P, Biehl V, Duschner H, Breme J, Hildebrand H (2002) Interactions between cells and titanium surfaces. Biomol Eng 19: 243–249
29. Ahmad M, Gawronski D, Blum J, Goldberg J, Gronowicz G (1999) Differential response of human osteoblast-like cells to commercially pure (cp) titanium grades 1 and 4. J Biomed Mater Res 46(1):121–131
30. Hazan R, Brener R, Oron U (1993) Bone growth to metal implants is regulated by their surface chemical properties. Biomaterials 14(8):570–574
31. Vinall RL, Gasser B, Richards RG (1995) Investigation of cell compatibility of titanium test surfaces to fibroblasts. Injury 26(1):21–27
32. Textor M, Sittig C, Frauchiger V, Tosatti S, Brunette DM (2001) Properties and biological significance of natural oxide films on titanium and its alloys. In: Brunette DM, Tengvall P, Textor M, Thomsen P (eds) Titanium in medicine. Springer, Heidelberg, pp 171-230
33. Morita M, Sasada T, Hayashi H, Tsukamoto Y (1988) The corrosion fatigue properties of surgical implants in a living body. J Biomed Mater Res 22(6):529–540

34. Hanawa T (2004) Metal ion release from metal implants. Mater Sci Eng 24:745–752
35. Hayes JS, Richards RG (2010) The use of titanium and stainless steel in fracture fixation. Expert Rev Med Devices 7(6):843–853
36. Degasne I, Basle MF, Demais V, Hure G, Lesourd M, Grolleau B, Mercier L, Chappard D (1999) Effects of roughness, fibronectin and vitronectin on attachment, spreading, and proliferation of human osteoblast-like cells (SaOs-2) on titanium surfaces. Calcif Tissue Int 64(6):499–507
37. Yang Y, Cavin R, Ong JL (2003) Protein adsorption on titanium surfaces and their effect on osteoblast attachment. J Biomed Mater Res A 67:344–349
38. Meyer AE, Baier RE, Natiella JR, Meenaghan MA (1988) Investigation of tissue/implant interactions during the first two hours of implantation. J Oral Implantol 14:363–379
39. Rapuano BE, MacDonald DE (2011) Surface oxide net charge of a titanium alloy: modulation of fibronectin-activated attachment and spreading of osteogenic cells. Colloids Surf B Biointerfaces 82(1):95–103
40. Howlett CR, Evans MD, Walsh WR, Johnson G, Steele JG (1994) Mechanism of initial attachment of cells derived from human bone to commonly used prosthetic materials during cell culture. Biomaterials 15:213–222
41. Horbett TA, Schway MB (1988) Correlations between mouse 3T3 cell spreading and serum fibronectin adsorption on glass and hydroxyethylmethacrylate-ethylmethacrylate copolymers. J Biomed Mater Res 22(9):763–793
42. Bauer TW, Schils J (1999) The pathology of total joint arthroplasty I. Mechanisms of implant fixation. Skeletal Radiol 28(8):423–432
43. Reyes CD, Petrie TA, Burns KL, Schwartz Z, García AJ (2007) Biomolecular surface coating to enhance orthopaedic tissue healing and integration. Biomaterials 28(21): 3228–3235
44. Yang Y, Kim KH, Ong JL (2005) A review on calcium phosphate coatings produced using a sputtering process an alternative to plasma spraying. Biomaterials 26(3):327–337
45. Rokkum M, Reigstad A, Johansson CB, Albrektsson T (2003) Tissue reactions adjacent to wellfixed hydroxyapatite-coated acetabular cups. Histopathology of ten specimens retrieved at reoperation after 0.3–5.8 years. J Bone Joint Surg Br 85:440–447
46. de Jonge LT, Leeuwenburgh SC, Wolke JG, Jansen JA (2008) Organic-inorganic surface modifications for titanium implant surfaces. Pharm Res 25(10):2357–2369
47. Schlegel P, Hayes JS, Frauchiger VM, Gasser B, Wieling R, Textor M, Richards RG (2009) An in vivo evaluation of the biocompatibility of anodic plasma chemical (APC) treatment of titanium with calcium phosphate. J Biomed Mater Res B Appl Biomater 90(1):26–34
48. O'Hare P, Meenan BJ, Burke GA, Byrne G, Dowling D, Hunt JA (2010) Biological responses to hydroxyapatite surfaces deposited via a co-incident microblasting technique. Biomaterials 31(3):515–522
49. O'Donoghue J, Haverty D (2008) Method of doping surfaces. PCT application no. WO2008/033867
50. Grzesik WJ, Robey PG (1994) Bone matrix RGD glycoproteins: immunolocalization and interaction with human primary osteoblastic bone cells in vitro. J Bone Miner Res 9(4): 487–496
51. Zreiqat H, Akin FA, Howlett CR, Markovic B, Haynes D, Lateef S (2003) Differentiation of human bone-derived cells grown on GRGDSP-peptide bound titanium surfaces. J Biomed Mater Res A 64(1):105–111
52. Rammelt S, Illert T, Bierbaum S, Scharnweber D, Zwipp H, Schneiders W (2006) Coating of titanium implants with collagen, RGD peptide and chondroitin sulfate. Biomaterials 27(32):5561–5571
53. Knight CG, Morton LF, Peachey AR, Tuckwell DS, Farndale RW, Barnes MJ (2000) The collagen-binding A-domains of integrins alpha(1) beta(1) and alpha(2) beta(1) recognize the same specific amino acid sequence, GFOGER, in native (triple-helical) collagens. J Biol Chem 275(1):35–40

54. Vlacic-Zischke J, Hamlet SM, Friis T, Tonetti MS, Ivanovski S (2011) The influence of surface microroughness and hydrophilicity of titanium on the up-regulation of TGFβ/BMP signalling in osteoblasts. Biomaterials 32(3):665–671
55. Lai M, Cai K, Zhao L, Chen X, Hou Y, Yang Z (2011) Surface functionalization of TiO(2) nanotubes with bone morphogenetic protein 2 and its synergistic effect on the differentiation of mesenchymal stem cells. Biomacromolecules 12(4):1097–1105
56. Liu Y, Enggist L, Kuffer AF, Buser D, Hunziker EB (2007) The influence of BMP-2 and its mode of delivery on the osteoconductivity of implant surfaces during the early phase of osseointegration. Biomaterials 28(16):2677–2686
57. Wikesjö UM, Qahash M, Polimeni G, Susin C, Shanaman RH, Rohrer MD, Wozney JM, Hall J (2008) Alveolar ridge augmentation using implants coated with recombinant human bone morphogenetic protein-2: histologic observations. J Clin Periodontol 35(11): 1001–1010
58. Cochran DL (1999) A comparison of endosseous dental implant surfaces. J Periodontol 70(12):1523–1539
59. Meredith DO, Eschbach L, Wood MA, Riehle MO, Curtis AS, Richards RG (2005) Human fibroblast reactions to standard and electropolished titanium and Ti-6Al-7Nb, and electropolished stainless steel. J Biomed Mater Res A 75:541–555
60. Hayes JS (2008) The efficacy of surface polishing clinically available internal fixation materials for ease of removal. PhD thesis, AO Research Institute, Davos/University of Wales, Cardiff
61. Brunette DM (1986) Spreading and orientation of epithelial cells on grooved substrata. Exp Cell Res 167:203–217
62. Boateng SY, Hartman TJ, Ahluwalia N, Vidula H, Desai TA, Russell B (2003) Inhibition of fibroblast proliferation in cardiac myocyte cultures by surface microtopography. Am J Physiol Cell Physiol 285(1):171–182
63. Lincks J, Boyan BD, Blanchard CR, Lohmann CH, Liu Y, Cochran DL, Dean DD, Schwartz Z (1998) Response of MG63 osteoblast-like cells to titanium and titanium alloy is dependent on surface roughness and composition. Biomaterials 19(23):2219–2232
64. Schmidt C, Ignatius AA, Claes LE (2001) Proliferation and differentiation parameters of human osteoblasts on titanium and steel surfaces. J Biomed Mater Res 54(2):209–215
65. Boyan BD, Bonewald LF, Paschalis EP, Lohmann CH, Rosser J, Cochran DL, Dean DD, Schwartz Z, Boskey AL (2002) Osteoblast-mediated mineral deposition in culture is dependent on surface microtopography. Calcif Tissue Int 71(6):519–529
66. Postiglione L, Di Domenico G, Ramaglia L, Montagnani S, Salzano S, Di Meglio F, Sbordone L, Vitale M, Rossi G (2003) Behavior of SaOS-2 cells cultured on different titanium surfaces. J Dent Res 82:692–696
67. Boyan BD, Lossdorfer S, Wang L, Zhao G, Lohmann CH, Cochran DL, Schwartz Z (2003) Osteoblasts generate an osteogenic microenvironment when grown on surfaces with rough microtopographies. Eur Cell Mater 6:22–27
68. Kieswetter K, Schwartz Z, Hummert TW, Cochran DL, Simpson J, Dean DD, Boyan BD (1996) Surface roughness modulates the local production of growth factors and cytokines by osteoblast-like MG-63 cells. J Biomed Mater Res 32(1):55–63
69. Brett PM, Harle J, Salih V, Mihoc R, Olsen I, Jones FH, Tonetti M (2004) Roughness response genes in osteoblasts. Bone 35:124–133
70. Arcelli D, Palmieri A, Pezzetti F, Brunelli G, Zollino I, Carinci F (2007) Genetic effects of titanium surface on osteoblasts: a meta-analysis. J Oral Sci 49:299–309
71. Hayes JS, Khan IM, Archer CW, Richards RG (2010) The role of surface microtopography in the modulation of osteoblast differentiation. Eur Cell Mater 20:98–108
72. Schneider GB, Perinpanayagam H, Clegg M, Zaharias R, Seabold D, Keller J, Stanford C (2003) Implant surface roughness affects osteoblast gene expression. J Dent Res 82:372–376
73. Isa ZM, Schneider GB, Zaharias R, Seabold D, Stanford CM (2006) Effects of fluoride-modified titanium surfaces on osteoblast proliferation and gene expression. Int J Oral Maxillofac Implants 21(2):203–211

74. Guo J, Padilla RJ, Wallace A, DeKoK IJ, Cooper LF (2007) The effect of hydrofluoric acid treatment of TiO2 grit blasted titanium implants on adherent osteoblast gene expression in vitro and in vivo. Biomaterials 28(36):5418–5425
75. Schneider GB, Zaharias R, Seabold D, Keller J, Stanford C (2004) Differentiation of preosteoblasts is affected by implant surface microtopographies. J Biomed Mater Res A 69(3):462–468
76. Verrier S, Peroglio M, Voisard C, Lechmann B, Alini M (2011) The osteogenic differentiation of human osteoprogenitor cells on anodic-plasma-chemical treated Ti6Al7Nb. Biomaterials 32(3):672–680
77. Hayakawa T, Yoshinari M, Nemoto K, Wolke JG, Jansen JA (2000) Effect of surface roughness and calcium phosphate coating on the implant/bone response. Clin Oral Implants Res 11(4):296–304
78. Pearce AI, Pearce SG, Schwieger K, Milz S, Schneider E, Archer CW, Richards RG (2008) Effect of surface topography on removal of cortical bone screws in a novel sheep model. J Orthop Res 26(10):1377–1383
79. Welton JL (2007) Master thesis: in vivo evaluation of defined polished surfaces to prevent soft tissue adhesion. AO Research Institute, Davos
80. Hayes JS, Seidenglanz U, Pearce AI, Pearce SG, Archer CW, Richards RG (2010) Surface polishing positively influences ease of plate and screw removal. Eur Cell Mater 19:117–126
81. Hayes JS, Vos DI, Hahn J, Pearce SG, Richards RG (2009) An in vivo evaluation of surface polishing of TAN intramedullary nails for ease of removal. Eur Cell Mater 18:15–26
82. Boyan BD, Schwartz Z (1999) Modulation of osteogenesis via implant surface design. In: Davies JE (ed) Bone engineering. EM Squared, Toronto, pp 232–239
83. Brunette DM (1988) The effects of implant surface topography on the behavior of cells. Int J Oral Maxillofac Implants 3(4):231–246
84. Lossdorfer S, Schwartz Z, Wang L, Lohmann CH, Turner JD, Wieland M, Cochran DL, Boyan BD (2004) Microrough implant surface topographies increase osteogenesis by reducing osteoclast formation and activity. J Biomed Mater Res A 70(3):361–369
85. Dalby MJ, Gadegaard N, Curtis AS, Oreffo RO (2007) Nanotopographical control of human osteoprogenitor differentiation. Curr Stem Cell Res Ther 2(2):129–138
86. Sjöström T, Dalby MJ, Hart A, Tare R, Oreffo RO, Su B (2009) Fabrication of pillar-like titania nanostructures on titanium and their interactions with human skeletal stem cells. Acta Biomater 5(5):1433–1441
87. Bjursten LM, Rasmusson L, Oh S, Smith GC, Brammer KS, Jin S (2009) Titanium dioxide nanotubes enhance bone bonding in vivo. J Biomed Mater Res A 92(3):1218–1224
88. Meirelles L, Currie F, Jacobsson M, Albrektsson T, Wennerberg A (2008) The effect of chemical and nanotopographical modifications on the early stages of osseointegration. Int J Oral Maxillofac Implants 23(4):641–647
89. Dalby MJ (2005) Topographically induced direct cell mechanotransduction. Med Eng Phys 27(9):730–742
90. Ingber DE (1993) Cellular tensegrity: defining new rules of biological design that govern the cytoskeleton. J Cell Sci 104:613–627
91. Forgacs G (1995) On the possible role of cytoskeletal filamentous networks in intracellular signalling: an approach based on percolation. J Cell Sci 108(6):2131–2143
92. Ingber DE (2003) Tensegrity I cell structure and hierarchical systems biology. J Cell Sci 116:1157–1173
93. Maniotis AJ, Chen CS, Ingber DE (1997) Demonstration of mechanical connections between integrins, cytoskeletal filaments, and nucleoplasm that stabilize nuclear structure. Proc Natl Acad Sci USA 4 94(3):849–854
94. Dahl KN, Kahn SM, Wilson KL, Discher DE (2004) The nuclear envelope lamina network has elasticity and a compressibility limit suggestive of a molecular shock absorber. J Cell Sci 117:4779–4786

95. Dalby MJ, Biggs MJ, Gadegaard N, Kalna G, Wilkinson CD, Curtis AS (2007) Nanotopographical stimulation of mechanotransduction and changes in interphase centromere positioning. J Cell Biochem 100(2):326–338
96. Ge C, Xiao G, Jiang D, Franceschi RT (2007) Critical role of the extracellular signal regulated kinase-MAPK pathway in osteoblast differentiation and skeletal development. J Cell Biol 176:709–718
97. Xiao G, Gopalakrishnan R, Jiang D, Reith E, Benson MD, Franceschi RT (2002) Bone morphogenetic proteins, extracellular matrix, and mitogen-activated protein kinase signaling pathways are required for osteoblast-specific gene expression and differentiation in MC3T3–E1 cells. J Bone Miner Res 17:101–110
98. You J, Reilly GC, Zhen X, Yellowley CE, Chen Q, Donahue HJ et al (2001) Osteopontin gene regulation by oscillatory fluid flow via intracellular calcium mobilization and activation of mitogen-activated protein kinase in MC3T3–E1 osteoblasts. J Biol Chem 276:13365–13371
99. Hata K, Ikebe K, Wada M, Nokubi T (2007) Osteoblast response to titanium regulates transcriptional activity of Runx2 through MAPK pathway. J Biomed Mater Res A 81(2): 446–452
100. Ge C, Xiao G, Jiang D, Franceschi RT (2007) Critical role of the extracellular signal-regulated kinase-MAPK pathway inosteoblast differentiation and skeletal development. J Cell Biol 176(5):709–718
101. Krause A, Cowles EA, Gronowicz G (2000) Integrin-mediated signaling in osteoblasts on titanium implant materials. J Biomed Mater Res 52(4):738–747
102. Hamamura K, Jiang C, Yokota H (2010) ECM-dependent mRNA expression profiles and phosphorylation patterns of p130Cas, FAK, ERK and p38 MAPK of osteoblast-like cells. Cell Biol Int 34(10):1005–1012
103. Kokubu E, Hamilton DW, Inoue T, Brunette DM (2009) Modulation of human gingival fibroblast adhesion, morphology, tyrosine phosphorylation, and ERK 1/2 localization on polished, grooved and SLA substratum topographies. J Biomed Mater Res A 91(3):663–670
104. Hamilton DW, Brunette DM (2007) The effect of substratum topography on osteoblast adhesion mediated signal transduction and phosphorylation. Biomaterials 28(10): 1806–1819
105. Schwartz Z, Lohmann CH, Vocke AK, Sylvia VL, Cochran DL, Dean DD, Boyan BD (2001) Osteoblast response to titanium surface roughness and 1a, 25-(OH)2D3 is mediated through the mitogen-activated protein kinase (MAPK) pathway. J Biomed Mater Res 56(3):417–426
106. Salasznyk RM, Klees RF, Williams WA, Boskey A, Plopper GE (2007) Focal adhesion kinase signaling pathways regulate the osteogenic differentiation of human mesenchymal stem cells. Exp Cell Res 313(1):22–37
107. Biggs MJP, Richards RG, Gadegaard N, Wilkinson ROC Oreffo CDW, Dalby MJ (2009) The use of nanoscale topography to modulate the dynamics of adhesion formation in primary osteoblasts and ERK/MAPK signalling in STRO-1 + enriched skeletal stem cells. Biomaterials 30:5094–5103
108. Fritz EA, Jacobs JJ, Glant TT, Roebuck KA (2005) Chemokine IL-8 induction by particulate wear debris in osteoblasts is mediated by NF-kappaB. J Orthop Res 23(6): 1249–1257
109. Beidelschies MA, Huang H, McMullen MR, Smith MV, Islam AS, Goldberg VM, Chen X, Nagy LE, Greenfield EM (2008) Stimulation of macrophage TNFalpha production by orthopaedic wear particles requires activation of the ERK1/2/Egr-1 and NF-kappaB pathways but is independent of p38 and JNK. J Cell Physiol 217(3):652–666
110. Osborn JF, Newesley H (1980) Dynamics aspects of the bone-implant interface. In: Heimke G (ed) Dental implants–material and systems. Hanser, Munich, pp 111–123
111. Davies JE (1998) Mechanisms of endosseous integration. Int J Prosthodontics 11(5): 391–400

112. Orr RD, deBruijn JD, Davies JE (1991) Scanning electron microscopy of the bone interface with titanium alloy and hydroxyapatite. Cells Mater 2:241–251
113. Davies JE, Baldan N (1997) Scanning electron microscopy of the bone–bioactive implant interface. J Biomed Mat Res 36:429–440
114. Zhou H, Chemecky R, Davies JE (1994) Deposition of cement at reversible lines in rat femoral bone. J Bone Miner Res 9:367–374
115. Schwartz Z, Shani J, Soskolne WA, Touma H, Amir D, Sela J (1993) Uptake and biodistribution of technetium-99 m-MD32P during rat tibial bone repair. J Nuc Med 34:104–108
116. Albreksston T, Hansson HA (1986) An ultrastructural characterisation of the interface between bone and sputtered titanium or steel surfaces. Biomaterials 7:201–205
117. Klinger MM, Rahemtulla F, Prince CW, Lucas LC, Lemons JE (1998) Proteoglycans at the bone-implant interface. Crit Rev Oral Biol Med 9(4):449–463

Studying Cell–Surface Interactions In Vitro: A Survey of Experimental Approaches and Techniques

Stefanie Michaelis, Rudolf Robelek and Joachim Wegener

Abstract A better understanding of the interactions of animal (or human) cells with in vitro surfaces is the key to the successful development, improvement and optimization of biomaterials for biomedical or biotechnological purposes. State-of-the-art experimental approaches and techniques are a prerequisite for further and deeper insights into the mechanisms and processes involved in cell–surface adhesion. This chapter provides a brief but not complete survey of optical, mechanical, electrochemical and acoustic devices that are currently used to study the structural and functional properties of the cell–surface junction. Each technique is introduced with respect to the underlying principles before example data are discussed. At the end of the chapter all techniques are compared in terms of their strengths, limitations and technical requirements.

Keywords Cell–surface interactions · Cell–surface junction · Cell adhesion · Cell migration · Cell spreading · Electric cell–substrate impedance sensing · Micromotion · Surface plasmon resonance · Quartz crystal microbalance

Abbreviations

ECIS	Electric Cell–Substrate Impedance Sensing
ECM	Extracellular Matrix
RICM	Reflection Interference Contrast Microscopy
SPR	Surface Plasmon Resonance
TIRF	Total Internal Reflection Fluorescence
TIRAF	Total Internal Reflection Aqueous Fluorescence
QCM	Quartz Crystal Microbalance

S. Michaelis · R. Robelek · J. Wegener (✉)
Institut für Analytische Chemie, Chemo- und Biosensorik,
Universität Regensburg, Universitätsstr. 31,
93053 Regensburg Germany
e-mail: Joachim.Wegener@chemie.uni-regensburg.de

Contents

1 Cell–Surface Interactions from Two Perspectives 34
2 Hallmarks of Cell Adhesion on In Vitro Surfaces 35
3 Molecular Architecture of Specific Cell–Surface Interactions.. 37
4 Experimental Techniques for Studying Cell–Surface Interactions................................... 39
    4.1 Optical Methods for Studying Cell–Surface Interactions 40
    4.2 Mechanical Methods for Studying the Stability of Cell–Surface Interactions 46
    4.3 Electrochemical Approaches for Studying Cell–Surface Interactions..................... 49
    4.4 Acoustic Techniques for Studying Cell–Surface Interactions 55
5 Synopsis... 64
References... 64

1 Cell–Surface Interactions from Two Perspectives

Cells interact with their environment in many different ways: (1) they generate and withstand mechanical stress; (2) they secrete and sense individual signal molecules or molecular cocktails; and (3) they can establish and perceive electrical signals. In a multi-cellular organism all of these or just a subset may provide important clues for a cell to differentiate into one specific phenotype with site-specific functionalities that are important for the organism as a whole. The interactions of cells with their extracellular environment mediated by cell-surface receptors are of paramount importance. Besides their mechanical importance for processes like cell migration during development and wound healing, they provide the basis for inside-out and outside-in signaling. The non-cellular extracellular environment, summarized as the extracellular matrix (ECM), is a complex multi-component mixture consisting of (glyco)-proteins, carbohydrates, low-molecular-weight compounds, electrolytes and water [1]. The macromolecules interact with each other forming a two- or three-dimensional fibrous network that plays a crucial role in tissue homeostasis, mechanics and functionality. Thus, from this biological perspective the interactions of cells with biomaterial surfaces within the organism are critically important for both the cell and the organism.

When cells are isolated from the organism and transferred to an in vitro environment for biomedical or biotechnological purposes, they may lose their specific differentiation and functions due to the absence of the three-dimensional tissue architecture and important molecular clues. In order to maintain the cellular phenotype in vitro for research, medical approaches or biotechnology applications it is important to provide a biocompatible environment. Besides the chemical composition of the growth medium, it is the surface of the cell culture vessels that is critical for cell survival and fate. In particular, the in vitro culture of anchorage-dependent cells, which undergo apoptosis unless they can find proper sites for cell adhesion, relies on tailored in vitro surfaces. Thus, from the perspective of biotechnology, there is a strong need to understand, develop and refine the properties of in vitro surfaces such that growth and differentiation of the cells is not affected—or even guided in a certain direction. There is not, however, one ideal surface that is well-suited for all kinds of cells and for all scenarios. The most appropriate material depends certainly on the cell type and also on the cellular property that the in vitro surface is supposed to support and to foster, such as cell adhesion, proliferation, differentiation, motility and expression of tissue-specific genes, to mention just a few.

Even though some general correlations between the physico-chemical properties of a given surface and its performance as a support for cell adhesion and growth have been established, there is no in-depth understanding of which surface features influence which cellular function. Many more systematic studies need to be done and they all rely on techniques capable of studying the cell–material interface from different perspectives. This chapter summarizes the established and some of the emerging techniques of analyzing the interface between cells and man-made surfaces. They comprise optical, electrochemical and acoustic approaches, which are compared and categorized at the end of the chapter.

2 Hallmarks of Cell Adhesion on In Vitro Surfaces

Cells do not interact directly with the surface of a man-made material but with a pre-adsorbed layer of extracellular biomolecules, mostly proteins from the ECM [1]. As a direct consequence, the adhesiveness of the surface for ECM-proteins is the first prerequisite for cytocompatibility and mainly determines the behavior and compliance of an in vitro surface in a physiological environment. However, when a man-made material is brought in contact with a biological fluid (e.g. blood, lymphatic fluid or cell culture medium), the surface initially encounters water molecules. These bind rapidly to the surface, establishing a water mono- or bi-layer. The specific arrangement of water molecules depends on the surface properties on the atomic level. Highly reactive surfaces lead to the dissociation of H_2O and form a hydroxylated, i.e. OH-terminated surface. Less reactive surfaces can interact with H_2O molecules by hydrogen bonding, leaving the water as intact, undissociated molecules. Surfaces that show either of these behaviors are termed *wetting* or *hydrophilic* surfaces. On the other hand, surfaces with a weak tendency for binding H_2O are termed *non-wetting* or *hydrophobic*. After the formation of this adsorbed water layer (adlayer), which occurs within nanoseconds, hydrated ions such as Cl^- and Na^+ get incorporated. The specific arrangement of these ions and their water shells is strongly influenced by the properties of the surface.

Subsequently, proteins from the biological fluid adsorb to the surface in a complex series of events, including initial adsorption, conformational changes and eventually replacement of smaller proteins by larger ones. In experiments in vitro these proteins originate from the serum-containing culture medium and/or they are synthesized and secreted by the cells themselves. Depending on the properties of the surface, the resulting mixture of proteins on the surface, their conformational

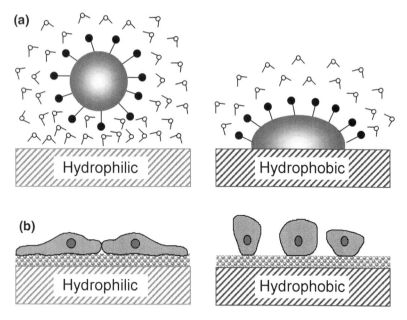

Fig. 1 Hydrophilic or hydrophobic in vitro surfaces in contact with a biological environment. **a** Protein adsorption, **b** cell adhesion (↶water molecules, ●┃polar amino acid side chains)

state and their orientation will be different [2]. Surface wettability is regarded as one of the most important surface parameters governing protein adsorption [3]. A common observation is that hydrophilic and hydrophobic surfaces bind proteins differently, i.e. proteins may adsorb intact or may undergo unfolding to minimize the free energy of the system (Fig.1a). Water-soluble proteins in a physiological environment commonly show a globular shape with a hydrophobic core and hydrophilic and charged amino acid side chains exposed to the solution. Thus, on hydrophilic surfaces protein adsorption occurs through polar and ionic interactions (Fig.1a, left panel). No conformational changes are induced and the proteins bind in their native conformation with intact water shells [4]. This leads to a rather weak, mostly reversible protein adsorption.

On hydrophobic surfaces the proteins are often irreversibly bound due to dehydration of the interface and the associated absence of intervening water shells. Dehydration of both the substrate and the protein surface provides an entropic driving force for the adsorption on hydrophobic surfaces. This leads inevitably to a significant rearrangement of the protein conformation with partial or total unfolding. The hydrophobic amino acids of the protein core are exposed to the substrate surface to allow for hydrophobic interactions with the surface [5]. Most of the polar and charged amino acid residues are oriented towards the aqueous solution (Fig.1a, right panel). The degree of the surface-induced conformational change mirrors the balance between the strength of protein–surface interactions and the internal conformational stability of the protein [6]. The adsorbed protein layer forms within the first few seconds of contact between the surface and the biological environment and determines the compliance of the surface with subsequent cell attachment and spreading.

After the adsorption of proteins, cell attachment and spreading are initiated by nonspecific interactions between the cells and the protein-decorated surface (Fig. 1b). These comprise electrostatic, electrodynamic, steric and entropic interactions. The first two are predominantly attractive in nature and based on the presence of fixed charges and dipoles on both the cell surface and the substrate surface. On the other hand, close adhesion between cell and surface requires the compression of the glycocalix decorating the cell membrane and the surface-attached protein layer which gives rise to steric and entropic repulsion. Once the balance of nonspecific interactions has provided sufficiently close proximity, specific interactions between cell–surface receptors and the surface immobilized proteins are established and provide mechanically stable substrate anchorage of the cells [7, 8]. The most prominent class of cell-surface receptors involved in cell adhesion and spreading is the integrin family, which will be discussed in more detail in the subsequent section.

Whether or not stable substrate anchorage occurs depends on (1) the expression of integrins with affinity for the extracellular proteins pre-adsorbed on the surface and (2) the composition of the surface attached protein layer and the conformation of the adsorbed proteins. These two conditions eventually determine the fate of cells settling upon an in vitro surface: the cells will start to attach firmly and spread, maximizing their interface with an adhesive surface (Fig.1b, left panel) or they will stay in a rounded morphology, loosely attached and unable to spread, when specific interactions cannot be formed (Fig. 1b, right panel). The latter will drive anchorage-dependent cells towards apoptosis. Generally speaking, while hydrophilic surfaces promote cell adhesion due to their coating with a native protein layer, hydrophobic surfaces covered with a layer of unfolded protein often counteract cell adhesion since specific recognition sequences within the extracellular proteins are not accessible to the cell-surface receptors.

3 Molecular Architecture of Specific Cell–Surface Interactions

The most prominent type of transmembrane receptors responsible for specific cell–substrate interactions are the integrins [9]. Integrins are a family of non-covalently associated, α,β-heterodimeric transmembrane glycoproteins that project from the cell membrane by roughly 20 nm [10]. To date, 24 different integrins have been identified, resulting from different combinations of 18 α- and 8 β-subunits [11].

Both subunits exhibit some structural similarity: each is composed of a long stalk-like extracellular segment with a globular domain at the N-terminus. The N-terminal domains of both subunits combine to form the specific ligand binding site that interacts with ECM proteins in the presence of divalent cations (Mg^{2+} or Ca^{2+}).

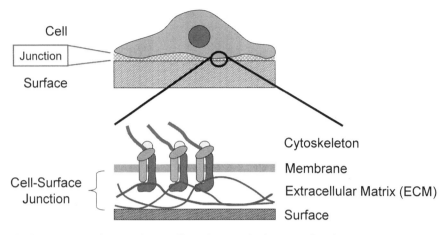

Fig. 2 Contact area between lower cell membrane and substrate surface forming the *cell–surface junction*. Adhesion is provided by cell-surface receptors that specifically bind to components of the ECM pre-adsorbed upon the substrate surface (adapted from [12])

Both subunits share a hydrophobic domain, which spans the cell membrane, and rather small cytosolic C-terminal domains. On the intracellular side, the β-subunit of the integrins is linked to the actin cytoskeleton [13, 14] by means of cytoskeletal adapter proteins (Fig. 2), such as talin, vinculin, α-actinin and paxillin. In contrast to all other integrins, the $\alpha_6\beta_4$ integrin associates with the keratin intermediate filaments (hemi-desmosomes). Thus, integrins interconnect the intracellular protein filaments of the cytoskeleton with the protein filaments of the ECM and serve as a transmembrane bridge between these two macromolecular networks (Fig. 2), providing mechanical stability to the cell–surface junction. The distribution of integrins in the plasma membrane of adherent cells is very often not homogeneous. After ligand binding, they tend to cluster locally, forming so-called focal adhesions or focal contacts [1, 15]. At these adhesion sites, the cells are believed to have the closest distance to the surface.

Depending on their subunit composition, integrins differ significantly with respect to their specificity for different ECM proteins. Molecular recognition and binding of individual ECM proteins are generally mediated by rather short amino acid sequences ($\sim 4 - 10$ amino acids) within the primary structure of ECM proteins. The most well-known amino acid sequence involved in integrin recognition is the tetrapeptide binding motif Arg-Gly-Asp-Ser (RGDS), a sequence found in many ECM ligands including the ECM proteins fibronectin and vitronectin [14,16]. Many integrins are multispecific receptors, meaning that they can bind several different ECM proteins as long as these carry a suitable recognition sequence. On the other hand, one particular ECM protein may interact with various integrins by carrying more than one recognition sequence in its primary structure.

Apart from their crucial functional role in cell adhesion and linkage of the cytoskeleton to the ECM [10], integrins act as bi-directional signaling receptors that mediate information transfer across the plasma membrane and thereby regulate various cellular processes. In *outside-in* signaling ligand binding is transmitted into the cell interior by conformational changes of the receptor carrying information that mediates cell growth, differentiation, proliferation, migration, morphology and survival [1, 13, 17–20]. In addition, integrins transfer signals from the cells to the ECM, a process termed *inside-out* signaling. This process is mainly involved in regulation of integrin conformation, ligand-binding affinity and ECM remodeling [19].

4 Experimental Techniques for Studying Cell–Surface Interactions

The constant improvement in biomaterial synthesis and surface modifications has provided various different synthetic materials that should be tested and screened for their ability to promote cell adhesion and to support or even induce cellular functionalities. A set of experimental approaches has been used in the past to evaluate the cyto-compatibility of a given surface. These established techniques cover a significant range of technical sophistication comprising low and high tech. On the low-tech side, for instance, the number of cells that has adhered to a surface under study within a given time is quantified by simple cell counting upon microscopic examination. For cell proliferation studies, this measurement is repeated at regular intervals. Apart from counting, the amount of cells adhering to a surface under study can be determined by photometry after intracellular uptake of membrane-permeable dyes or other methods of cell staining or biochemical assays. The colorimetric MTT assay is an established in vitro assay for evaluating the cyto-compatibility of biomaterials by measuring the metabolic activity of the cells in contact with the surface. It is based on the intracellular reduction of a colorless tetrazolium salt to colored formazan, which only occurs in metabolically active cells with sufficient supplies of reducing agents ($FADH_2$, NADH). The amount of the colored formazan is proportional to the number of vital cells and can easily be quantified by photometry.

More on the high-tech side are high-resolution microscopic techniques that are capable of imaging the morphology of cells in contact with a surface under study, such as scanning force microscopy (SFM) or scanning electron microscopy (SEM). Both approaches provide detailed images of the upper cell surface with a high spatial resolution. Other microscopic approaches can also be used to image the cells on the surface at different resolutions and contrast.

All the methods mentioned above do not have direct access to the interface between the lower cell membrane and the substrate that the cell is adhered to. We refer to this interface as the *cell–surface junction* (cf. Fig. 2). However, most of the processes important for adhesion, spreading or migration of cells are localized at this particular interface between cell and surface. Thus, all experimental techniques capable of reporting from this hidden area should be very useful for studying all aspects of cell–substrate adhesion. Moreover, most of the techniques mentioned above are invasive in nature and only provide an endpoint analysis. In order to get insight into the dynamics of cell–surface interactions, non-invasive approaches are required capable of recording the different steps of cell–material encounter with a reasonable time resolution.

In the following paragraphs we will highlight a few techniques and approaches that are particularly valuable for studying the cell–surface interface. Most of the techniques are non-invasive in nature; all of them report directly from the cell–surface junction. They are based on optical, mechanical, electrical or acoustical principles and are grouped accordingly. This survey does not claim completeness but picks the most valuable techniques in the authors' judgment.

4.1 Optical Methods for Studying Cell–Surface Interactions

4.1.1 Reflection Interference Contrast Microscopy

Reflection interference contrast microscopy (RICM) is capable of visualizing the contact area between living cells and a transparent substrate, providing something like the "footprints" of cells rather than their projections. It has been used extensively to study cell adhesion dynamics [21]. In RICM, cells are grown on a glass coverslip which is placed under an inverted microscope and is illuminated from below by monochromatic light using an objective with high illumination numerical aperture (INA). The RICM image results from light that is reflected at interfaces between media of different refractive indices like the glass/liquid and the liquid/cell membrane interfaces. When the incident light hits the transparent substrate at a cell-free area, a fraction of the incident light is reflected at the glass/liquid interface (Fig. 3a). The intensity of the reflected light depends on the difference in the refractive indices of the two adjacent media. Since this difference is more significant for the glass/liquid interface compared to any of the other interfaces of the sample, the reflection is relatively strong, which makes cell-free areas of the sample appear bright in RICM images (Fig. 3b). In cell-covered areas, the incident light is also reflected at the glass/liquid interface but here the reflected light is modulated by interference. The fraction of the incident light passing through the glass/liquid interface is reflected at the liquid/cell membrane interface. Due to the close proximity of these two interfaces (10–200 nm), both reflected light beams are partly coherent and interfere. Thus, the intensity of the reflected light—or the overall brightness of the image in cell-covered areas-depends on the optical path difference of the two reflected light beams. Reflections from interfaces deeper in the sample cannot modify the contrast of RICM images, as the condition of local coherence is not valid for points within the sample that are further away from the surface than approximately 100 nm. Taken together, the brightness of RICM images in cell-covered areas is a function of the distance between the lower cell membrane and the surface [22].

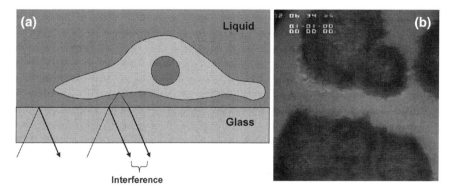

Fig. 3 **a** Image formation in RICM. Light reflected from the glass/liquid interface and the liquid/cell membrane interface is partly coherent and interferes. The image contrast depends on the optical path difference between the two light beams and, thus, the cell–substrate separation distance. **b** Typical RICM image of a cell after attachment and spreading on a glass coverslip

In the corresponding RICM image (Fig. 3b) the contact area of a cell with a substrate appears dark against a homogeneous grey and cell-free background, whereas the intensity of the dark regions themselves depends on the local distance of the lower cell membrane from the substrate surface. Besides providing static information about the cell–surface junction, this technique can also record dynamic processes, such as cell motion, due to its non-invasive nature [23]. However, it is very tricky to extract absolute distances between cell membrane and surface from RICM images, as the refractive indices of the different layers of the sample are very critical parameters during analysis but hard to measure with sufficient precision.

4.1.2 Fluorescence Interference Contrast Microscopy

Another microscopic technique imaging the cell–surface junction is fluorescence interference contrast microscopy (FLIC), which was introduced by Braun and Fromherz in 1997. This technique is capable of quantifying the exact cell–substrate separation distance [24]. Cells are grown on silicon substrates with steps made from silicon dioxide on their surface. The steps have at least four different, known heights ranging between 20 and 200 nm (Fig. 4a). After attachment and spreading of cells on a FLIC substrate, the cell membranes are stained with a lipophilic fluorescent dye and the sample is examined in an upright fluorescence microscope. FLIC microscopy is based on the effect that the fluorescence intensity of the fluorophore in the substrate-facing membrane is modulated by the silicon/silicon dioxide interface which behaves like a mirror.

During illumination, standing waves of the incident light are formed with a node at the silicon surface. Thus, the intensity of fluorophore *excitation* is dependent on the distance between fluorophore (cell membrane) and silicon. The fluorescent light emitted by the fluorophore upon excitation is collected from the objective lens of

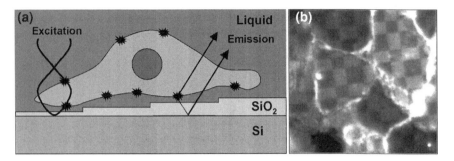

Fig. 4 a Schematic illustrating image formation in FLIC microscopy (adapted from [24]). **b** Fluorescence micrograph of a cell grown on a FLIC substrate

the microscope either directly or after reflection at the silicon/silicon dioxide interface. Since both the direct and the reflected fluorescent light are partly coherent, interference occurs so that the intensity of the fluorescence emission is also modulated by the optical path difference between membrane and silicon surface. Taken together, the intensity of fluorophore excitation and the intensity of the resulting fluorescence light are a function of the cell–substrate separation distance. However, the relationship between the relative fluorescence intensity and the distance of the fluorophore to the silicon substrate surface is not unique but a damped periodic function. The four different steps of silicon dioxide, serving as well-defined spacers between the cell membrane and the reflecting silicon surface, are used to provide four data pairs. The four different fluorescence intensities (cp. Fig. 4b) are analyzed using an optical theory, providing a distinct cell–substrate separation distance with unprecedented precision of 1 nm [24, 25]. However, FLIC is not a label-free method and the cells may experience phototoxicity when repeated experiments are performed to follow dynamic processes at the cell–surface junction.

4.1.3 Total Internal Reflection (Aqueous) Fluorescence Microscopy

Total internal reflection fluorescence microscopy (TIRF) [26] and total internal reflection aqueous fluorescence microscopy (TIRAF) [27] are additional microscopic techniques for visualizing the cell–surface junction of living cells as long as they are grown on a transparent substrate. Both TIRF and TIRAF are subsumed under the generic term evanescent field microscopy. In contrast to RICM and similarly to FLIC, either the cell membrane (TIRF) or the incubation fluid (TIRAF) requires fluorescent labeling. The cells under study are grown on a transparent substrate that is illuminated from below with a laser beam. The laser beam is aligned in such a way that it strikes the glass/liquid interface at an angle bigger than or equal to the critical angle of total internal reflection θ_{crit} (Fig. 5a). Due to diffraction phenomena at the interface between an optically thicker and an optically thinner medium, an evanescent electric field is generated at the surface facing the liquid. Fluorophores attached to some component of the cell (TIRF) or added to the bathing fluid (TIRAF) are excited by the evanescent field. The evanescent field decays exponentially with the distance from the substrate surface. The penetration depth is rather short and is in the order of 100 nm. Fluorophores residing deeper inside the sample than the penetration depth of the evanescent field are not excited. Thus, only fluorophores quite close to the surface contribute to TIRF and TIRAF images.

In TIRF microscopy, transmembrane proteins such as integrins are commonly labeled by a fluorescent tag so that their distribution within the cell–surface junction can be analyzed. For TIRAF microscopy, the extracellular fluid is stained with a water-soluble fluorescent dye instead of staining the cell membrane. When the cells attach and spread, the cellular bodies displace the aqueous phase with the dyes from areas of close cell-to-substrate adhesion [27]. Consequently, cell-covered areas appear dark in TIRAF images, in contrast to TIRF images. Figure 5b shows a typical TIRAF image from the cell–surface junction of an adherent fibroblast. The image shows a non-uniform adhesion along the contact area.

4.1.4 Surface Plasmon Resonance

Surface plasmon resonance (SPR) spectroscopy is another experimental approach to study cell–surface interactions and it is also based on evanescent electric fields. As with the other evanescent wave techniques, the penetration depth of SPR using visible light is below 200 nm. Thus, the sensitivity is confined to the interface between cell and substrate whereas the technique is blind to processes that occur deeper in the sample. It is therefore ideally suited to monitoring cell–substrate interactions, in particular when time-resolved measurements are required [28]. SPR is an emerging technique as far as cell–substrate interactions are concerned but it has a long history as a transducer in biomolecular interaction analysis [29].

As explained before for TIRF and TIRAF, the surface plasmon resonance technique is also based on the phenomenon of total internal reflection and the generation of an evanescent electric field (cf. Figs. 5, 6). In SPR the latter is, however, used to excite surface plasmons (i.e. electron density fluctuations) in a thin layer of a noble metal (most often gold) that is coated on the interface at which total internal reflection occurs. However, surface plasmons are only excited if the resonance condition is precisely met [30]. The resonance condition depends on the angle of incidence, the wavelength of the incident light and the refractive index close to the metal surface. With constant instrument parameters, SPR measures the changes in refractive index of thin layers of inorganic, organic and biological material adsorbed on the thin noble metal surface. As such it has become a very versatile surface-sensitive technique with a myriad of applications.

For a given wavelength of incident light, the excitation of surface plasmons is seen as a dip in intensity of reflected light at a specific angle of incidence (Fig. 6a). This fact opens a whole field of possible optical configurations by which the relationship between reflected light intensity, incident angle and excitation wavelength can be exploited to result in label-free spectroscopic and microscopic

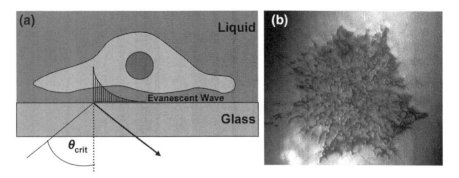

Fig. 5 a Schematic illustration of image formation in TIRF/TIRAF microscopy. **b** Typical TIRAF image of the cell–substrate junction (adapted from [27])

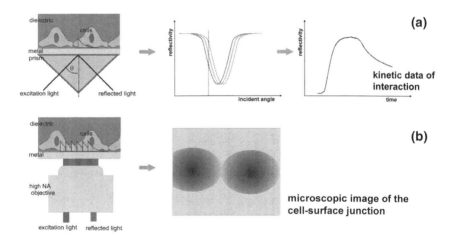

Fig. 6 a Kretschmann configuration for performing integral SPR analysis of the cell–surface junction. Surface plasmon excitation is recorded as a dip of the reflectivity as a function of incident angle. During kinetic measurements the changes in reflectivity are acquired at a constant angle of incidence. **b** Setup for SPR imaging based on a high numerical aperture objective

approaches for sensitive and time-resolved analyses of cell–substrate interactions. Most of these approaches are based on the so-called Kretschmann configuration in which the excitation light is coupled into the metal surface via a high refractive index prism for spectroscopic (Fig. 6a) or a high numeric aperture objective for microscopic applications (Fig. 6b). Spectroscopic analyses are performed by either measuring the changes in reflectivity at a constant (monochromatic) excitation wavelength but variable angle of incidence (angle-dependent mode) or by measuring the reflectivity at a constant angle of incidence but with polychromatic excitation and subsequent spectral analysis of the reflected light (wavelength-dependent mode) [29]. In order to perform SPR-based microscopy, a collimated monochromatic light path is used to excite the whole field of view and the reflected light is recorded by a CCD chip [31]. In this configuration, local changes in refractive index close to the sensor surface are visualized as a microscopic picture (Fig. 6b).

Taken together, there are two ways of studying cell–substrate interactions by SPR: (1) Changes in the refractive index averaged over the entire illumination spot are recorded as a single parameter that integrates over all processes that occur within the evanescent field. (2) SPR-generated reflectivity differences are sampled with lateral resolution and converted into microscopic pictures. Initial SPR studies addressing cell–substrate interactions were conducted in the spectroscopic mode (1) by Yanase et al. in 2007 [32], reporting on their pioneering experiments to grow adherent cells and immobilize suspended cells on SPR sensors. In a subsequent report the group correlated the cell-induced changes in SPR signals and light microscopic images, providing the first correlation between SPR signal strength and the area of cell–surface adherence. The refractive index as an integral parameter that changes when cells—or parts of cells—enter or leave the evanescent field or simply change their morphology was extensively discussed [33]. Cuerrier and Chabot applied SPR successfully to a label-free, time-resolved analysis of changes in cell–cell and cell–surface interactions of human embryonic kidney (HEK) cells when these were stimulated with toxins or physiological agonists or antagonists of cell-surface receptors [34, 35]. Phase-contrast microscopy was used to support a direct correlation of the SPR signal and the cellular reaction which led to changes in cell–surface interactions. The studies clearly showed that SPR detects changes in cell–cell and cell–surface interactions with significantly more sensitivity than phase-contrast micrographs. All of these studies emphasize the pros and cons of the limited penetration depth of the evanescent field. On the one hand the limited decay length of the evanescent field shields off contributions to the signal that do not originate from the cell–surface junction but at the same time it provides a sensitivity problem for cells that do not adhere tightly to their growth surface. This problem has been overcome by the novel concept of FTIR–SPR, which was introduced by Golosovsky et al. [36]. Exciting the surface plasmons with infrared light results in a substantially higher penetration depth of up to 2.5 µm, as the penetration depth corresponds approximately to half the wavelength of the incident light. This setup is capable of conducting a more flexible but still sensitive real-time monitoring of the different phases of the formation of cell–substrate interactions during cell adhesion. Due to the novel quality of the SPR data recorded via infrared excitation, temporal fine structures during adhesion were observed that have not been revealed by other analytical techniques so far [37].

Only a few important studies of surface plasmon resonance microscopy (SPRM) have been published to date. Giebel et al. [31] used SPRM to study cell–substrate interactions of primary goldfish glial cells. Besides the qualitative information obtained from the recorded SPR micrographs about leading and tailing lamellipodia during cell migration, the average distance between surface and different parts of the cell bodies was extracted from the raw data. Comparing SPR data to that from other state-of-the-art microscopic techniques identified the SPRM

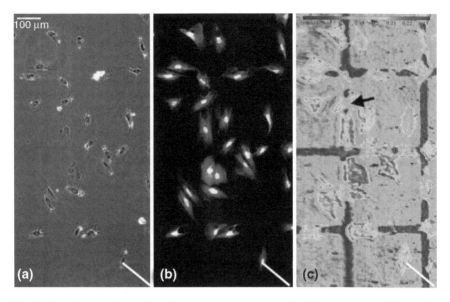

Fig. 7 Vascular smooth muscle cells grown on a substrate with 300 μm squares of fibronectin. The cells had been fixed prior to imaging. **a** Phase-contrast micrograph. **b** Fluorescence micrograph after Texas Red staining. **c** SPR-based micrograph (Adapted from [38])

as a suitable and in some respects even more powerful method for the visualization and quantification of cell–substrate interactions.

Only recently the group of Peterson et al. [38, 39] described an SPRM-based analysis of remodeling processes of the ECM performed by vascular smooth muscle cells during adherence, spreading and migration. By using a sophisticated optical setup and a growth surface carrying fibronectin patterns, the group could simultaneously collect data about the cell density and distance from the matrix as well as the amount of protein that was deposited or removed from the ECM, respectively. Figure 7c compares the SPRM image with conventional phase-contrast microscopy (Fig. 7a) and fluorescence microscopy (Fig. 7b) of the same field of view.

While the sensitivity of the system is remarkably high (~ 20 ng/cm^2), the lateral resolution of the micrographs is still low (~ 2 μm) compared to other microscopic techniques. Even though SPRM has not been used extensively to study cell–surface adhesion, its unique technical features and readouts may drive further applications.

4.2 Mechanical Methods for Studying the Stability of Cell–Surface Interactions

The mechanical stability of cell–surface interactions can be determined from so-called *detachment assays*. In these assays, substrate-anchored cells are exposed to mechanical forces that aim to detach the cells from the surface under study. The

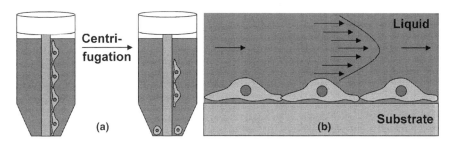

Fig. 8 Detachment assays used to determine the mechanical resistivity in cell–substrate interactions. **a** Centrifugation assay. **b** Hydrodynamic flow assay

more mechanical stress the cells can take without detachment, the more stable are the cell–substrate interactions with the growth surface. Depending on the assay, the mechanical stress is more or less defined and can be gradually increased to find the critical shear force for detachment.

The simplest approach to study cell adhesion strength requires seeding of suspended cells upon a substrate of interest. After a pre-defined incubation time, the surface is rinsed with a physiological washing buffer. Weakly attached cells adhering with a force smaller than the shear forces generated by the flow of buffer are washed from the substrate surface. The number of cells that remain attached to the surface is counted and serves as a measure for the adhesive interactions between surface and cells under study. Although these wash-off assays provide semi-quantitative information about the mechanical stability of cell–substrate interactions, the applied shear forces are ill-defined, difficult to control and of limited use [40].

Thus, more precise assays have been developed that use well-defined shear forces to probe the stability of cell–surface interactions in mature cell populations. According to the type of force application, these assays are classified as centrifugation assays [41, 42] and hydrodynamic shear force assays [43, 44].

4.2.1 Centrifugation Assay

The centrifugation assay (Fig. 8a) applies the shear forces necessary to probe the stability of cell adhesion by centrifugation—as the name implies. After cells are allowed to fully attach and spread upon a surface under study, the cell-covered surfaces are placed into centrifuge tubes filled with culture medium and are centrifuged at a preset angular velocity. The centrifugal force acts as a well-defined shear force parallel to the surface, generating tangential mechanical pull on the cell bodies. After centrifugation the cells which resisted the mechanical stress and remained attached to the substrate surface are counted using microscopic techniques. Repeated runs with increasing angular velocity provide the critical shear force that characterizes the mechanical stability of cell–substrate interactions for this particular pair of cells and substrate. A quantitative indicator that has been used to describe the strength of cell–substrate interactions is the centrifugal force necessary to detach 50% of the initial cell population. The centrifugation assay is, however, a low-throughput assay as only a single force can be applied per experiment. Moreover, in some cases the strength of cellular adhesion may exceed the forces that can be applied in these assays with reasonable effort.

Channavajjala et al. [41] applied the centrifugation assay to quantify the adhesion strength of tumor cells to immobilized HIV-1 Tat-protein, containing the amino acid sequence RGD, in comparison to other specific ECM proteins such as fibronectin and vitronectin. HIV-1 Tat was shown to mediate cell adhesion, but, unlike the ECM proteins, the interaction between cells and the surface-immobilized protein was mechanically weak.

4.2.2 Hydrodynamic Flow Experiment

In hydrodynamic flow experiments the cell-covered surface is placed in a laminar flow channel and the cells are challenged with increasing flow velocities of the fluid. As in the centrifugation assay, the flow of liquid generates mechanical forces on the cell body tangential to the surface, which may lead to detachment (Fig. 8b). Most frequently a parallel-plate flow apparatus is used. Here the opposing side of the channel is a parallel plate that moves with a preset velocity. A laminar shear flow is generated over the cell surface by viscous coupling of the liquid. In order to guarantee laminar flow the gap height between the parallel plates has to be small compared to the length of the flow path. In a parallel-plate flow chamber the shear stress is constant and depends on the flow rate and the gap between the two plates. Thus, the applied shear stress can be easily adjusted by altering one of these two parameters. After exposing the cells to laminar flow for a definite period of time, the number of adherent cells is counted and compared to the number before the onset of flow, providing the fraction of adherent cells that were capable of resisting a given laminar shear stress.

Using a laminar flow assay Xiao and Truskey [44] analyzed the adhesion strength of endothelial cells grown on a glass substrate that had been pre-coated with linear or cyclic RGD peptides as well as fibronectin. The critical shear stress, defined as the shear stress required to detach 50% of the cells from the coated substrate surface, was determined to be (59 ± 13) dyne/cm^2 for cyclic and (39 ± 4) dyne/cm^2 for linear RGD peptide. The value for fibronectin-coated surfaces was lower than those for the peptide coatings.

By mounting the flow chamber on the stage of an inverted phase-contrast microscope and using time-lapse video microscopy, the dynamics of cell detachment can be monitored simultaneously. However, a major limitation of these flow systems is that the detachment forces are usually non-uniform along the cell surface and cannot be calculated without simplifying assumptions about cell shape.

4.3 Electrochemical Approaches for Studying Cell–Surface Interactions

Experimental approaches based on electrochemical impedance analysis—also referred to as impedimetric approaches—are emerging and very versatile research tools for studying cell–substrate interactions in real time. The basic principle of this technique was introduced by Giaever and Keese in 1984 and has been continuously optimized ever since. In the initial publication the technique was named *electric cell–substrate impedance sensing* or short *ECIS* and it has paved the way for several modifications that are all based on the ECIS principles [45]. In ECIS, adherent cells are grown on the surface of planar gold film electrodes, which are pre-deposited on the bottom of a cell culture dish by thin-film technology. The gold films serve as growth substrate for the cells and, at the same time, as electrodes for the electrochemical measurement. With the cells adhering essentially to the electrode surface, there is only a gap of 20–200 nm between the cell bodies and the measurement probe. By virtue of this arrangement the measurement is particularly sensitive to changes that occur within the cell–surface junction in unprecedented detail.

In the most commonly used configuration the measurement system contains two electrodes: a small working electrode ($5 \times 10^{-4} cm^2$) and a substantially larger (~ 500-fold) counter electrode (Fig. 9). The electric circuit is completed by the cell culture medium on top of the cell layer. ECIS is based on measuring changes in the electrochemical impedance of the gold film electrodes at different AC frequencies (alternating current). As the cells behave essentially like insulating particles, they force the current to flow through the cells or around the cell bodies. Both situations result in an increase of the measured impedance due to the presence of the cell bodies on the electrode surface compared to a cell-free electrode.

When initially suspended cells are seeded on an ECIS electrode, it is possible to follow attachment and spreading of the cells upon the electrode surface from time-resolved impedance readings due to the gradual constriction of current flow. ECIS recordings are therefore particularly well-suited to follow the kinetics of cell spreading. Once the cells form a confluent monolayer the impedance becomes stationary as long as the cells do not change their shape. In this situation ECIS is capable of monitoring all experimental challenges that are mirrored by a change in cell shape. As many chemical, physical or biological stimuli result in minute changes of cell morphology, the technique is widely applicable in many research areas that will not be addressed here, such as cytotoxicity screening, GPCR-mediated signal transduction or stem cell differentiation. ECIS readings rely only on small amplitude currents and voltages such that the cells are not affected in any way by the electric field used for the measurement. It is considered to be a non-invasive approach even if the experiment spans several days or even weeks.

The method was originally developed using gold as the electrode material, and gold is by far the most widely used material due to its inertness, its biocompatibility,

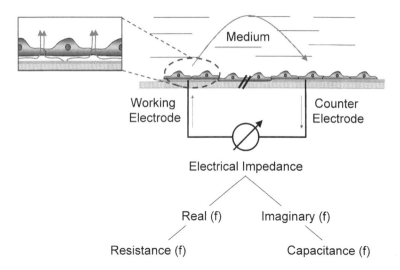

Fig. 9 Schematic of the ECIS principle, indicating AC current flow between the small working electrode and the larger counter electrode via the culture medium. The arrows, which indicate AC current flow, are drawn unidirectional only for the sake of clarity. The size of the electrodes is not drawn to scale with respect to cell size

its electrochemical properties and the well-established concepts for surface modifications. However, the method has been successfully transferred to other electrode materials such as platinum or indium tin oxide (ITO). The latter cases will not be covered in the following paragraphs, which will deal only with ECIS using gold film electrodes. The gold electrodes can be coated with individual components of the ECM or reconstituted preparations of native matrices without losing any sensitivity. Due to their inertness they can be modified by simple adsorption with almost any compound that has to be tested for its impact on cell–substrate adhesion. Moreover, there is an enormous tool box available for modifying the gold surfaces covalently by self-assembly reactions with compounds that carry thiol moieties as functional groups. ECIS is, however, of no use for polymeric or ceramic coatings as these are non conducting and they cannot be coated on the electrodes without losing the ability to perform ECIS readings.

Moreover, ECIS is very well suited to study cell–surface interactions on topographically structured substrates with specific topographical features. For this, inert substrates that carry the topography of interest have to be coated with a thin gold film in a well-defined electrode layout such that impedance readings can be performed with sufficient sensitivity. Characterizing the kinetics of cell attachment and spreading upon topographically structured in vitro surfaces might be of interest before these materials can be used as biomaterials in biosensors or implants.

4.3.1 Monitoring Cell Spreading Kinetics with High Time Resolution

The AC frequency used for ECIS recordings is an extremely important parameter as it determines the current pathway across the cell layer and, thus, the processes that are mirrored in the time-resolved signal. For the electrode size discussed here, the presence of the cells on the electrode surface alters the impedance of the electrode in the frequency range between 10 Hz and 100 kHz. Within this frequency range the current can flow along two different current pathways: (1) *around* the cell bodies via the cell–surface junctions and the cell–cell junctions into the bulk (*paracellular pathway*) or (2) *across* the plasma membranes and directly through the cell bodies (*transcellular pathway*). The first case describes approximately the flow of AC current for frequencies f < 10 kHz whereas the latter case describes the current pathway for frequencies f > 10 kHz. Thus, selecting the frequency determines where the current flows and as a consequence which part of the cell or which cellular processes are actually probed. A rule of thumb says that whenever morphological changes of the cells are the focus of interest, the measurement should be made sensitive for changes in the paracellular current pathway (f < 10 kHz). When coverage of the electrode is of interest—as in spreading and migration experiments—the measurement should be performed in the transcellular frequency regime (f > 10 kHz).

According to this rule of thumb, cell attachment and spreading is usually recorded in the high-frequency regime (>10 kHz). At these frequencies, the main part of the current passes capacitively through the cells, passing the basal and the apical cell membrane. For a more detailed analysis of cell spreading kinetics, not the impedance but the capacitive part of the complex impedance (cf. Fig. 9) is followed at a sampling frequency of 40 kHz. When the dielectric cell bodies attach and spread on the electrode surface, they decrease the equivalent capacitance of the electrode at 40 kHz *proportionally* to the fraction of the area they cover [46]. Measuring the capacitance of the system at 40 kHz as a function of time is therefore the most direct approach to monitor the coverage of the electrode surface with time, thus providing the spreading kinetics.

The following examples illustrate the analytical performance of the device. Figure 10a shows the kinetics of cell spreading for epithelial MDCK (Madin–Darby canine kidney) cells seeded on ECIS electrodes that were pre-coated with different ECM proteins [46]. The time courses of the individual electrode capacitances at a sampling frequency of 40 kHz show clear differences in the time to confluence on these different ECM proteins. The electrode capacitance decreases as the cells spread out on the electrode surface. Whereas cell attachment and spreading is fastest on a fibronectin-coated electrode, spreading on the non-adhesive serum albumin (BSA) takes significantly longer. Thus, the individual spreading kinetics provide quantitative information on the interaction of the cells under study with this particular protein coating.

Two parameters can be extracted to describe the adhesion and spreading kinetics on the different proteins quantitatively: the parameter $t_{1/2}$ provides the time required for half-maximal cell spreading and the parameter s stands for the

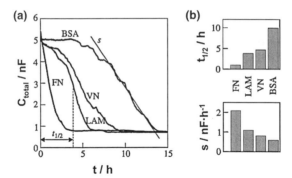

Fig. 10 Monitoring attachment and spreading of suspended cells by ECIS recordings. **a** Time course of the electrode capacitance at 40 kHz during attachment and spreading of initially suspended MDCK cells seeded at time point zero on ECIS electrodes pre-coated with different ECM proteins (FN = fibronectin, VN = vitronectin, LAM = laminin, BSA = bovine serum albumin). **b** Half-times $t_{1/2}$ and spreading rates s as determined from the data in a

spreading rate. The spreading rate is deduced from the slope of the curve at $t=t_{1/2}$ (Fig. 10a). It is directly proportional to the adhesion energy of the cells with a given surface composition [47]. $t_{1/2}$ values for the different protein coatings clearly mirror the time courses of the electrode capacitance, identifying BSA as the least adhesive protein with the highest $t_{1/2}$ value. However, the spreading rate s for BSA is close to the values for LAM and VN, indicating a similar adhesion energy for BSA compared to these two proteins (Fig. 10b).

This apparent discrepancy is readily explained by the onset of ECM production in the cells that were seeded on BSA-coated electrodes. The absence of adhesion sites on the surface triggers ECM production and secretion.

4.3.2 Monitoring Micromotion within Confluent Monolayers

Besides the kinetics of cell attachment and spreading, ECIS is also capable of recording metabolically driven cell shape fluctuations that have been referred to as micromotion [48]. Micromotion recordings integrate over transient and minute changes in cell–cell contacts, cell–substrate contacts, cell volume and cell membrane invaginations. The scale of these cell shape fluctuations can be in the sub-nanometer range and still be visible in ECIS measurements. Micromotion has been electrically recorded as small and rapid fluctuations of the impedance of cell-covered ECIS electrodes, when the impedance is tracked as a function of time at a single frequency sensitive for these movements. According to the rule of thumb described in the preceding paragraph, micromotion recordings are typically performed at intermediate frequencies, most often at 4 kHz. The higher the time resolution of the measurement the more of the inherent dynamics of the cell layer is revealed. Thus, micromotion recordings provide direct and indirect information

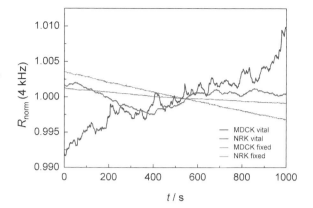

Fig. 11 Cellular micromotion as revealed in time-resolved measurements of the normalized resistance of an ECIS electrode covered with a confluent monolayer of MDCK (*blue*) or NRK cells (*red*) before (*thick lines*) and after fixation (*thin lines*) with formaldehyde. The resistance was measured at 4 kHz each time

about the cytocompatibility of a given surface—via direct contributions from the cell–surface junction to the observed impedance fluctuations, and indirectly since micromotion is a general indicator of cell viability [49].

Figure 11 compares the micromotion of two different cell lines before (thick lines) and after (thin lines) the cells were fixed with formaldehyde. Different cell lines show individual and characteristic "fingerprints" of their motility whereas dead cells no longer induce any significant resistance fluctuations anymore. Analysis of micromotion for different electrode coatings may provide valuable insights into the interactions between cells and a given surface.

4.3.3 Monitoring Cell Migration

Cell–surface interactions play an important role in the ability of a cell to migrate, as, for instance, during wound healing or embryonic development. The easiest assay for assessing cell migration on a given ECM in vitro is called the *wound healing scratch assay*: a confluent cell monolayer grown on the surface under study is mechanically wounded by scratching the tip of a pipette or a needle through the cell layer. The size of the lesions depends on the size of the needle. Cells from the periphery of the scratch migrate into the center of the wound and this process can be documented and analyzed microscopically over time. A weakness of this assay is the time-consuming analysis of the micrographs and the fact that the applied mechanical wounds are often hard to reproduce. The assay becomes significantly more reproducible and more convenient when the cell layer is established on ECIS electrodes such that the electrodes can be used to apply a lethal electric field and thereby wound those cells residing on the electrode surface. When the conditions of the electric wounding pulse are properly selected, all the cells on the electrode surface die but not those in the periphery of the electrode (Fig. 12). Migration of viable cells from the electrode periphery to the center of the electrode (*wound healing*) can be followed quantitatively by time-resolved ECIS

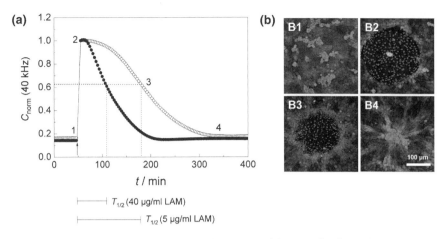

Fig. 12 ECIS-based cell migration assay. **a** Time course of the normalized capacitance measured at an AC frequency of 40 kHz along a complete wound healing/migration assay with NRK cells grown on ECIS electrodes pre-coated with 40 μg/mL (*filled circle*) or 5 μg/mL (*open circle*) laminin. Numbers 1–4 indicate the time points at which a vital stain of the cells on the electrode was performed (cf. Fig. 12b). $T_{1/2}$ is the time to reach half-maximal repopulation of the electrode. Capacitance data was normalized to the first value after electrical wounding (cell-free electrode). **b** Fluorescence micrographs of NRK cells stained with calcein AM and ethidium homodimer-1 at different points of the wound healing/migration assay (green = vital cells, red = dead cells; the staining was performed **1**: before, **2**: immediately after the wounding pulse, **3**: after 50% wound healing, and **4**: after complete wound healing)

readings [50]. In contrast to the mechanical scratch assay, the ECIS-based wound healing assay provides well defined and highly reproducible wounds, as they correspond to the size of the electrode.

Similar to cell attachment and spreading studies, the capacitance at a sampling frequency of 40 kHz is the most useful indicator for monitoring the repopulation of the electrode surface by cells migrating in from the periphery. Figure 12a shows the time course of the normalized capacitance at 40 kHz after wounding a confluent layer of NRK cells that have been grown on ECIS electrodes pre-coated with laminin (LAM) in different concentrations. The arrow at position 1 marks the time point when the invasive electric field was applied to the cells, killing the cells on the electrode. Immediately after pulse application the capacitance increases from its minimum value, which mirrors a confluent cell layer, to a typical reading for an open, cell-free electrode. The electrical permeabilization of the cell membranes allows the current to flow freely through the dead cells without a measurable capacitance contribution from the cell membranes. As time progresses C_{norm} continuously decreases back to the pre-pulse values as the electrode is gradually repopulated by cells that migrate in from the periphery. The rate of the capacitance decrease depends on the LAM concentration. Eventually, the capacitance reaches the stationary level of a cell-covered electrode again, indicating that the healing process is completed (closure of the wound). To characterize the assay readout for different electrode coatings by a single parameter, it is useful to

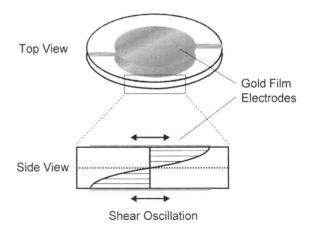

Fig. 13 *Top* and *side* views of a shear wave resonator as used in QCM-based experiments. The quartz resonator is sandwiched between two gold film electrodes used to drive the resonant oscillation and to read the resonance frequency. Under resonance conditions a standing acoustic wave is established between the crystal's surfaces. Resonance parameters are very sensitive to adsorption or desorption processes at the surface

determine the time $T_{1/2}$ needed to attain half-maximal capacitance decrease after wounding and, thus, half-maximal repopulation of the electrode (= wound healing). The slope of the capacitance versus time curves mirrors the migration velocity.

Fluorescence microscopic observation of the ECIS-based wound healing assay (Fig. 12b) provides images of the different stages of the wound healing process. A fluorescence-based viability assay based on ethidium homodimer-1 (EthD-1; red fluorescence) and calcein acetoxymethylester (CaAM; green fluorescence) was used to discriminate live and dead cells in the micrographs at the times indicated in Fig. 12a. The DNA-intercalating dye EthD-1 is a marker for membrane integrity as it is non-membrane-permeable and can only access the nuclei after membrane permeabilization. Calcein AM (CaAM) is essentially non-fluorescent but membrane-permeable. Intracellular esterases inside living cells hydrolyze CaAM to the membrane-impermeable calcein, which emits a green fluorescence. Before the high-field application, all cells exhibit a green cytoplasmic fluorescence, which thus indicates vital cells. After the elevated field is applied, all cells residing on the electrode are selectively wounded as indicated by their EthD-1 stained cell nuclei (red), while the cells surrounding the electrode remain vital, showing a green cytoplasmic fluorescence (Fig. 12b2). Figure 12b3 shows a fluorescence image of an electrode covered with NRK cells after half-maximal wound healing. A radial growth pattern in the cell layer near the electrode periphery can be observed as the cells have migrated inward, suggesting a re-alignment of the cells during the migration process. This pattern is even more pronounced for the image recorded after wound closure (Fig. 12b4).

4.4 Acoustic Techniques for Studying Cell–Surface Interactions

Several acoustic approaches have been described that are capable of providing valuable information about the formation and modulation of cell–surface interactions. By far the most widely known device is the quartz crystal microbalance

(QCM) that has a long track record as a mass-sensitive tool to study adsorption reactions at the solid–liquid interface. It operates non-invasively and with a superb time resolution that is much better than necessary for most cell-related studies. The core component of this technique is a thin, disk-shaped piezoelectric (AT-cut) quartz crystal sandwiched between two gold film electrodes. When an oscillating potential difference is applied between the surface electrodes, the piezoelectric resonator is excited to perform mechanical shear oscillations parallel to the crystal faces at the resonator's resonance frequency (Fig. 13). This mechanical oscillation is highly sensitive to changes that occur at the resonator surface, so that adsorption or desorption processes can be followed by readings of the resonance frequency f [51] or by analyzing the shear oscillation of the resonator using principles of impedance analysis [52–54].

For many years the QCM technique was used as an established and accepted tool for studying deposition processes of thin material films in the gas phase or in vacuum. As long as the adlayer film is rigid and homogeneous, the resonance frequency decreases in proportion to the amount of deposited mass [55], providing a balance with nanogram sensitivity. Recent progress in designing better oscillator circuits to determine the resonance frequency or alternative readout approaches has paved the way to monitor adsorption processes even in an aqueous environment—a prerequisite for the detection of protein adsorption or cell adhesion processes under physiological conditions.

4.4.1 Monitoring Attachment and Spreading on Protein-Coated Resonators

The most sensitive operational mode for a quartz resonator is the *active oscillator mode*. Here, the quartz resonator is integrated as the frequency-controlling element in an oscillator circuit and the resonance frequency of the crystal is recorded with high sensitivity and a time resolution of less than 1 s. The oscillator circuit only compensates for energy losses and maintains the quartz resonator at its resonance frequency. The general applicability of the QCM technique in the active oscillator mode for studying cell adhesion has been demonstrated by various authors addressing a variety of bioanalytical issues. Gryte et al. [56], Redepenning et al. [57] and Wegener et al. [58] monitored the attachment and spreading of initially suspended mammalian cells on the resonator surface in real time by readings of the resonance frequency. They showed that the attachment and spreading of mammalian cells upon the resonator surface induced a decrease in the resonance frequency that was proportional to the fraction of the surface area covered with cells (Fig. 14). Thus, time-resolved measurements of the resonance frequency mirror the kinetics of cell attachment and spreading on the resonator surface. To illustrate the quality of the data, Fig. 14a shows the time course of the resonance frequency shift when increasing amounts of MDCK cells are seeded on the resonator surface at time zero. After a transient slight increase of the resonance frequency due to warm-up of the medium, Δf decreases, reporting on the

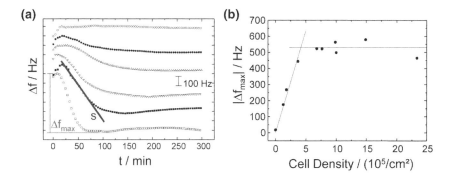

Fig. 14 a Shift in resonance frequency during attachment and spreading of initially suspended MDCK cells. From the upper to the lower curve seeding densities (in $10^5 cm^{-2}$) increase (*open circle* 0; *filled circle* 1.3; *triangle* 1.8; *filled down triangle* 3.7; *filled square* 7.7; *open square* 15). Δf_{max} indicates the maximum frequency shift observed for a given seeding density. **b** Maximum frequency shift Δf_{max} as a function of the cell density seeded on the resonator surface at time point zero. T = 37 C

continuous progress in cell attachment and spreading. The total shift in resonance frequency (Δf_{max}) increases with increasing seeding density up to a threshold value. This becomes obvious when the total frequency shift is plotted against the number of seeded cells, providing a saturation curve (Fig. 14b).

For low seeding densities the frequency shift is proportional to the fractional surface coverage with cells. However, when all adhesion sites on the surface are occupied by cells, the frequency shift does not further increase even though the number of seeded cells is further raised. This observation confirms that only those cells in direct contact with the resonator surface contribute to the QCM signal.

Wegener et al. [58], who studied the adhesion of different mammalian cell types, have additionally found that confluent monolayers of different cell types produce individual shifts in resonance frequency, possibly reflecting individual molecular architectures of their cell–substrate contacts. With a more detailed understanding of this cell-type-specific QCM readout, unprecedented information about the interactions of cells with in vitro surfaces will become available. This next step requires methodological improvements with more observables than just the resonance frequency. The resonance frequency of the quartz resonator is an integral parameter sensitive to both mass deposition and changes in the density or viscosity of the material in contact to the resonator surface [59]. Discrimination between these two contributions is not possible from readings of the resonance frequency alone. This, however, becomes important when the resonator is loaded with a complex material that neither behaves like a rigid mass nor has uniform contact with the surface. Cells are viscoelastic bodies for which the linear relationship between adsorbed mass and change in frequency is not valid [60, 61]. Rodahl et al. [61, 62] developed an extension of the traditional QCM technique,

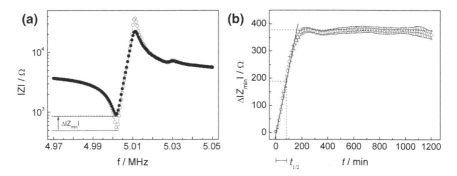

Fig. 15 a Impedance spectra near resonance of a cell-free (*open circle*) and a cell-covered (*filled circle*) 5-MHz quartz resonator. The arrow indicates the change of the minimal impedance Δ|Z$_{min}$|, which was used to follow cell attachment and spreading upon the resonator surface. **b** Time course of the minimal impedance Δ|Z$_{min}$| during the attachment and spreading of NRK cells upon the quartz resonator. Changes of the load parameters are given relative to a cell-free but medium-loaded resonator. t$_{1/2}$ is the time required to induce a half-maximal change of Δ|Z$_{min}$|. Mean ± SEM, n = 5; T = 37 C

the so-called QCM-D technique, which provides measurements of the dissipation factor D in addition to the common measurement of the resonance frequency f. The dissipation factor quantifies the damping of the quartz oscillation and is defined as the ratio of the dissipated energy to the energy that is elastically stored during one period of oscillation. Simultaneous f and D measurements are performed by periodically switching off the driving power to the quartz oscillator and subsequently recording the free decay of the quartz oscillation. The measured decay time of the damped sine wave is expressed as the dissipation factor, to which it is inversely proportional. Readings of D indicate whether the recorded shift in resonance frequency arises from dissipative processes, as occur with any viscous loading of the resonator surface [61]. Thus, simultaneous f and D measurements facilitate the detailed interpretation of QCM experiments.

Fredriksson et al. [63] and Nimeri et al. [64] used this approach to characterize the attachment and spreading of mammalian cells and the interaction of neutrophils with a protein-coated surface. Reiss et al. [65] used functionalized (biotin-doped) lipid vesicles and studied their adhesion onto a resonator surface coated with specific ligands (avidin) as a simple model system to mimic cell attachment and spreading. The adhesion of mammalian cells and liposomes gave rise to very similar shifts in resonance frequency, whereas the viscous energy dissipation was at least one order of magnitude higher when cells made contact with the resonator. As a conclusion of these studies it was obvious that the viscous properties of the cell–surface junction and the cell body had a great impact on QCM readings.

The second operational mode for QCM experiments is termed the *passive mode*. Here, the quartz resonator is not oscillating freely but a sinusoidal voltage is applied to the surface electrodes and the crystal is thereby forced to oscillate at frequencies determined by the frequency of the applied AC voltage. The quartz is excited sequentially along a narrow frequency band around its fundamental resonance and the electrical impedance of the system is recorded. Figure 15a shows a typical impedance spectrum for a medium-loaded resonator compared to the same resonator covered with a confluent monolayer of NRK cells.

The presence of cells on the resonator surface is most obviously expressed by a significant damping of the shear oscillation as indicated by shifts of the minimal ($|Z_{min}|$) and the maximal impedance ($|Z_{max}|$). Attachment of the cells to the resonator induces only a minor shift of the impedance spectrum along the frequency axis towards lower frequencies. This confirms that the cell monolayer induces primarily dissipation of motional energy and only a negligible storage of elastic energy.

Attachment and spreading of initially suspended cells to the resonator surface is followed over time by continuously recording impedance spectra such as the one shown in Fig. 15a. The gradual coverage of the surface is mirrored in the change of the minimal impedance magnitude $\Delta|Z_{min}|$, which is directly extracted from the raw data (blue horizontal lines in Fig. 15a), as a function of time (Fig. 15b). Figure 15b shows the time course of the minimal impedance $\Delta|Z_{min}|$ during the attachment and spreading of initially suspended NRK cells seeded to confluence upon the resonator surface, expressed relative to the value for the medium-loaded resonator. Immediately after cell inoculation (t = 0), $\Delta|Z_{min}|$ shows a characteristic steep increase, reaching a stationary value once spreading is complete. Since the cells were seeded to confluence into the measuring chamber, i.e. they already cover the entire quartz surface after sedimentation and adhesion without any need for further cell proliferation, the final change in $\Delta|Z_{min}|$ corresponds to a confluent monolayer of cells on the surface. The initial increase of $\Delta|Z_{min}|$ with time describes the kinetics of cell attachment and spreading, which is quantified by two parameters, $t_{1/2}$ and s. The quantity $t_{1/2}$ describes the time that is needed to reach the half-maximal change of $\Delta|Z_{min}|$, which corresponds to half-maximal surface coverage. The slope s of the attachment curve quantifies the spreading rate and is directly proportional to the adhesion energy of the cells.

With the help of these different operational modes it is possible to unravel several key features of any QCM-based analysis of cell–surface junctions:

1. QCM readings only report on specific, integrin-mediated cell adhesion to the resonator surface. Sedimentation and loose attachment of cells to the resonator surface via nonspecific interactions do not influence the QCM readout [58, 66].
2. QCM readings are only sensitive to those parts of the cellular body that are involved in making cell–substrate contacts and that are close to the resonator surface [66]. Thus, sensitivity is confined to the cell–surface junction.
3. The presence of a confluent cell layer on top of the resonator surface leads to a significant increase in viscous energy dissipation, usually many times (2- to 10-fold) larger than the increase in the stored energy [67]. The impact of cells on energy dissipation was shown to be cell-type-dependent reflecting individual acoustic properties.

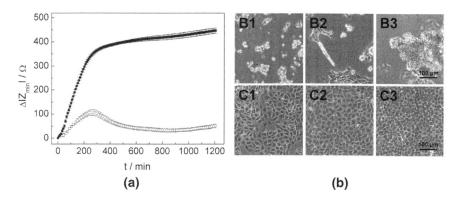

Fig. 16 **a** Time course of $\Delta|Z_{min}|$ after seeding equal numbers of initially suspended NRK cells on a polystyrene-coated quartz resonator at time point zero. The PS film on the resonator was either left unmodified (*open circle*) or had been exposed to an argon plasma (*filled circle*) prior to inoculation. The value of $|Z_{min}|$ at the beginning of the experiment was set to zero. (Mean ± SEM, n ≥ 5; T = 37 C). **b** Phase-contrast micrographs of NRK cells 10 h (**1**), 24 h (**2**) and 48 h (**3**) after inoculation on unmodified PS (*B1, B2, B3*) and plasma-treated PS (*C1, C2, C3*)

4. A major contributor to the QCM signal of adherent cells is the actin cytoskeleton [66]. Taken together with point (2) above, it is the part of the actin cytoskeleton associated with cell–surface junctions that contributes to the QCM signal and can be probed by QCM experiments.

4.4.2 Biocompatibility Testing of Polymer Surfaces

In contrast to the electrochemical approaches described in the preceding section, which are limited to conductive coatings of the electrodes with proteins or other biomolecules, the QCM-based analysis of cell–surface interactions is also applicable to polymeric, metallic or ceramic coatings of the resonator. Any of these materials can be coated as a thin film on the resonator surface and then resonance analysis is performed just as described above. The only requirements with respect to material deposition are homogeneity and rigidity of the coating such that no significant acoustic loss occurs within the adlayer, as this would reduce the inherent sensitivity of the resonator for subsequent cell adhesion studies. Figure 16a shows the time course of attachment and spreading of NRK cells after a single-cell suspension was seeded to confluence on a quartz resonator coated with a thin layer of polystyrene (PS). Spreading kinetics are again mirrored in the time course of $\Delta|Z_{min}|$. When initially suspended NRK cells are seeded on an unmodified PS surface (open symbols), the minimal impedance $\Delta|Z_{min}|$ shows only a minor and transient increase, indicating that specific interactions between cells and the underlying surface are not allowed to form. Apparently, the cells are not able to attach and spread properly on this surface. By contrast, after a selective hydrophilization of the PS surface by short-term exposure to an argon plasma, cell attachment and spreading occurs immediately (filled symbols in Fig. 16a). $\Delta|Z_{min}|$ increases to a transient maximum within 350 min after cell inoculation and even continues to increase to a final maximum. These results were confirmed by phase-contrast micrographs recorded 10, 24 and 48 h after seeding of the NRK cell suspension on both polymer surfaces (Fig. 16b).

On the unmodified PS surface only a small fraction of the seeded cells has spread on the surface within 10 h (Fig. 16b1), whereas the predominant fraction remains spherically shaped and forms clump-like aggregates floating in the culture medium. 24 h after cell inoculation, small cell islets have formed (Fig. 16b2) showing individual cells with a quite untypical, elongated morphology. 48 h after cell inoculation, no adherent cells can be observed on the unmodified PS surface but large aggregates of cell fragments and apoptotic cells are floating in the culture medium (Fig. 16b3). By contrast, 10 h after cell inoculation upon the plasma-treated surface, the biggest fraction of the surface is covered with spread cells (Fig. 16c1). As time progresses a confluent cell monolayer is established with a gradual increase in the cell number per unit area (Fig. 16c2, c3).

Compared to microscopic studies, an inherent advantage of the QCM approach is that it directly provides quantitative and time-resolved data of cell spreading kinetics without any need for complex and time-consuming image analysis. The inability of cells to attach and spread upon unmodified, hydrophobic PS is in good agreement with the literature [68–70] and might be related to the pre-adsorbed protein layer which influences the behavior of cells approaching the surface. On hydrophobic surfaces like unmodified PS, the soluble proteins are thought to adsorb under conformational rearrangement (denatured conformation) on the polymer surface, making the binding sites within the adsorbed proteins inaccessible to cell–surface receptors. Consequently, specific cell–substrate contacts cannot be established and cell adhesion is almost completely inhibited. On the other hand, PS surfaces after argon plasma treatment generally allow the proteins to adsorb without unfolding, rendering the surface cytocompatible. Consequently, the integrin binding sites within the adsorbed proteins should be accessible for the cell-surface receptors of the arriving cells. This single example shows the general applicability of the QCM approach to study the biocompatibility of different biomaterials, no matter whether they are polymers, metals or ceramics.

4.4.3 Probing the Cell–Surface Junction Under Shear Stress

In the preceding paragraphs the QCM-device has been described as a *sensor* to monitor the mechanical interactions between cells and surface. It can also be used as an *actuator* capable of disturbing or even dissolving molecular recognition at the solid–liquid interface, such as cell–surface contacts. In the *actuator mode*, the resonator is used at an elevated lateral shear amplitude. The latter is controlled by the driving voltage applied to the surface electrodes to excite the crystal. Heitmann and Wegener [71] studied the impact of elevated lateral oscillation amplitudes on the adhesion kinetics of different mammalian cells. By gradually increasing the

Table 1 Analytical techniques capable of probing cell–surface interactions

Technique	Information provided	Substrate	Strength	Limitation
RICM Microscopic/Optical	Images of contact area between living cells and growth substrate	transparent glass substrate	• non-invasive • high time resolution • readout confined to cell–surface junction	• areas with very thin cells or cell protrusions cannot be analyzed • cell–substrate separation distance cannot be quantified reliably
FLIC Microscopic/Optical	Distance between lower cell membrane and surface with high precision	silicon substrate with steps of silicon dioxide	• most precise determination of cell–substrate separation distance	• requires membrane staining by lipophilic membrane dye; • requires Si-substrate with steps of SiO_2
TIRF/TIRAF Microscopic/Optical	TIRF: identification of molecular components close to the growth surface after specific labeling TIRAF: images of contact area between living cells and growth substrate	transparent glass substrate	• readout confined to cell–surface junction • detailed image of cell–surface junction	• TIRF: requires fluorescent membrane staining • TIRAF: requires fluorescent staining of extracellular buffer
SPR Optical	Refractive index of material within hundreds of nm from the surface; used as integral readout parameter for cell coverage	gold film of precise thickness on glass substrate with specified refractive index	• label-free and non-invasive • high time resolution • readout confined to cell–surface junction	• interpretation of integral refractive index still difficult • limited parallelization
Flow-Induced Mechanical	Quantitative assessment of the shear resistance of cell–surface interactions	all substrates	• applicable to all substrates • direct mechanical information	• invasive • no time-resolved information • requires cell counting

(continued)

Table 1 (continued)

Technique	Information provided	Substrate	Strength	Limitation
ECIS Electrochemical	Cell spreading kinetics, micromotility of adherent cells, migration velocity in wound healing assay	planar gold film electrodes	• label-free and non-invasive • 96 samples in parallel • automated assay • real time approach	• restricted to conducting and electrochemically well-behaved electrode materials • sensitive to cell morphology as competing signal contribution
QCM Acoustic	Cell spreading kinetics, micromechanics of cell–surface junction, shear resistance of cell–surface interactions	piezoelectric quartz crystal sandwiched between two gold film electrodes	• label-free and non-invasive • real time approach • resonator can be coated with any thin and rigid material	• limited parallelization • sensitive to cellular micromechanics as competing signal contribution

amplitude of the resonator's shear displacement during cell attachment, the authors were able to determine a threshold lateral oscillation amplitude of ~ 20 nm in the center of the resonator (driving voltage: > 5 V) beyond which cell adhesion to the quartz surface was retarded or even entirely blocked. A maximum shear amplitude of 35 nm (driving voltage: 10 V) was sufficient to completely inhibit cell adhesion for all cell lines under study. However, shear oscillations of similar amplitudes were unable to disrupt cell–surface interactions within established cell layers. The cells could not be displaced from the surface once they had formed mature adhesions. In the long run, this method might develop into a new approach to probe the mechanical shear resistance of cell–surface interactions. As it is possible to coat the resonator with different biomaterials without losing its sensor function, the assay can be used to characterize a wide variety of materials.

5 Synopsis

Table 1 summarizes the various approaches and techniques for studying cell–surface interactions as they have been described in this survey. We have included the individual strengths and limitations of each technique according to our personal perspective and judgment. The survey is not exhaustive but focuses on those assays and concepts that have given or are expected to give significant input to a better understanding of cell–surface interactions.

Acknowledgments The authors would like to express their gratitude to the Kurt-Eberhard Bode Stiftung (Germany) for financial funding.

References

1. Davies JA (2001) Encyclopedia of Life Sciences. Wiley, Chichester www.els.net
2. Ziats NP, Miller KM, Anderson JM (1988) Biomaterials 9:5
3. Lee JH, Khang G, Lee JW, Lee HB (1998) J Colloid Interface Sci 205:323
4. Vogler EA (1998) Adv Colloid Interface Sci 74:69
5. Norde W, Haynes CA (1995) In: Horbett TA, Brash JL (eds) Proteins at interfaces II: fundamentals and applications, American Chemical Society, Washington, p 26
6. Roach P, Farrar D, Perry CC (2005) J Am Chem Soc 127:8168
7. Bongrand P (1998) J Dispersion Sci Technol 19:963
8. Pierres A, Benoliel AM, Bongrand P (2002) Eur Cell Mater 3:31
9. Arnaout MA, Mahalingam B, Xiong J-P (2005) Annu Rev Cell Dev Biol 21:381
10. Hynes RO (1992) Cell 69:11
11. Barczyk M, Carracedo S, Gullberg D (2010) Cell Tissue Res 339:269
12. Wegener J (2006) Wiley encyclopedia of biomedical engineering, vol 6. Wiley, Hoboken, p 1
13. Krissansen GW, Danen EH (2006) Encyclopedia of life sciences. Wiley, Chichester www.els.net
14. Krissansen GW, Danen EH (2007) Encyclopedia of life sciences. Wiley, Chichester www.els.net

15. Wozniak MA, Modzelewska K, Kwong L, Keely PJ (2004) Biochim Biophys Acta 1692:103
16. Pierschbacher MD, Ruoslahti E (1984) Nature 309:30
17. Bokel C, Brown NH (2002) Dev Cell 3:311
18. Cox EA, Huttenlocher A (1998) Microsc Res Tech 43:412
19. Hynes RO (2002) Cell 110:673
20. Luo BH, Carman CV, Springer TA (2007) Annu Rev Immunol 25:619
21. Verschueren H (1985) J Cell Sci 75:279
22. Curtis ASG (1964) J Cell Biol 20:199
23. Schindl M, Wallraff E, Deubzer B, Witke W, Gerisch G, Sackmann E (1995) Biophys J 68:1177
24. Braun D, Fromherz P (1997) Appl Phys A-Mater Sci Process 65:341
25. Lambacher A, Fromherz P (1996) Appl Phys A-Mater Sci Process 63:207
26. Truskey GA, Burmeister JS, Grapa E, Reichert WM (1992) J Cell Sci 103(Pt 2):491
27. Geggier P, Fuhr G (1999) Appl Phys A-Mater Sci Process 68:505
28. Su YD, Chiu KC, Chang NS, Wu HL, Chen SJ (2010) Opt Express 18:20125
29. Robelek R (2009) Bioanal Rev 1:57
30. Knoll W (1998) Annu Rev Phys Chem 49:569
31. Giebel KF, Bechinger C, Herminghaus S, Riedel M, Leiderer P, Weiland U, Bastmeyer M (1999) Biophys J 76:509
32. Yanase Y, Suzuki H, Tsutsui T, Uechi I, Hiragun T, Mihara S, Hide M (2007) Biosens Bioelectron 23:562
33. Yanase Y, Suzuki H, Tsutsui T, Hiragun T, Kameyoshi Y, Hide M (2007) Biosens Bioelectron 22:1081
34. Chabot V, Cuerrier CM, Escher E, Aimez V, Grandbois M, Charette PG (2009) Biosens Bioelectron 24:1667
35. Cuerrier CM, Chabot V, Vigneux S, Aimez V, Escher E, Gobeil F, Charette PG, Grandbois M (2008) Cell Mol Bioeng 1:229
36. Golosovsky M, Lirtsman V, Yashunsky V, Davidov D, Aroeti B (2009) J Appl Phys 105
37. Yashunsky V, Lirtsman V, Golosovsky M, Davidov D, Aroeti B (2010) Biophys J 99:4028
38. Peterson AW, Halter M, Tona A, Bhadriraju K, Plant AL (2009) BMC Cell Biol 10
39. Peterson AW, Halter M, Tona A, Bhadriraju K, Plant AL (2010) Cytometry Part A 77A:895
40. Sagvolden G, Giaever I, Pettersen EO, Feder J (1999) Proc Natl Acad Sci USA 96:471
41. Channavajjala LS, Eidsath A, Saxinger WC (1997) J Cell Sci 110(Pt 2):249
42. Thoumine O, Ott A, Louvard D (1996) Cell Motil Cytoskeleton 33:276
43. Truskey GA, Pirone JS (1990) J Biomed Mater Res 24:1333
44. Xiao Y, Truskey GA (1996) Biophys J 71:2869
45. Giaever I, Keese CR (1984) Proc Natl Acad Sci USA 81:3761
46. Wegener J, Keese CR, Giaever I (2000) Exp Cell Res 259:158
47. Frisch T, Thoumine O (2002) J Biomech 35:1137
48. Giaever I, Keese CR (1991) Proc Natl Acad Sci USA 88:7896
49. Lo CM, Keese CR, Giaever I (1993) Exp Cell Res 204:102
50. Keese CR, Wegener J, Walker SR, Giaever L (2004) Proc Natl Acad Sci USA 101:1554
51. Schuhmacher R (1990) Angew Chem 102:347
52. Buttry DA, Ward MD (1992) Chem Rev 92:1355
53. Kipling AL, Thompson M (1990) Anal Chem 62:1514
54. Yang MS, Thompson M (1993) Anal Chem 65:1158
55. Sauerbrey G (1959) Zeitschrift Fur Physik 155:206
56. Gryte DM, Ward MD, Hu WS (1993) Biotechnol Progr 9:105
57. Redepenning J, Schlesinger TK, Mechalke EJ, Puleo DA, Bizios R (1993) Anal Chem 65:3378
58. Wegener J, Janshoff A, Galla HJ (1998) Eur Biophys J Biophys Lett 28:26
59. Kanazawa KK, Gordon JG (1985) Anal Chem 57:1770
60. Janshoff A, Wegener J, Sieber G, Galla HJ (1996) Eur Biophys J 25:93

61. Rodahl M, Hook F, Fredriksson C, Keller CA, Krozer A, Brzezinski P, Voinova M, Kasemo B (1997) Faraday Discuss:229
62. Rodahl M, Kasemo B (1996) Rev Sci Instrum 67:3238
63. Fredriksson C, Kihlman S, Rodahl M, Kasemo B (1998) Langmuir 14:248
64. Nimeri G, Fredriksson C, Elwing H, Liu L, Rodahl M, Kasemo B (1998) Colloids Surf B-Biointerfaces 11:255
65. Reiss B, Janshoff A, Steinem C, Seebach J, Wegener J (2003) Langmuir 19:1816
66. Wegener J, Seebach J, Janshoff A, Galla HJ (2000) Biophys J 78:2821
67. Reiß B (2004) Westfälische Wihelms-Universität, Münster
68. Mitchell SA, Poulsson AHC, Davidson MR, Bradley RH (2005) Colloids Surf B-Biointerfaces 46:108
69. Mitchell SA, Poulsson AHC, Davidson MR, Emmison N, Shard AG, Bradley RH (2004) Biomaterials 25:4079
70. Welle A, Gottwald E (2002) Biomed Microdevices 4:33
71. Heitmann V, Wegener J (2007) Anal Chem 79:3392

Adv Biochem Engin/Biotechnol (2012) 126: 67–104
DOI: 10.1007/10_2011_107
© Springer-Verlag Berlin Heidelberg 2011
Published Online: 6 October 2011

Harnessing Cell–Biomaterial Interactions for Osteochondral Tissue Regeneration

Kyobum Kim, Diana M. Yoon, Antonios G. Mikos and F. Kurtis Kasper

Abstract Articular cartilage that is damaged or diseased often requires surgical intervention to repair the tissue; therefore, tissue engineering strategies have been developed to aid in cartilage regeneration. Tissue engineering approaches often require the integration of cells, biomaterials, and growth factors to direct and support tissue formation. A variety of cell types have been isolated from adipose, bone marrow, muscle, and skin tissue to promote cartilage regeneration. The interaction of cells with each other and with their surrounding environment has been shown to play a key role in cartilage engineering. In tissue engineering approaches, biomaterials are commonly used to provide an initial framework for cell recruitment and proliferation and tissue formation. Modifications of the properties of biomaterials, such as creating sites for cell binding, altering their physicochemical characteristics, and regulating the delivery of growth factors, can have a significant influence on chondrogenesis. Overall, the goal is to completely restore healthy cartilage within an articular cartilage defect. This chapter aims to provide information about the importance of cell–biomaterial interactions for the chondrogenic differentiation of various cell populations that can eventually produce functional cartilage matrix that is indicative of healthy cartilage tissue.

Keywords Biomaterials · Cartilage · Cells · Tissue engineering

Abbreviations

ADSC	Adipose-derived stem cell
BMP	Bone morphogenetic protein
CS	Chondroitin sulfate
EB	Embryoid body
ECM	Extracellular matrix
ESC	Embryonic stem cell
FGF	Fibroblast growth factor
HA	Hyaluronic acid
hDF	Human dermal fibroblast

K. Kim · D. M. Yoon · A. G. Mikos · F. K. Kasper (✉)
Department of Bioengineering, Rice University, MS-142, P.O. Box 1892,
Houston TX 77251-1892, USA
e-mail: kasper@rice.edu

GAG	Glycosaminoglycan
GMP	Gelatin microparticle
IGF	Insulin-like growth factor
MDSC	Muscle-derived stem cell
MMP	Matrix metalloproteinase
MSC	Mesenchymal stem cell
OPF	Oligo(poly(ethylene glycol) fumarate)
PCL	Poly(ε-caprolactone)
PDSC	Periosteum-derived stem cell
PEG	Poly(ethylene glycol)
PHA	Poly(hydroxyalkanoate)
PLGA	Poly(L,D-lactic-*co*-glycolic acid)
RGD	Arg-Gly-Asp
mRNA	Messenger RNA
TGF	Transforming growth factor

Contents

1 Introduction	69
2 Background	69
3 Cartilage	70
3.1 Chondrocytes	70
3.2 Extracellular Matrix	71
3.3 Architecture	72
4 Chondrocyte–ECM Interactions	73
5 Cell Sources for Cartilage Tissue Engineering	74
5.1 Stem/Progenitor Cell Sources	74
5.2 Fibroblasts	79
6 Chondrogenesis in Biomaterials	80
6.1 Cell Homing	80
6.2 Hydrogels	81
7 Modification of Biomaterials	82
7.1 RGD Peptide Incorporation	82
7.2 Growth Factor Incorporation	84
7.3 Gene Delivery	86
7.4 Modulation of the Intrinsic Properties of Biomaterials	87
8 Osteochondral Tissue Regeneration	90
8.1 Zonal Cartilage Engineering	90
8.2 Bilayered Hydrogels	91
9 Summary	93
References	93

1 Introduction

Biomaterials have been used in the medical field for several decades to address a variety of tissue defects and diseases. Initially, biomaterials applied for tissue defects were generally intended to be biologically inert so as to prevent an immune response. More recently, biomaterials have evolved to be bioactive and degradable by design so they can temporarily fill a defect space and serve as a conduit for tissue repair. Specifically, in the field of tissue engineering/regenerative medicine, biomaterials are commonly employed as cell transplantation vehicles or as conduits to support the infiltration of host cells for tissue formation. Various progenitor cells can differentiate into distinctive lineages and are responsible for tissue maintenance. Biomaterials can act as a template and support structure for cell differentiation, cell proliferation, and the production of proteins and extracellular matrix (ECM). Additionally, biomaterials can be modified to promote cell adhesion or delivery of growth factors necessary to enhance tissue growth and maintenance. The field of tissue engineering has investigated the use of cells, growth factors, and biomaterials for regeneration of tissues from a variety of systems, such as cardiovascular, dental, endocrine, gastrointestinal, maxillofacial, nervous, ophthalmologic, and orthopedic. In this review, cell–biomaterial interactions will be highlighted within the context of cartilage tissue engineering applications.

2 Background

Articular cartilage is a tissue that is located on the surface of articulating joints and consequently may experience a significant amount of force and trauma. Since native cartilage tissues have a limited capacity to heal naturally, cartilage may accumulate damage over time, which can lead to discomfort and pain. As a result, surgical procedures are often applied to aid in repairing the injured cartilage tissue. Current surgical approaches for cartilage repair, including arthroscopic lavage/debridement, autologous chondrocyte implantation, microfracture, and osteochondral grafting, provide some pain relief and improved joint function but fail to fully restore cartilage that has the same biomechanical function as healthy cartilage. Accordingly, tissue engineers are investigating alternative strategies that involve the incorporation of cells, biomaterials, and/or biologically active factors to create a suitable initial support system for cartilage repair. To successfully realize a tissue engineering approach for cartilage regeneration, it is imperative to first understand the organization of native cartilage and the way in which the tissue components, especially cells and ECM molecules, interact with each other to allow proper cartilage tissue function. By studying this communication in native cartilage, researchers can determine the key components that are necessary to successfully recapitulate articular cartilage in repair strategies. This chapter will briefly review the biological aspects and structure of cartilage and discuss how cell–biomaterial interactions can be harnessed to aid in chondrogenesis and regeneration of cartilage.

3 Cartilage

Cartilage is located throughout the human body in sites such as the ears, elbows, knees, intervertebral discs, nose, and ribs. Articular cartilage, or hyaline cartilage, is primarily located on the surface of load-bearing articulating joints (e.g., ankles, elbows, knees, wrists) and enables the movement of the joints to occur smoothly. Articular cartilage, unlike most tissue, is alymphatic, aneural, and avascular. Consequently, the primary mechanism by which cells access nutrients and waste products are removed is by diffusion through the synovial fluid [1]. Accordingly, the transport of various molecules to and from cartilage is facilitated by the high water content of the tissue. Water composes approximately 60–85% of the total wet weight of cartilage [2]. When the tissue undergoes compression, some water is expelled from the tissue, providing lubrication to the articulating surface and allowing the joints to move with low friction. The remainder of the cartilage weight comprises cells and the ECM. This section will briefly review the components and architecture of cartilage.

3.1 Chondrocytes

Chondrocytes are the primary cell type in cartilage and compose 1–10% of the total articular cartilage volume [3, 4]. They originate from embryonic mesodermal cells, which are responsible for limb development. Mesenchymal stem cells (MSCs) derive from the mesodermal cells and can differentiate into a variety of cell types, including chondrocytes. Chondrogenic differentiation occurs by a process known as cellular condensation. The process begins with degradation of the local ECM to allow the MSCs to aggregate and enhance the number of cell–cell interactions, which are necessary for chondrogenesis to occur [5–8]. Early differentiated chondrocytes are metabolically active, exhibit a high proliferation rate, and start to secrete ECM proteins. During this process, the chondrocytes delineate into two different zones: peripheral and central. In the peripheral zone, the characteristic zonal architecture of cartilage begins to develop, whereas in the central zone, endochondral ossification occurs [9]. The mature chondrocytes have a rounded morphology close to the subchondral bone, but at the cartilage surface, their shape is flatter and more discoidal. Mature chondrocytes have limited proliferative ability and are commonly surrounded by pericellular matrix [9, 10]. Chondrocytes interact with surrounding ECM through receptors, and they are primarily involved in maintaining the cartilage tissue by providing a balance between degradation and synthesis of ECM molecules.

3.2 Extracellular Matrix

The ECM makes up approximately 95% of the total cartilage volume and thus has a significant influence on the physical properties of cartilage [11]. The ECM comprises collagen, proteoglycans, and glycoproteins, with the dry weight of cartilage containing 60%, 25–35%, and 15–20% of these molecules, respectively [9]. There are multiple types of collagen such as types II, VI, and IX–XI [12, 13]. Type II collagen is the predominant collagen in cartilage (90–95% of collagen dry weight) [14]. The other collagens are fibrillar and tend to interweave throughout the ECM to create a framework that imparts tensile strength to the cartilage tissue [9]. Proteoglycans contain a protein core, which accounts for 5% of the molecule, with the rest being composed of branched glycosaminoglycans (GAGs) stemming from the protein core. [3]. GAGs are composed of long, nonrepeating disaccharides with negatively charged carboxylate/sulfate groups, such as hyaluronic acid (HA), decorin, biglycan, keratin sulfate, chondroitin sulfate (CS), dermatan sulfate, and heparin sulfate [9, 10]. Proteoglycans are found within the pericellular matrix and are known to aid in chondrocyte attachment to other matrix components [15]. Large proteoglycan monomers tend to aggregate, which results in the formation of aggrecan containing keratan sulfate and CS. Aggrecans can interact noncovalently with hyaluronan, which is stabilized by link proteins [16, 17]. The negative charge of GAGs and proteoglycans imparts the tissue with a high degree of hydrophilicity, and repulsion of these molecules has been linked to the compressive properties of cartilage [18]. Additionally, these aggregated molecules can prevent a large displacement of the cartilage matrix when undergoing compression, which confers cartilage with its resiliency and durability. Glycoproteins are simply polypeptides that are covalently attached to a carbohydrate group, such as link protein, fibronectin, laminin, vitronectin, thrombospondin, and tenascin-C, and are distributed throughout the ECM [12, 15]. Fibronectin contains specific amino acid sequences that are responsible for cell binding such as Arg-Gly-Asp (RGD), Arg-Gly-Asp-Ser, Leu-Asp-Val, and Arg-Glu-Asp-Val [19]. Glycoproteins also contain adhesion sites for chondrocytes and multiple binding sites that are able to stabilize collagens and proteoglycans in the tissue [15]. Superficial chondrocytes produce lubricin, also known as proteoglycan 4 or superficial zone protein, which is another glycoprotein that has been found within the synovial fluid to facilitate joint lubrication [20].

3.3 Architecture

Cartilage has a zonal architecture with four main compartments: superficial/tangential, middle/transitional, deep/basal, and calcified. Overall, when looking from the superficial zone to the deep zone, one finds that the water content decreases, oxygen pressure decreases, compressive strength increases, and tensile strength increases [9, 21]. The superficial/tangential zone is located at the articulating surface and composes 10–20% of the total cartilage volume [22]. In the superficial/tangential zone, chondrocytes are flat and disc-shaped with collagen fibrils densely packed and parallel to the tissue surface. A limited number of proteoglycans are located in this area with high amounts of fibronectin [3]. The superficial/tangential zone receives the highest tension, compression, shear, and hydrostatic pressures [22]. A large amount of superficial zone protein has been isolated primarily in this zone [20]. The middle/transitional zone accounts for 40–60% of the total cartilage volume [22]. In this zone, chondrocytes are more rounded in shape and the proteoglycan content is greater compared with the superficial zone, with collagen fibrils being more dispersed throughout the tissue. There is no specific orientation of the cells and ECM in this zone. Within the deep/basal zone (30% of the total tissue volume), chondrocytes are rounded, clustered, and aligned into columns that are perpendicular to the subchondral bone, and the cells are more proliferative compared with those in the other zones [22, 23]. Additionally, the ECM molecules are aligned perpendicular to the joint surface. The ECM contains more collagen and GAG than in the superiorly oriented zones [21]. The collagen is larger in diameter in this zone and transcends the tidemark which delineates the deep zone from the calcified zone. The calcified zone is a thin layer that is located above the surface of the subchondral bone. Here, chondrocytes are rounded but smaller than in the other zones, and the cells are surrounded by calcified ECM [9]. The quantity of proteoglycans and collagen fibrils is lower in this zone, with the collagen fibrils anchored into the underlying calcified bone [9].

Chondrocytes are also surrounded by pockets of ECM proteins that are classified into three different regions defined as the pericellular matrix, territorial matrix, and interterritorial matrix. The pericellular and territorial matrices allow the chondrocytes to bind to matrix proteins and protect them during mechanical loading [10]. A chondrocyte surrounded by the pericellular matrix is known as a chondron, and the thickness of pericellular matrix is approximately 2 μm [12]. The pericellular matrix contains minimal collagen fibrils, with the exception of collagen type VI, and there are abundant proteoglycans, such as aggrecan, hyaluronan, decorin, and biglycan, and glycoproteins such as fibronectin, link protein, and laminin [10]. The territorial and interterritorial matrices function primarily for load bearing. The territorial matrix surrounds the chondrocytes 2–5 μm from the membrane surface [12]. In this area, type II collagen is present with high amounts of aggrecan containing CS [12]. All these proteins wrap around the chondrocytes to protect them from the mechanical stresses. Collagen fibers from the territorial matrix adhere to the pericellular matrix. The interterritorial matrix is the largest portion of cartilage (more than 5 μm from the surface of chondrocytes) and is responsible for the zonal architecture of cartilage [9, 12]. This area contains the most type II collagen and the lowest amount of aggrecan [12].

4 Chondrocyte–ECM Interactions

Cell–cell and cell–matrix interactions play a key role in the differentiation of MSCs into chondrocytes during limb development. Initially, before condensation occurs, MSCs produce type I and type II collagen, hyaluronan, fibronectin, and tenascin C. These ECM components hinder the movement of the MSCs. However during condensation, the progenitor cells produce enzymes that breakdown local ECM components [24, 25]. The lower ECM density allows the cells to directly interact with each other via cell adhesion molecules, such as neural cadherin and neural cell adhesion molecule [26, 27]. Whereas neural cadherin and neural cell adhesion molecule need to be expressed during condensation, their expression must be lowered during the process of differentiation in order for chondrogenesis to successfully occur [12, 28]. During chondrogenesis, interactions of cells with the ECM once again become crucial for proper cartilage formation. One important ECM component that is not modulated during both the condensation and the chondrogenesis process is type II collagen [12]. The expression of the transcription factor *SOX9* by the cells allows the continual synthesis of type II collagen and is considered a major constituent in chondrogenesis, as the absence of *SOX9* gene expression has been shown to result in lack of cartilage development [29, 30]. Additionally, *L-SOX5* and *SOX6* have been linked with *SOX9* to further promote chondrogenesis and can upregulate cartilage matrix production of type II collagen and aggrecan [31, 32].

Mature chondrocytes are able to maintain healthy cartilage as they balance the degradation and synthesis of ECM components. The signals that chondrocytes receive from the surrounding environment help to define what is necessary to maintain the tissue. Chondrocytes interact with the ECM via receptors that are classified as non-integrin and integrin. Two common non-integrin receptors are annexin V/anchorin CII and CD44. Type II collagen binds to chondrocytes via the annexin V/anchorin CII receptor [33]. CD44 is a cell-surface glycoprotein that has a high affinity for hyaluronan in cartilage [34]. Integrins themselves are glycoproteins that function as heterodimeric transmembrane receptors with α and β subunits. Different types of α and β subunits can noncovalently associate to form receptors with a high affinity for various ligands. β_1 integrins with α_1, α_2, α_3, or α_5 have been found to influence chondrocyte attachment to type II collagen [2]. Chondrocytes interact with type VI collagen by $\alpha_1\beta_1$ integrin and NG2/human melanoma proteoglycan receptors [35, 36]. The $\alpha_3\beta_1$ and $\alpha_5\beta_1$ integrins can mediate the binding of fibronectin [37].

5 Cell Sources for Cartilage Tissue Engineering

Chondrocytes are the primary cell type in cartilage and are an obvious cell source to be explored for cartilage regeneration. Indeed, a clinical product approved by the US Food and Drug Administration for articular cartilage repair, Carticel®, involves autologous chondrocyte implantation. The surgical procedure associated with the application of Carticel® involves isolation of chondrocytes from a non-load-bearing portion of cartilage, cell expansion, and subsequent implantation of the autologous cells into the defect area. Recent studies have reported that 68% of patients treated with Carticel® had graft failure, delamination, or tissue hypertrophy [38]. One possible cause of these complications is the chondrocytes themselves. In healthy tissue, chondrocytes are able to produce and breakdown the ECM in a balanced manner. However, in damaged cartilage, chondrocytes are generally unable to repair the tissue. The chondrocytes that are placed into the defect site might not be integrating with the surrounding tissue and may therefore hinder cartilage repair. Additionally, the limited quantity of chondrocytes in cartilage requires in vitro culture to expand the cell number to clinically viable levels. The simplest method to culture chondrocytes is to passage them in a monolayer, but this may alter the morphology and dedifferentiate the chondrocytes [39]. The cells tend to adopt a fibroblastic morphology and the levels of phenotypic expression markers such as aggrecan and type II collagen decrease with an increase in the level of type I collagen, indicating a fibroblastic phenotype [39]. Transformation of the chondrocytes in this manner may result in fibrocartilage tissue formation. Other studies have found that chondrocytes entrapped in a 3D hydrogel such as alginate or agarose are able to retain a spherical morphology and have shown their ability to redifferentiate [40, 41]. A disadvantage of this approach is that mature chondrocytes tend not to proliferate readily. Additionally, 3D constructs have diffusion limitations, with lower levels of nutrient and waste exchange occurring in the center of the scaffold. Consequently, the size of the construct is limited, which in the end affects how many cells can be entrapped within the system. Further, it is difficult to achieve clinically relevant cell numbers in 3D construct cultures as quickly as with monolayer cultures. Although understanding chondrocyte function is important to the development of approaches for cartilage repair, major challenges are associated with the application of chondrocytes in strategies for cartilage regeneration. These limitations have led to the investigation of alternative cell sources for promoting cartilage repair.

5.1 Stem/Progenitor Cell Sources

A variety of stem cell types can undergo chondrogenic differentiation and have been investigated for application in cartilage tissue engineering. Stem cells are used because they are able to differentiate along multiple cell lineages, can proliferate readily, and can be easily harvested.

5.1.1 Bone-Marrow-Derived Stem Cells

The bone marrow contains both hematopoietic stem cells and MSCs. The cloning abilities of MSCs along with their multilineage capacity suggest their potential for use in tissue engineering applications [42, 43]. The method used to culture MSCs and the associated culture conditions can affect the phenotype of the cells. Therefore, to identify a homogenous population of MSCs, the Mesenchymal and Tissue Stem Cell Committee of the International Society of Cellular Therapy proposed minimal criteria for human designation of MSCs as follows: (1) MSCs must adhere to tissue culture plastic; (2) MSCs must be positive for CD105, CD73, and CD90 and negative for CD45, CD34, CD14, or CD11b, CD79a or CD19, and HLA-DR; (3) MSCs must be capable of differentiating into osteoblasts, adipocytes, and chondroblasts under standard in vitro conditions [44]. The ability of MSCs to differentiate into chondroprogenitor cells as well as to form cartilage in vivo has led to continued efforts to utilize MSCs for cartilage engineering [45–47].

As previously discussed, the process of condensation drives cell–cell interactions in vivo and initiates chondrogenesis. Therefore, MSCs have been cultured as pellets, aggregates, or spheroids to promote chondrogenic cell–cell interactions. MSCs are able to express type II collagen and aggrecan with low expression levels of type I collagen when they are in pellet culture [48, 49]. Cell culture conditions, such as the culture method and the presence of chondrogenic factors, play a key role in directing MSCs towards a chondrogenic lineage. For example, MSCs exposed to dexamethasone, an anti-inflammatory agent for MSC chondrogenesis, and transforming growth factor (TGF)-β_1/TGF-β_3 together in pellet culture demonstrated an increased level of aggrecan and type II collagen expression compared with the presence of the individual chondrogenic factors [50, 51]. MSCs in pellet culture in the presence of dexamethasone and TGF-β_1/TGF-β_3 have also shown increased levels of type X collagen expression and alkaline phosphatase activity, which indicates the pellet culture may induce hypertrophic chondrocyte differentiation under certain conditions [49, 52]. However, when MSCs we co-cultured with mature chondrocytes in a pellet culture with dexamethasone and TGF-β_3, type II collagen expression was found to be significantly higher than with culture of chondrocytes alone [53]. Human MSC pellets cultured with dexamethasone and TGF-β_1 along with conditioned medium from human chondrocyte pellets induced type II collagen expression and lowered type X collagen expression, when compared with MSC pellets without conditioned medium from chondrocytes [51]. These co-culture systems provide a mechanism to expose MSCs to cartilage proteins and ECM molecules produced by chondrocytes, which may facilitate the initiation of chondrogenesis. Additionally, exposure of MSC pellets to 10 ng/mL parathyroid-hormone-related peptide has been shown to reduce type X collagen expression, which is responsible for inhibiting chondrocytes from transiting from a prehypertrophic to a hypertrophic state during condensation [32, 51]. Overall, these findings suggest that chondrogenic differentiation of MSCs promoted by cell–cell contact during in vitro culture can be enhanced by exposure to chondrogenic or cartilage-like factors.

The microenvironment has also been shown to greatly influence chondrogenic differentiation of MSCs. For example, when MSCs are entrapped in alginate gels, they are able to maintain a rounded morphology and express type II collagen in the presence of TGF-β_1, and type X collagen expression was found to be lower that for MSCs grown in pellet culture [48, 54]. A lower level of type X collagen has been shown to be expressed by MSCs when ECM components are present, such as native cartilage-derived matrix with TGF-β_3 and/or bone morphogenetic protein (BMP)-6, versus culture in alginate alone [55], indicating the importance of cell–matrix interactions in MSC differentiation. Exposure of MSCs to methacrylated HA has been found to upregulate type II collagen expression and, by also incorporating TGF-β_3, higher levels of type II collagen, *SOX9*, and aggrecan are observed by 14 days of culture [56]. When the methacrylated HA constructs were implanted subcutaneously in mice, the presence of TGF-β_3 with MSCs in the constructs resulted in better neocartilage expression than observed with MSCs in poly(ethylene glycol) (PEG) hydrogels. Other work has shown that HA in poly(ethylene oxide) diacrylate hydrogels allows entrapped MSCs to express type II collagen and cartilage proteoglycans, but incorporation of TGF-β_3 was able to lower the expression of type I collagen [57]. Fibrin hydrogels with heparinized nanoparticles releasing TGF-β_3 were able to induce higher levels of type II collagen, aggrecan, and *SOX9* expression compared with fibrin hydrogels, fibrin hydrogels with TGF-β_3, and fibrin hydrogels with nanoparticles alone. In a rabbit full-thickness articular cartilage defect, the MSCs in the fibrin construct containing the nanoparticles and TGF-β_3 showed the best neocartilage formation [58]. Overall, interactions of cells with growth factors and ECM molecules are essential for chondrogenic differentiation of MSCs.

5.1.2 Adipose-Derived Stem Cells

Adipose-derived stem cells (ADSCs) are part of the embryonic mesoderm and can be isolated from fat tissue, which is commonly removed from the body by liposuction. ADSCs have a fibroblastic morphology and are positive for cell markers CD90 and CD105 and negative for CD14, CD34, and CD45, which is similar to the criteria suggested for selection of human MSCs [59]. Additionally, ADSCs can be expanded for long periods of time with low levels of senescence. Cartilage nodules containing sulfated proteoglycan-rich matrix and type II collagen have been observed when ADSCs are cultured in a micromass culture [60]. The chondrogenic culture medium with dexamethasone containing TGF-β_1, BMP-6, and/or TGF-β_3 has been found to be important to facilitate the differentiation of ADSCs into chondrocytes [55, 61].

ADSCs in direct contact with porous native articular cartilage ECM showed increased expression levels of type II collagen and aggrecan, with decreased levels of type X collagen in the absence of chondrogenic growth factors in the culture medium. These results indicate the importance of cell–matrix interactions for chondrogenic differentiation of ADSCs [62]. Additionally, the interaction of ADSCs with biomaterials can further influence chondrogenesis. For instance, biomaterials such as alginate and agarose, which naturally lack attachment sequences, allow ADSCs to maintain a rounded morphology. However, in fibrin or gelatin hydrogels, which contain cell binding sites, ADSCs tend to spread and become more fibrochondrogenic [63]. ADSCs entrapped within elastin-like polypeptides with repeating sequences of Val-Pro-Gly-Xaa-Gly, where Xaa is any amino acid except proline, were able to synthesize high levels of type II collagen, with low type I collagen formation in the absence of chondrogenic medium for at least 2 weeks [64]. Human ADSCs cultured on silanized hydroxypropylmethyl-cellulose hydrogels demonstrated higher type II collagen, *COMP*, and *SOX9* expression in 3D culture versus 2D culture without chondrogenic medium. However, to form cartilage matrix in vivo, culture of ADSCs in chondrogenic medium was necessary [65]. The preculture of ADSCs with chondrogenic medium has been found in other studies to be effective in the growth of cartilage ECM in vivo. When ADSCs are grown in a monolayer with chondrogenic medium and then subcutaneously implanted with alginate, an increase in type II collagen, *COMP*, aggrecan, and *SOX9* expression was exhibited after 20 weeks, but there were low levels of type I and type X collagen [66]. ADSCs in micromass culture with TGF-β_1 have been grown in an atelocollagen honeycomb-shaped scaffold and placed into an osteochondral rabbit defect [67]. Although cartilage-like tissue formed in the scaffold with and without ADSCs, larger amounts of cartilage tissue and the histological scoring of the tissue indicated that the presence of ADSCs results in improved cartilage repair.

5.1.3 Embryonic Stem Cells

Human embryonic stem cells (ESCs) are pluripotent cells that are derived from the inner mass of the embryonic blastocyst. ESCs are self-maintaining and are able to proliferate indefinitely [68]. Even though the use of these cells is subject to ethical debate, there have been initial promising signs of their ability to differentiate into chondrocytes. Successful chondrogenesis of ESCs was initially thought to require the formation of embryoid bodies (EBs) [69]. Enzymatically dissociated EB cells cultured in 2% agarose with chondrogenic medium without any exogenous growth factors showed higher collagen and sulfated GAG content than native EBs [70]. Exposing ESCs to BMP-2, BMP-4, and BMP-7 has been found to aid in the expression of cartilage matrix, but TGF-β_1 can hinder chondrogenic differentiation of these cells [71–74].

Directed differentiation of ESCs to chondrocytes has also been accomplished by culturing the cells with exogenous growth factors that may allow them to first differentiate into primitive streak mesendoderm [fibronectin matrix, WNT3A, activin A, fibroblast growth factor (FGF)-2, BMP-4], then to differentiate into a mesoderm population (fibronectin matrix, FGF-2, BMP-4, neurotrophin-4, follistatin) [75]. Culturing the mesoderm cells with fibronectin and gelatin, and weaning them off BMP-4 while supplying FGF-2, neurotrophin-4, and growth differentiation factor 5, led the cells to express the chondrogenic marker *SOX9* as well as to produce type II collagen and sulfated GAGs [75]. Additionally, new approaches have begun to investigate how to bypass the formation of an EB, because of the lack of control of EB size and the associated cell number. Cell–cell interactions promoted by pellet or micromass culture of ESCs in combination with growth factors can further enhance the formation of type II collagen [73, 74, 76]. Co-culture of ESCs with chondrocytes can also aid in chondrogenic differentiation in vitro and in vivo [77, 78]. ESCs were initially co-cultured with irradiated chondrocytes and TGF-β_3. Co-culturing these cells with fresh ESCs in Hyaff-11, a hyaluronan gel, and TGF-β_1 showed positive alician blue–van Gieson staining for collagen and GAGs [78]. In addition, human ESCs cultured in RGD-modified PEG hydrogels showed an increase in synthesis of GAG and collagen as well as a stimulated gene expression level of link protein and type II collagen versus cells cultured in unmodified PEG hydrogels [79].

5.1.4 Other Stem Cells

Stem cells can be isolated from other tissues as well, such as the muscle and periosteum. These stem cells have shown the potential to become chondroprogenitor cells. Muscle-derived stem cells (MDSCs) are located in the muscle tissue, have been shown to be multipotent, and can proliferate quickly with limited senescence. New cartilage was able to form in full-thickness osteochondral defects in rats after they had been treated for 5 weeks with a type I collagen gel scaffold containing MDSCs [80]. The presence of 10 μg BMP-2 in a MDSC pellet culture in a diffusion chamber resulted in expression of type II collagen and aggrecan [81]. Additionally, when this pellet was placed in an in vivo rat patellar groove defect, the newly formed tissue covered the defect area with GAGs and collagen, as seen by histological sections stained with hematoxylin and eosin as well as with toluidine blue. MDSCs retrovirally transduced with a BMP-4 gene have been found to express type II collagen in vitro when cultured in chondrogenic medium supplemented with TGF-β_1 [82]. Injection of the MDSCs with acellular fibrin glue into a rat osteochondral defect resulted in glossy cartilage being formed after 24 weeks.

The periosteum is located on the surface of the bone cortex and contains two distinct layers: a fibrous outer layer and the cambium. Chondroprogenitor cells have been isolated in the thin, inner, cambium layer, which is located adjacent to the bone surface [83]. Surface markers that are present for MSCs have also been found in periosteum-derived stem cells (PDSCs) [84]. However, further research will be necessary to properly identify and locate chondrocyte precursors within the bulk PDSCs. The exposure of PDSCs to growth factors such as TGF-β_1, TGF-β_3, FGF-2, and insulin-like growth factor (IGF)-1 significantly influences chondrogenesis [85–87]. PDSCs and periosteum explants have been successfully cultured in alginate, agarose, and atelocollagen gels, but the presence of growth factors is necessary for successful neocartilage formation with these cells under these conditions [87, 88].

5.2 Fibroblasts

As an alternative to stem cells, fibroblasts are being investigated for cartilage engineering. Almost every organ in the human body contains fibroblasts, which originate from the mesoderm. An abundant source of fibroblasts can be isolated from skin biopsies and can be used for potential autologous transplantation. Fibroblasts have properties similar to those of MSCs, such as being able to adhere to tissue culture plastic, being positive for CD73 and CD105, and being negative for hematopoietic markers (e.g., CD14, CD34, CD45) [89]. Human dermal fibroblasts (hDFs) are able to produce cartilage-like matrices containing components such as chondroitin 4-sulfate and keratin sulfate when cultured on a collagen-sponge-demineralized bone matrix composite [90]. Treating adult hDFs with IGF-1 before culturing them on aggrecan led to production of type II collagen as observed by immunohistochemical staining [91]. Additionally, exposing entrapped neonatal hDFs in alginate beads to 5% oxygen with 100 ng/mL BMP-2 under 3 weeks of hydrostatic compression (1 Hz for 4 h/day) increased collagen production and aggrecan gene expression compared with static culture conditions [92]. Dermis-isolated, aggrecan-sensitive cells, which are isolated by culturing dermal fibroblasts in a monolayer on aggrecan, can be cultured as a micromass in an agarose gel [93]. The dermis-isolated, aggrecan-sensitive cells have been shown to produce a cartilage-like ECM. These results underscore the importance of cell–ECM interactions. Additionally, cell–cell interactions have been shown to be a significant parameter for fibroblasts to progress along a chondrogenic lineage. Co-culturing hDFs with porcine articular chondrocytes on poly(lactic acid)/poly(glycolic acid) as well as a micromass of hDFs with lactic acid has shown that these cells are able to synthesize type II collagen [94, 95]. Additionally, fibroblasts reprogrammed to progress towards a chondrogenic lineage, such as BMP-7-transduced fibroblasts cultured on collagen hydrogels, retrovirally *SOX9* induced fibroblasts, and fibroblasts induced by cartilage-derived morphogenetic protein 1, have shown promising in vitro and in vivo results for the production of cartilage-associated ECM proteins [96–98].

6 Chondrogenesis in Biomaterials

6.1 Cell Homing

Biomaterials can be utilized as progenitor cell transplantation vehicles as well as to provide moieties that can aid in cartilage regeneration. Tissue engineering strategies may also involve leverage of biomaterials for the homing of progenitor cells, such as MSCs, from the host to the construct to facilitate cartilage repair. In general, the recruitment of endogenous host cells from a cell storage niche, such as the bone marrow, to an anatomic compartment is considered cell homing [99]. MSC homing is also specifically defined as a MSC population that is arrested within the vasculature of a tissue and transmigrated across the endothelium [100].

Natural healing and regeneration in defect tissues involves mobilization, homing, and subsequent reparative actions at the injured sites [101]. MSCs released from a cell storage niche first circulate in response to signals from distal injured tissues (mobilization), and vasculature arrestment as well as transendothelial migration (homing) occurs where MSCs will develop into mature healthy tissue [100, 101]. One of the early studies to investigate the origin and function of the progenitor cells involved in the repair of full-thickness defects of articular cartilage demonstrated that the repair was mediated entirely by proliferation and chondrogenic differentiation of primitive MSCs from the bone marrow [102]. It was also indicated by autoradiography after labeling with ^{3}H-thymidine and ^{3}H-cytidine that the chondrocyte population from the residual adjacent articular cartilage was not fully involved in defect repair. Therefore, this study emphasized the importance of bone marrow as a progenitor cell reservoir for articular cartilage regeneration and osteochondral tissue repair.

Owing to the potential drawbacks of clinical cell delivery, including undesired immune responses, pathogen transmission, and technical barriers associated with regulatory approval, cell homing to recruit MSCs from surrounding host tissues has been suggested [99]. Nonetheless, there are a limited number of publications investigating cell homing strategies for articular cartilage repair and chondrogenic tissue regeneration [101]. Owing to the lack of vasculature in a cartilage tissue, MSC homing strategies for cartilage tissue repair may need to target cell storage sources such as bone marrow and synovial fluid. One recent study has shown the potential use of synovium stem cells as a host cell source as well as the feasibility of a biomaterial-based cell homing strategy to induce articular surface regeneration [99]. The scaffold for the entire articular surface of the synovial joint was anatomically customized with poly(ε-caprolactone) (PCL) and 20% hydroxyapatite powder based on the surface morphology of a rabbit forelimb joint. An acellular scaffold containing interconnected microchannels (200–400 μm in diameter) with and without TGF-β_3 infusion was implanted in a rabbit model. TGF-β_3 was envisioned to serve as a cell homing molecule to accelerate functional recovery and hyaline cartilage regeneration. Compared with the scaffolds in the absence of TGF-β_3, scaffolds infused with TGF-β_3 recruited a greater number of cells to the injured site, exhibited mechanical properties similar to those of native cartilage, and contained higher concentrations of type II collagen and aggrecan. In addition, the microchannels in the scaffold functioned as conduits for cell homing, diffusion, histogenesis, and angiogenesis, thereby promoting tissue regeneration. In this study, it was proposed that some of the endogenous cells were recruited from the synovium, bone marrow, adipose tissue, and vasculature. The porous surface of the proximal end of the implanted scaffold and the presence of the microchannels may enable access to synovium stem cells and bone marrow progenitor cells, while providing a conduit for the migration of these cells. Therefore, this study demonstrated the possibility of utilizing a cell homing strategy to regenerate the entire articular cartilage surface, beyond focal defect healing, with the aid of growth-factor-loaded scaffolds.

6.2 Hydrogels

For cartilage tissue engineering, hydrogels are the most extensively investigated class of scaffold materials, as they provide a variety of advantages, including high water content and elastic properties that mimic native cartilage tissue, technical capacity for cell encapsulation, and effective transport of water and nutrients owing to their high equilibrium swelling [103]. Hydrogels may be tuned to mimic the function and architecture of native cartilage, which is composed of chondrocytes and ECM proteins [104, 105]. Therefore, success criteria for a functional hydrogel material for the delivery of cells and growth factors for cartilage regeneration include (1) biocompatibility, (2) an ability to encapsulate cells and support their viability and proliferation, (3) a capacity for sufficient hydration to provide effective diffusion of molecules in physiological conditions, (4) suitable mechanical properties, and (5) possible degradation to provide space for neotissue regeneration and remodeling. Naturally derived polymers such as alginate [106], fibrin [58, 107], silk [108, 109], chitosan [110, 111], and blends of these components [112, 113] as well as ECM components in native cartilage, including collagen, HA [56, 114], and CS [115], have been used to fabricate functional hydrogels for cartilage regeneration [116]. Although hydrogels comprising naturally derived materials generally exhibit excellent cytocompatibility and cell adhesion, synthetic polymers have the potential to overcome some of the limitations of naturally derived hydrogel materials, especially the general lack of tunability and insufficient mechanical strength. Synthetic polymers have been utilized in the fabrication of functional hydrogel systems owing, in large part, to their ability to modulate key physical properties of the hydrogels, such as the mechanical properties (e.g., elasticity and injectability), as well as the degradation kinetics. Surface properties of synthetically derived polymeric biomaterials can also be modulated to increase hydrophilicity, mobilize ECM-derived proteins, and fabricate micropatterned and nanopatterned surfaces [117]. In addition, it is also possible to fabricate hybrid composite gels by combining natural and synthetic polymers [110, 118] or by incorporating other molecules into the polymer structure [119]. Modification of hydrogel materials may be harnessed to provide enhanced functionality to stimulate encapsulated cell responses, particularly chondrogenic differentiation and matrix formation.

7 Modification of Biomaterials

Although synthetic polymeric materials generally lack biological moieties on the surface to interact with cells, an advantage of synthetic polymers is their tunable properties, which may be leveraged to conjugate a series of biologically active peptides. Other biological modifications of materials include delivery of exogenous growth factors through substrate materials or carrier molecules. Genetic modification by delivering genes encoding therapeutic growth factors could also be introduced to stimulate the production of growth factors from the seeded cell population without further modulation in a system. Moreover, intrinsic properties of biomaterials, including surface charge, roughness, topology, and scaffold architecture, could also be key parameters to modulate the cell–biomaterial interaction and influence chondrogenic differentiation of progenitor cells.

7.1 RGD Peptide Incorporation

Modification of biomaterials by utilizing biomimetic and bioactive motifs, such as short ligands, could improve the interaction between cells and materials [120]. Many studies have investigated a variety of bioactive peptides to improve cell adhesion, osteoinductivity of biomaterials, and maintenance of chondrocyte phenotype in a synthetic scaffold, including laminin-derived YIGSR [121], elastin-derived VPGIG [122], proteoglycan-binding peptide FHRIKA [123], osteopontin-derived peptide [124], and matrix metalloproteinase (MMP)-derived peptide [125]. Among various peptides, adhesive peptides containing RGD sequences are known to function as a binding domain for cell integrins [117]. Many studies have demonstrated that the immobilized RGD peptides on the surface of scaffolds facilitate cell–biomaterial interactions, specifically cell binding and adhesion. Recent investigations have shown that the presence of RGD peptides on biomaterial surfaces or the proper bulk incorporation of RGD peptides within a 3D matrix can improve the chondrogenic differentiation of stem cells to facilitate the process of cartilage repair [126, 127]. As a cell-binding sequence, RGD peptides present on poly(hydroxyalkanoate) (PHA) scaffolds may improve cell survival and motility [126]. PHA scaffolds coated with PHA granule binding protein PhaP and fused with RGD peptide showed increased cell spreading, adhesion, proliferation, and chondrogenic differentiation of human MSCs [127]. This RGD-modified scaffold exhibited higher expression levels of chondrogenic differentiation markers, including type II collagen, aggrecan, and *SOX9*, as well as increased production of sulfated GAG and total collagen than a PhaP scaffold without RGD and a blank scaffold.

RGD peptides immobilized on macroporous alginate scaffolds have also been shown to stimulate chondrogenic differentiation of human MSCs [128]. Specifically, the positive effect of RGD modification of scaffolds upon chondrogenesis was illustrated in the observation that the TGF-induced Smad signaling pathway involved in chondrogenic differentiation of MSCs was more activated in the presence of RGD peptide under the conditions studied [128]. A western blot and its densitometric analysis showed that the phosphorylation of both SMAD2 and ERK1/2 was significantly higher in the RGD-incorporating scaffolds than in the control alginate scaffolds. This result is consistent with the observed upregulation of chondrogenic maker gene expression, including type II collagen and *SOX9*. A study using human articular chondrocytes also demonstrated that bioactive RGD incorporation on a copolymer substrate of polystyrene/poly(L-lysine)/PEG improved GAG production and type II collagen messenger RNA (mRNA) expression compared with a blank polystyrene substrate [129]. Another study sought to mimic native RGD release by combing MMP-13 cleavage sites and demonstrated the importance of temporal regulation of integrin-binding peptides in chondrogenic differentiation [130]. In this study, human MSCs were encapsulated in PEG hydrogels with either an uncleavable RGD tether (CRGDSG) or a cleavable RGD tether (CPENFFGRGDSG). Both tethers were designed with MMP-13-specific cleavage sites. Once MMP-13 had been produced by encapsulated cells, RGD was released from the hydrogel system via the MMP-13 enzymatic cleavage of the tether. Released RGD induced greater chondrogenic differentiation of the MSCs than gels with uncleavable sequences, as indicated by higher GAG deposition and type II collagen staining. It has been suggested that RGD also functions as a mechanotransducer [131]. Under the mechanical loading environment, RGD ligands stimulated cartilage-specific gene expression and ECM protein synthesis. When dynamic compressive strains were applied to bovine chondrocytes encapsulated in a PEG hydrogel, the chondrocyte phenotype index (the expression ratio of collagen II and collagen I) and the proteoglycan synthesis were enhanced in RGD-incorporating gels relative to those without RGD incorporation.

ADSCs obtained from rats have also been used to investigate the influence of the integrin-binding peptides on chondrogenic differentiation [132]. RGD-chimeric protein with a cellulose binding domain in alginate beads resulted in increased gene expression of type II collagen, *SOX9*, aggrecan, and fibronectin, as well as the accumulation of chondrogenic matrix during TGF-β_3-induced differentiation. The results also demonstrated that the mechanism of RGD-chimeric protein stimulation of chondrogenic differentiation might be associated with suppressed RohA activity in the early differentiation stage. Furthermore, human ESC-derived cells could also be encapsulated in RGD-incorporating hydrogels [79, 133]. Human ESC-derived MSCs were positive for MSC surface markers including CD29, CD44, CD109, and platelet-derived growth factor α [79]. These cells exhibited in vitro neocartilage formation with basophilic ECM deposition and upregulated chondrogenic gene expression in RGD-modified PEG diacrylate hydrogels. An in vivo study using predifferentiated human ESC-derived MSCs demonstrated the potential application of this cell type for osteochondral tissue repair, indicated by a smooth articular cartilage surface and architecture of repaired cartilage similar to that of normal cartilage [133].

Although RGD-incorporating hydrogel systems have shown the capacity to induce in vitro chondrogenesis and possible approaches to stimulate in vivo articular cartilage repair, there have also been some controversial results with use of RGD-modified materials [134, 135]. For example, bovine MSCs encapsulated in unmodified alginate gels have been shown to enhance chondrogenic gene expression and matrix accumulation, whereas this observation was not found in RGD-modified alginate gels [134, 135]. However, it was found that the inhibition of sulfated GAG synthesis was stimulated by increasing RGD density [134, 135]. Another study also reported apoptosis of chondrocytes and synovial cells could be induced by RGD peptides [135].

7.2 Growth Factor Incorporation

Chondrogenic differentiation of progenitor cells into mature chondrocytes, cartilage-specific ECM deposition, and articular cartilage regeneration require a dynamic interaction of various growth factors as soluble signaling molecules to initiate, stimulate, and maintain differentiated cell function. A number of studies have investigated the functions of specific growth factors with a given progenitor cell population in a synthetic environment as well as the influence of combinations of growth factors in a dynamic fashion. Members of the TGF-β superfamily, such as certain TGF-βs and BMPs, have been found to aid in the upregulation of type II collagen, *SOX9*, and aggrecan expression. Specific molecules that have shown promising results for chondrogenesis include TGF-β_1, TGF-β_2, and TGF-β_3 [49, 52, 136–141], IGF-1 [142–146], BMP-2, BMP-4, BMP-6, and BMP-9 [55, 147–151], and FGF [152–158]. Combinational and dynamic effects of these growth factors on chondrogenic differentiation and articular cartilage repair have also been investigated [144, 158–160]. Effective delivery of growth factors is of importance to induce the chondrogenesis of progenitor cells and enhance articular cartilage regeneration. In addition to selection of the growth factor and the combination of a series of growth factors, other parameters such as the dose and release kinetics are also of importance to augment cartilage repair. To this end, sustained, controlled growth factor delivery to defect sites by using various delivery vehicles has also been investigated.

7.2.1 Gelatin Microparticles

Among a number of available methods to incorporate a growth factor into a scaffold or matrix material, gelatin microparticles (GMPs) have been intensively investigated because of their ability to form electrostatic complexes with charged growth factors under physiological conditions (pH 7.4), depending upon the charge of the growth factor and the gelatin [161]. Through enzymatic degradation of gelatin, incorporated growth factor(s) can be released in a controlled manner by regulating tunable fabrication parameters, such as the extent of cross-linking. For instance, controlled release of TGF-β_3 from GMPs stimulated chondrogenic matrix production by MSCs in pellet culture [162].

Simultaneously, degradation of gelatin can also result in a porous inner morphology in a 3D matrix or hydrogel containing GMPs. Growth-factor-loaded GMPs can be incorporated with injectable hydrogels, such as systems based on oligo(poly(ethylene glycol) fumarate) (OPF) [163]. GMPs made of acidic gelatin with an isoelectric point of 5.0 can electrostatically complex with appropriately charged protein in aqueous solution under physiological conditions [164]. An initial in vitro release study revealed that the release profile of TGF-β_1 from GMPs in OPF hydrogels and degradation of the hydrogel composites could be modulated by altering key fabrication parameters, including the amount of loaded GMPs, the cross-linking extent of GMPs, and the molecular weight of OPF [165]. A study of TGF-β_1 release in the presence of the enzyme collagenase demonstrated that this OPF–GMP hydrogel system could be applicable to the cartilage wound healing environment [166]. In addition, dual growth factor loading using TGF-β_1 and IGF-1 was also evaluated for controlled and localized release [160]. To demonstrate the capability of this composite hydrogel system as a delivery vehicle for cells and growth factors, primary chondrocytes from the condyle of calf femurs [167] and rabbit bone marrow MSCs were embedded in the hydrogel [168]. Encapsulated chondrocytes in OPF hydrogels exhibited higher levels of proliferation and GAG production in the presence of GMPs loaded with TGF-β_1 for 28 days of in vitro culture compared with control OPF gels as well as OPF gels with unloaded GMPs [167]. Chondrogenic differentiation of bone marrow MSCs derived from rabbit femurs was also investigated because of the potential application of MSCs as progenitor cells for cartilage tissue regeneration as mentioned previously [168]. Quantitative reverse transcription polymerase chain reaction data indicated that chondrogenic differentiation of bone marrow MSCs was upregulated with a medium dose of TGF-β_1 incorporation (10 ng/mL) at low cell seeding density (ten million cells per milliliter of gel). Dual growth factor delivery of TGF-β_1 and IGF-1 affected in vitro chondrogenic differentiation of encapsulated rabbit MSCs in the OPF–GMP composite hydrogels [145]. In this study, the dual-growth-factor-loaded group showed a significantly higher expression level of collagen type II and aggrecan on day 14, as determined by quantitative reverse transcription polymerase chain reaction, compared with the group with single growth factor incorporation. Moreover, the molecular weight of OPF was suggested to be a tunable parameter to potentially modulate the chondrogenic differentiation of MSCs, owing to the higher swelling ratio and larger mesh sizes of surrounding OPF hydrogels [169]. Therefore, in vitro chondrogenic differentiation of encapsulated bone-marrow-derived MSCs could be upregulated by modulating the formulation of OPF–GMP composite hydrogels.

7.2.2 Polymeric Microspheres

In addition to GMPs, polymeric microparticles could also be utilized as a delivery vehicle for growth factors and therapeutic agents. Poly(L,D-lactic-*co*-glycolic acid) (PLGA) is one of the most widely used polymeric materials for the fabrication of microsized carriers for growth factors [170]. Pharmacologically active microcarriers fabricated with fibronectin-coated PLGA microspheres have been applied to deliver TGF-β_3 in a controlled and sustained fashion for chondrogenesis [171]. Adsorbed fibronectin induced MSC adhesion onto the surface of the microspheres, and released TGF-β_3 stimulated chondrogenic differentiation of MSCs. In vitro differentiation of adherent MSCs was dependent on the amount of TGF-β_3, as indicated by upregulated mRNA expression levels for aggrecan, type IIB collagen, and type X collagen. Qualitative histological image analysis of subcutaneous implantation of TGF-β_3/PLGA microcarriers with adherent MSCs onto the surface in SCID mice also indicated high levels of type II collagen and aggrecan production as well as the formation of neotissue surrounding microspheres.

PLGA microspheres have also been applied for the delivery of other therapeutic factors, such as dexamethasone [172, 173]. Porous PLGA microspheres loaded with dexamethasone were incorporated within HA (4% w/v) based hydrogels and subcutaneously implanted in nude mice [172]. After 4 weeks, gene expression levels for cartilage-specific markers, including type II collagen and *SOX9*, were significantly higher in porous PLGA microspheres than in nonporous microspheres and the control treatment (bulk dexamethasone loading without PLGA carrier). Histological and immunohistochemical analyses also showed higher levels of GAG staining, as well as type II collagen synthesis, in the porous microsphere group. In addition, for the dual delivery of growth factor and dexamethasone, a more complex delivery system was introduced [173]. Dexamethasone-coated PLGA microspheres were conjugated with heparinized TGF-β_3, and this complex allowed dual release of dexamethasone and growth factor from the surface of the microspheres [173, 174]. Rabbit MSCs were cultured with these complexes and injected subcutaneously into the backs of nude mice. The dual delivery complex exhibited higher gene expression levels for chondrogenic differentiation markers, including type II collagen and aggrecan, than dexamethasone-coated PLGA microspheres as well as the blank PLGA control. In the presence of both dexamethasone and TGF-β_3, the accumulation of proteoglycans and polysaccharides in the microspheres was observed.

7.3 Gene Delivery

In addition to exogenous growth factor incorporation during the hydrogel fabrication for localized and sustained delivery, gene transfer has also been explored to stimulate the encapsulated cell population to produce various chondrogenic growth factors [175–178]. Delivery of growth factor to the cartilage defect site in an animal model might be limited because of rapid clearance from the joint tissues [175]. Therefore, as an alternative means of delivery of exogenous growth factor, it has been hypothesized that overexpression of growth factors by transplanted chondrocytes or MSCs might enhance the repair of articular cartilage defects. Rabbit articular chondrocytes transfected with plasmid vectors encoding human IGF-1 complementary DNA by using nonviral and nonliposomal lipid formulations (FuGENE 6) demonstrated in vitro IGF-1 secretion from cells in the alginate construct [176]. In addition to the in vitro IGF secretion from transfected cells in alginate over a prolonged period (36 days), in vivo transplantation of the constructs led to enhanced articular cartilage repair and subchondral bone formation compared with what was seen in groups that were transfected with the *lacZ* reporter gene. This lipid-based transfection was also used for combined gene transfer for both human IGF-1 and human FGF-2 in vivo [177]. This study demonstrated that combined overexpression of both growth factors from NIH 3T3 cells encapsulated in alginate could accelerate the repair of full-thickness osteochondral cartilage defects, compared with the group receiving IGF alone and *lacZ* implants. In addition to nonviral gene delivery, viral vectors, including adenovirus, lentivirus, and retrovirus, could also be used as gene transfer agents [178–180]. In a study using adenovirus-mediated TGF-β_1 gene transfer to human MSCs, improved cartilage repair was observed 12 weeks after osteochondral implantation in a rat model [178]. FGF-2 has also been produced successfully by adeno-associated-virus-delivered transgene and improved in vivo cartilage tissue repair [179, 180]. Moreover, direct implantation of vector-laden, coagulated bone marrow aspirates (i.e., gene plug) has also been developed [175]. Typical ex vivo gene transfection techniques require a series of processes including expansion of cell number, transfection of cells with target genes, fabrication of cell/scaffold constructs, and surgical implantation [175]. By use of this alternative ex vivo protocol, implantation of bovine bone marrow aspirate transduced with adenoviral vector to deliver TGF-β_1 resulted in improved cartilage repair in partial-thickness defects of the medial condyle in mature sheep models.

7.4 Modulation of the Intrinsic Properties of Biomaterials

Other than the incorporation of biological moieties such as peptides and growth factors into hydrogel systems, there have been various investigations to enhance the level of cell–material interaction by altering surface properties of scaffolds, by modulating mechanical stimulation, and by changing the architecture of biomaterial scaffolds. In addition to promotion of cell–material interactions by addition of biologically active stimuli, modulation of the intrinsic properties of biomaterial substrates can also influence the chondrogenic differentiation of cell populations as well as the in vivo tissue responses.

7.4.1 Surface Properties

Wetability

Surface hydrophilicity/hydrophobicity of the biomaterials can be one of the important parameters to regulate a series of cellular functions, including attachment, migration, cytoskeletal organization, and differentiation [181]. Although it is known that hydrophobic surface characteristics favor protein adsorption from the aqueous surrounding solution, a hydrophilic surface is also necessary to initiate cell attachment [117]. Therefore, controlling the optimum level of hydrophilicity/hydrophobicity of the material surface could induce positive cell responses. Technical methods to increase the hydrophilicity of hydrophobic synthetic polymeric biomaterials include grafting hydrophilic polymer through copolymerization [182], plasma treatment to increase the number of oxygen-containing groups such as –OH and –C=O [183, 184], and photooxidation to introduce peroxide groups onto the material surface with the aid of UV treatment [117]. For instance, increasing hydrophilicity in copolymeric hydrogels by increasing the hydrophilic PEG content relative to the hydrophobic PCL content resulted in higher proliferation of primary rabbit chondrocytes [182]. In addition, more hydrophilic gels (e.g., 14 wt% PEG and 6 wt% PCL) could be optimum to induce chondrogenic differentiation, as determined by stimulated gene expression levels of type II collagen, aggrecan, *SOX9*, and *COMP*. Plasma-treated electrospun PCL nanofibers also showed higher chondrocyte adhesion and proliferation than untreated hydrophobic PCL surfaces [184].

Roughness and Topography

Modulation of the hydrophilicity of the material surface can usually be related to changes in topography and roughness [181, 184–186]. Cell adhesion of both human MSCs and porcine chondrocytes mediated by integrin β was influenced by different topologies and surface roughness of PLGA-, PLA-, and PCL-coated plates [187]. In addition to initial cell adhesion, topographical changes of the material surface also affected chondrocyte aggregation (i.e., mesenchymal condensation) [188] and the osteoblastic signaling pathway [186]. Another study using a composite bone scaffold also demonstrated the related changes in increasing hydrophilicity and roughness by incorporation of hydroxyapatite nanoparticles into a hydrophobic poly(propylene fumarate) scaffold [185]. In this study, a mineral particle content of 20 wt% resulted in higher hydrophilicity and roughness, and the changes in the physicochemical properties of the composite material influenced osteogenic signal expression of rat bone marrow stromal cells.

7.4.2 Scaffold Architecture

Architectural Design for Cartilage Tissue Engineering

In addition to altering surface properties of biomaterials to improve cell–material interactions, modulation of the structure and architecture of a scaffold can also influence the cellular behaviors, since architectural changes, including pore size, porosity, interconnectivity, and morphology of the substrate surface, can affect subsequent cellular behaviors [189–192]. Architectural design is of importance in fabrication of bone-tissue-engineered scaffolds to induce osteoblastic differentiation [190, 191], but the influence of structural parameters can also be observed in altering chondrocyte behaviors and differentiation for cartilage tissue engineering. One recent study demonstrated that morphological changes in four different cell-graft systems, including hyaluronan web, collagen fleece, collagen gel, and collagen sponge, affected chondrocyte distribution, morphology, and cell–scaffold interactions [192]. Passive chondrocyte distribution throughout the inner region of porous scaffolds might depend on porosity and structure, whereas changes in cell morphology may be correlated to fiber size. In addition, adhesion might be influenced by material composition through membrane receptors and adhesive matrix molecules.

Nanofiber Mesh Scaffold

Another example of architectural changes for cartilage tissue engineering scaffolds is a nanofiber mesh scaffold. Fiber meshes are commonly fabricated via an electrospinning technique [193, 194]. A potential advantage of electrospun nanofibrous scaffolds is the similarity of the fiber diameter to that of native collagen fibrils, which may provide an appropriate microenvironment for chondrogenic cell responses [194]. However, there are also some limitations to nanofiber scaffolds, such as insufficient control of pore size, inherent planar structure, and subsequent limited cell infiltration into the inner region of scaffolds [194]. Nevertheless, many researchers have shown that nanofiber mesh scaffolds can support chondrogenic differentiation of MSCs seeded on the scaffold [195–198] as well as multilineage differentiation, including osteogenesis and adipogenesis [199–201]. Composite fibrous scaffolds can also be fabricated by the dispersion of nanoparticles (e.g., hydroxyapatite minerals) in polymeric solution for electrospinning [197]. Some in vivo studies using nanofibrous scaffolds with MSCs revealed a promising method to repair cartilage defects [202, 203]. Six months after implantation, PCL nanofibrous scaffolds with both allogeneic chondrocytes and xenogeneic human MSCs in a swine model exhibited higher articular tissue regeneration over acellular PCL scaffold and the no-implant control [203]. Another in vivo study using periosteal cells from skeletally mature New Zealand White rabbits was also reported [202].

Varying the diameter of fibers in fibrous scaffolds could provide different architectures and morphologies of the substrate for cell interaction. Subsequent architectural changes such as specific surface area could be related to chondrocyte phenotype, ECM protein synthesis, and chondrogenic differentiation [192, 195, 198]. It has been demonstrated that a rounded morphology and disorganized cytoskeletal structure of cells were observed in nanosized fiber meshes, whereas well-spread chondrocytes with organized cytoskeletons were seen in microfibers [195]. This observation is correlated to another report demonstrating that rounded cell shape was retained when chondrocytes attached on fibers with a smaller diameter than the size of the cell [192]. In addition to morphological changes caused by altering the fiber diameter, GAG production and qualitative immunostaining for type II/IX collagen, aggrecan, and cartilage proteoglycan link protein were higher in nanofibrous poly(lactic acid) scaffolds than in microfiber scaffolds [195]. Moreover, type II collagen gene expression in a PCL fibrous scaffold with a diameter of 500 nm was also higher than in a PCL fibrous scaffold with a diameter of 1,000 nm [198]. Therefore, it can be speculated that changes in fiber diameter and subsequent modulation of architecture in nanofibrous scaffolds can regulate the cell–material interaction and chondrocyte behavior can be optimized.

8 Osteochondral Tissue Regeneration

8.1 Zonal Cartilage Engineering

As previously described, articular cartilage is an avascular tissue with a single cell population and dense ECM that has a zonal organization. Each zone has a unique distribution of chondrocytes, biochemical composition, and mechanical properties [118, 204]. To closely mimic the native phenotype and formation across articular cartilage tissues, chondrocyte subpopulations in two or more distinct layers of hydrogel have been engineered.

Superficial and deep zone chondrocytes from bovine articular cartilage have been encapsulated in photopolymerized bilayered poly(ethylene oxide) diacrylate hydrogels [204]. In this bilayer co-culture system, deep zone cells produced more collagen and proteoglycan than superficial cells after 6 weeks of in vitro culture. In addition to the inhomogeneity of ECM production, deep zone cells also exhibited higher shear and compressive strength than the homogeneous cell control. This research showed the heterotrophic cell interaction and modulated biological/mechanical properties of engineered cartilage tissues. Another study also demonstrated that engineered agarose hydrogels containing zonal chondrocytes exhibited depth-varying cellular and mechanical inhomogeneity similar to that of native tissue [205]. Following 42 days of in vitro culture, the data indicated that production of GAG and collagen from superficial and middle/deep zone chondrocytes was enhanced when they were layered with the other subpopulation in a bilayered construct. One of the recent approaches to mimic the highly organized zonal architecture of articular cartilage investigated a variety of hydrogel formulations with a combination of CS, MMP-sensitive peptides, and HA in PEG hydrogels [118]. This study demonstrated that the unique mechanical properties and ECM composition of each formulation might direct a zonal-phenotype-specific chondrogenic differentiation. The result indicated that the PEG/CS/MMP-sensitive peptide group corresponded to the superficial zone, the PEG/CS group corresponded to the middle zone, the PEG/HA group corresponded to the deep zone, and the CS group corresponded to the calcified zone.

8.2 Bilayered Hydrogels

In addition to zonal engineering of articular cartilage, implantable bilayered hydrogel systems have also been investigated for osteochondral tissue regeneration with two distinct sublayers. Osteochondral tissue contains an articular cartilage surface at the top and subchondral bone underneath the cartilage tissue that provides mechanical support [206–208]. Therefore, biphasic layers with distinct biomechanical properties could simultaneously mimic the chondrogenic surface as well as the bony subchondral tissue. For the full-thickness joint defect, an approach using a bilayered structure of implantable hydrogel or scaffold could induce osteochondral tissue regeneration.

A composite bilayered hydrogel of OPF matrix and growth factor-incorporating GMPs has been studied as a functional model for osteochondral tissue regeneration [209–211]. An early in vivo trial with this composite hydrogel demonstrated the support of healthy tissue growth in New Zealand White rabbit osteochondral defects by showing hyaline cartilage improvement in the chondral region and bone filling in the subchondral region at 14 weeks [211]. In this study, hydrogel composites of 3-mm diameter and 3-mm thickness were implanted in the full-thickness defect of a rabbit knee joint. The TGF-β_1-loaded chondral layer exhibited a significant improvement in morphology of neoformed surface tissues among various histological scoring criteria compared with gels encapsulating blank GMPs without the growth factor. Additionally, the in vitro influence of the differentiation stage of an encapsulated cell population in this hydrogel system was also investigated [210]. A combination of MSCs with TGF-β_1-loaded GMPs in the chondral (top) layer and osteogenically induced (6 days of culture in medium with osteogenic supplements) MSCs with blank GMPs in the subchondral (bottom) layer showed significantly higher mRNA expression levels of collagen type II and aggrecan compared with both nonosteogenic cells and TGF-β_1-free groups. This study demonstrated that osteogenic cells in the subchondral region might produce chondrogenic-signaling molecules to induce chondrogenic differentiation of MSCs in the chondral layer. This observation was only seen in the presence of TGF-β_1, which indicated the importance of the additional effect of growth factor on MSC differentiation in a hydrogel system. Similarly, precultured MSCs in medium with osteogenic supplements in the subchondral layer with the aid of TGF-β_3 induced a significantly higher level of in vitro chondrogenesis of MSCs in a chondral layer after 28 days of culture compared with groups without TGF-β_3 [209].

Although many in vitro studies have demonstrated that incorporation of growth factor within various cartilage scaffolds enhances chondrogenesis of progenitor cell populations and results in minor improvement after subcutaneous implantation in vivo, functional in vivo tissue regeneration in an articular cartilage surface remains a challenge. Some of the studies have shown only partial repair of cartilage defects and a minor level of improvement. Dual growth factor delivery using a bilayer osteochondral hydrogel was also investigated to demonstrate the interaction of growth factors in cartilage repair [159]. In this study, GMPs with TGF-β_1, IGF-1, and both of them were incorporated in a chondral (top) layer of OPF hydrogels and the gels were implanted in rabbits. In vivo analysis indicated that single IGF-1 delivery showed minor chondral repair with GAG and cell content of the cartilage compared to other groups, whereas single TGF-β_1 and dual delivery did not show any improved tissue repair. A lack of any synergistic effect of dual growth factor delivery suggests the complexity of the dynamic process of cartilage repair and the existence of other parameters to investigate beyond a simple combination of growth factors.

In addition to growth factor delivery, functional remodeling of osteochondral tissue by MSC delivery remains a target for investigation. Despite various in vitro studies that indicate successful chondrogenic differentiation of encapsulated MSC populations in hydrogels with the aid of exogenous growth factor delivery, in vivo cartilage regeneration with complete cartilage repair remains a challenge. Rabbit MSCs in OPF hydrogels with or without TGF-β_1 incorporation did not show any significant improvement in cartilage tissue regeneration [212]. Both reduced cartilage thickness and improved surface regularity were observed with MSC-loaded gels. It might be hypothesized that faster subchondral bone formation in OPF/MSC groups provided sufficient mechanical support to the articular surface region and resulted in smoother articular surfaces. A smoother surface could also be obtained by the participation of implanted MSCs in cartilaginous matrix secretion and remodeling. A similar MSC/growth factor delivery in a rabbit in vivo model using a composite hydrogel made from the self-assembling peptide sequences (RADA)$_4$ and (KLDL)$_3$ showed inconsistent results with in vitro chondrogenesis and chondrocyte phenotypes [213]. Neither the addition of dexamethasone as well as the chondrogenic growth factors TGF-β_1 and IGF-1 nor the combinational incorporation of these growth factors and bone-marrow-derived MSCs in a hydrogel led to any beneficial effect on cartilage repair. Fibrous tissue formation was even observed in the MSC/growth factor/hydrogel group. This study demonstrated a possibility to direct a single stem cell population to different zonal phenotypes within a 3D structure with multiple layers.

Other in vivo studies using tricopolymer scaffolds with gelatin, chondroitin 6-sulfate, and sodium hyaluronate demonstrated TGF-β_1 release could help articular cartilage repair in the full-thickness defect (4 mm in diameter and 3 mm in thickness) in rabbits [214, 215]. An amount of 0.8 ng of TGF-β_1 released from embedded GMPs induced chondrogenic differentiation of autologous MSCs loaded onto scaffolds [214]. Histological observation and semiquantitative scoring data indicated that a controlled TGF-β_1 release using GMPs might be superior to stimulate cartilage repair to the absence of growth factor delivery to the cells. A hybrid PLGA scaffold with this tricopolymer also showed better in vivo cartilage regeneration [215]. After 24 weeks of implantation, histological grading revealed that cell morphology/matrix staining, surface regularity, and subchondral bone reconstruction were significantly better in the tricopolymer-incorporated PLGA scaffolds than in blank PLGA scaffolds.

9 Summary

The microenvironment plays a key role in engineering tissue. Therefore, special care and attention are necessary to create a successful combinatorial approach for tissue regeneration that involves the integration of cells, growth factors, and biomaterials. A variety of stem cells as well as fibroblasts have been investigated for cartilage regeneration. The process of chondrogenic differentiation requires the interaction of cells with growth factors and biomaterials. Biomaterials can be modified to control the release of bioactive molecules as well as to aid in cell adhesion to enhance cartilage formation. Ultimately, biomaterials can be used to recapitulate the cartilage architecture, which can enhance cellular function to successfully tissue-engineer cartilage as well as to potentially regenerate full-thickness osteochondral defects.

Acknowledgements Work in the area of biomaterials science and cartilage tissue engineering is supported by the US National Institutes of Health (R01-AR048756, A.G.M. and F.K.K.; R01-AR057083, A.G.M.; and R21-AR056076, A.G.M.).

References

1. Freemont AJ, Hoyland J (2006) Lineage plasticity and cell biology of fibrocartilage and hyaline cartilage: its significance in cartilage repair and replacement. Eur J Radiol 57(1):32–36
2. Yoon DM, Fisher JP (2006) Chondrocyte signaling and artificial matrices for articular cartilage engineering. Adv Exp Med Biol 585:67–86
3. Temenoff JS, Mikos AG (2000) Review: tissue engineering for regeneration of articular cartilage. Biomaterials 21(5):431–440
4. Archer CW, Francis-West P (2003) The chondrocyte. Int J Biochem Cell Biol 35(4): 401–404
5. Janners MY, Searls RL (1970) Changes in rate of cellular proliferation during the differentiation of cartilage and muscle in the mesenchyme of the embryonic chick wing. Dev Biol 23(1):136–165
6. Fell HB (1925) The histogenesis of cartilage and bone in the long bones of the embryonic fowl. J Morphol Physiol 40(3):417–459
7. Thorogood PV, Hinchliffe JR (1975) An analysis of the condensation process during chondrogenesis in the embryonic chick hind limb. J Embryol Exp Morphol 33(3):581–606
8. Summerbell D, Wolpert L (1972) Cell density and cell division in the early morphogenesis of the chick wing. Nat New Biol 239(88):24–26

9. Buckwalter JA, Mankin HJ (1998) Articular cartilage: tissue design and chondrocyte-matrix interactions. Instr Course Lect 47:477–486
10. Poole CA (1997) Articular cartilage chondrons: form, function and failure. J Anat 191 (Pt 1):1–13
11. Ulrich-Vinther M, Maloney MD, Schwarz EM, Rosier R, O'Keefe RJ (2003) Articular cartilage biology. J Am Acad Orthop Surg 11(6):421–430
12. Bobick BE, Chen FH, Le AM, Tuan RS (2009) Regulation of the chondrogenic phenotype in culture. Birth Defects Res C Embryo Today 87(4):351–371
13. Buckwalter JA (1983) Articular cartilage. Instr Course Lect 32:349–370
14. Eyre DR, Wu JJ, Apone S (1987) A growing family of collagens in articular cartilage: identification of 5 genetically distinct types. J Rheumatol 14(Spec No):25–27
15. Rosso F, Giordano A, Barbarisi M, Barbarisi A (2004) From cell–ECM interactions to tissue engineering. J Cell Physiol 199(2):174–180
16. Heinegard D, Hascall VC (1974) Aggregation of cartilage proteoglycans. 3. Characteristics of the proteins isolated from trypsin digests of aggregates. J Biol Chem 249(13):4250–4256
17. Franzen A, Bjornsson S, Heinegard D (1981) Cartilage proteoglycan aggregate formation. Role of link protein. Biochem J 197(3):669–674
18. Newman AP (1998) Articular cartilage repair. Am J Sports Med 26(2):309–324
19. Kao WJ (1999) Evaluation of protein-modulated macrophage behavior on biomaterials: designing biomimetic materials for cellular engineering. Biomaterials 20(23–24):2213–2221
20. Khalafi A, Schmid TM, Neu C, Reddi AH (2007) Increased accumulation of superficial zone protein (SZP) in articular cartilage in response to bone morphogenetic protein-7 and growth factors. J Orthop Res 25(3):293–303
21. Coates EE, Fisher JP (2010) Phenotypic variations in chondrocyte subpopulations and their response to in vitro culture and external stimuli. Ann Biomed Eng 38(11):3371–3388
22. Buckwalter JA, Mow VC, Ratcliffe A (1994) Restoration of injured or degenerated articular cartilage. J Am Acad Orthop Surg 2(4):192–201
23. Wong M, Wuethrich P, Eggli P, Hunziker E (1996) Zone-specific cell biosynthetic activity in mature bovine articular cartilage: a new method using confocal microscopic stereology and quantitative autoradiography. J Orthop Res 14(3):424–432
24. Toole BP, Jackson G, Gross J (1972) Hyaluronate in morphogenesis: inhibition of chondrogenesis in vitro. Proc Natl Acad Sci USA 69(6):1384–1386
25. Kulyk WM, Upholt WB, Kosher RA (1989) Fibronectin gene expression during limb cartilage differentiation. Development 106(3):449–455
26. Oberlender SA, Tuan RS (1994) Expression and functional involvement of N-cadherin in embryonic limb chondrogenesis. Development 120(1):177–187
27. Widelitz RB, Jiang TX, Murray BA, Chuong CM (1993) Adhesion molecules in skeletogenesis: II. Neural cell adhesion molecules mediate precartilaginous mesenchymal condensations and enhance chondrogenesis. J Cell Physiol 156(2):399–411
28. Yoon YM, Oh CD, Kim DY, Lee YS, Park JW, Huh TL et al (2000) Epidermal growth factor negatively regulates chondrogenesis of mesenchymal cells by modulating the protein kinase C-alpha, Erk-1, and p38 MAPK signaling pathways. J Biol Chem 275(16):12353–12359
29. Wright E, Hargrave MR, Christiansen J, Cooper L, Kun J, Evans T et al (1995) The Sry-related gene *SOX9* is expressed during chondrogenesis in mouse embryos. Nat Genet 9(1):15–20
30. Ng LJ, Wheatley S, Muscat GE, Conway-Campbell J, Bowles J, Wright E et al (1997) *SOX9* binds DNA, activates transcription, and coexpresses with type II collagen during chondrogenesis in the mouse. Dev Biol 183(1):108–121
31. Lefebvre V, Li P, de Crombrugghe B (1998) A new long form of *SOX5* (*L-SOX5*), *SOX6* and *SOX9* are coexpressed in chondrogenesis and cooperatively activate the type II collagen gene. EMBO J 17(19):5718–5733
32. de Crombrugghe B, Lefebvre V, Behringer RR, Bi W, Murakami S, Huang W (2000) Transcriptional mechanisms of chondrocyte differentiation. Matrix Biol 19(5):389–394

33. Mollenhauer J, Bee JA, Lizarbe MA, von der Mark K (1984) Role of anchorin CII, a 31,000-mol-wt membrane protein, in the interaction of chondrocytes with type II collagen. J Cell Biol 98(4):1572–1579
34. Culty M, Miyake K, Kincade PW, Sikorski E, Butcher EC, Underhill C (1990) The hyaluronate receptor is a member of the CD44 (H-CAM) family of cell surface glycoproteins. J Cell Biol 111(6 Pt 1):2765–2774
35. Loeser RF (2002) Integrins and cell signaling in chondrocytes. Biorheology 39(1–2):119–124
36. Midwood KS, Salter DM (2001) NG2/HMPG modulation of human articular chondrocyte adhesion to type VI collagen is lost in osteoarthritis. J Pathol 195(5):631–635
37. Salter DM, Hughes DE, Simpson R, Gardner DL (1992) Integrin expression by human articular chondrocytes. Br J Rheumatol 31(4):231–234
38. Wood JJ, Malek MA, Frassica FJ, Polder JA, Mohan AK, Bloom ET et al (2006) Autologous cultured chondrocytes: adverse events reported to the United States food and drug administration. J Bone Joint Surg 88(3):503–507
39. Darling EM, Athanasiou KA (2005) Rapid phenotypic changes in passaged articular chondrocyte subpopulations. J Orthop Res 23(2):425–432
40. Hauselmann HJ, Fernandes RJ, Mok SS, Schmid TM, Block JA, Aydelotte MB et al (1994) Phenotypic stability of bovine articular chondrocytes after long-term culture in alginate beads. J Cell Sci 107(Pt 1):17–27
41. Benya PD, Shaffer JD (1982) Dedifferentiated chondrocytes reexpress the differentiated collagen phenotype when cultured in agarose gels. Cell 30(1):215–224
42. Becker AJ, Mc CE, Till JE (1963) Cytological demonstration of the clonal nature of spleen colonies derived from transplanted mouse marrow cells. Nature 197:452–454
43. Friedenstein AJ, Chailakhyan RK, Gerasimov UV (1987) Bone marrow osteogenic stem cells: in vitro cultivation and transplantation in diffusion chambers. Cell Tissue Kinet 20(3):263–272
44. Dominici M, Le Blanc K, Mueller I, Slaper-Cortenbach I, Marini F, Krause D et al (2006) Minimal criteria for defining multipotent mesenchymal stromal cells. The International Society for Cellular Therapy position statement. Cytotherapy 8(4):315–317
45. Caplan AI (1991) Mesenchymal stem cells. J Orthop Res 9(5):641–650
46. Ashton BA, Allen TD, Howlett CR, Eaglesom CC, Hattori A, Owen M (1980) Formation of bone and cartilage by marrow stromal cells in diffusion chambers in vivo. Clin Orthop Relat Res 151:294–307
47. Pittenger MF, Mackay AM, Beck SC, Jaiswal RK, Douglas R, Mosca JD et al (1999) Multilineage potential of adult human mesenchymal stem cells. Science 284(5411):143–147
48. Yang IH, Kim SH, Kim YH, Sun HJ, Kim SJ, Lee JW (2004) Comparison of phenotypic characterization between "alginate bead" and "pellet" culture systems as chondrogenic differentiation models for human mesenchymal stem cells. Yonsei Med J 45(5):891–900
49. Johnstone B, Hering TM, Caplan AI, Goldberg VM, Yoo JU (1998) In vitro chondrogenesis of bone marrow-derived mesenchymal progenitor cells. Exp Cell Res 238(1):265–272
50. Derfoul A, Perkins GL, Hall DJ, Tuan RS (2006) Glucocorticoids promote chondrogenic differentiation of adult human mesenchymal stem cells by enhancing expression of cartilage extracellular matrix genes. Stem Cells 24(6):1487–1495
51. Fischer J, Dickhut A, Rickert M, Richter W (2010) Human articular chondrocytes secrete parathyroid hormone-related protein and inhibit hypertrophy of mesenchymal stem cells in coculture during chondrogenesis. Arthritis Rheum 62(9):2696–2706
52. Winter A, Breit S, Parsch D, Benz K, Steck E, Hauner H et al (2003) Cartilage-like gene expression in differentiated human stem cell spheroids: a comparison of bone marrow-derived and adipose tissue-derived stromal cells. Arthritis Rheum 48(2):418–429
53. Tsuchiya K, Chen GP, Ushida T, Matsuno T, Tateishi T (2004) The effect of coculture of chondrocytes with mesenchymal stem cells on their cartilaginous phenotype in vitro. Mater Sci Eng C Biomim Supramol Syst 24(3):391–396

54. Ma HL, Hung SC, Lin SY, Chen YL, Lo WH (2003) Chondrogenesis of human mesenchymal stem cells encapsulated in alginate beads. J Biomed Mater Res A 64(2):273–281
55. Diekman BO, Rowland CR, Lennon DP, Caplan AI, Guilak F (2010) Chondrogenesis of adult stem cells from adipose tissue and bone marrow: induction by growth factors and cartilage-derived matrix. Tissue Eng Part A 16(2):523–533
56. Chung C, Burdick JA (2009) Influence of three-dimensional hyaluronic acid microenvironments on mesenchymal stem cell chondrogenesis. Tissue Eng Part A 15(2):243–254
57. Sharma B, Williams CG, Khan M, Manson P, Elisseeff JH (2007) In vivo chondrogenesis of mesenchymal stem cells in a photopolymerized hydrogel. Plast Reconstr Surg 119(1):112–120
58. Park JS, Yang HN, Woo DG, Jeon SY, Park KH (2011) Chondrogenesis of human mesenchymal stem cells in fibrin constructs evaluated in vitro and in nude mouse and rabbit defects models. Biomaterials 32(6):1495–1507
59. Zuk PA, Zhu M, Ashjian P, De Ugarte DA, Huang JI, Mizuno H et al (2002) Human adipose tissue is a source of multipotent stem cells. Mol Biol Cell 13(12):4279–4295
60. Zuk PA, Zhu M, Mizuno H, Huang J, Futrell JW, Katz AJ et al (2001) Multilineage cells from human adipose tissue: implications for cell-based therapies. Tissue Eng 7(2):211–228
61. Estes BT, Diekman BO, Gimble JM, Guilak F (2010) Isolation of adipose-derived stem cells and their induction to a chondrogenic phenotype. Nat Protoc 5(7):1294–1311
62. Cheng NC, Estes BT, Awad HA, Guilak F (2009) Chondrogenic differentiation of adipose-derived adult stem cells by a porous scaffold derived from native articular cartilage extracellular matrix. Tissue Eng Part A 15(2):231–241
63. Awad HA, Wickham MQ, Leddy HA, Gimble JM, Guilak F (2004) Chondrogenic differentiation of adipose-derived adult stem cells in agarose, alginate, and gelatin scaffolds. Biomaterials 25(16):3211–3222
64. Betre H, Setton LA, Meyer DE, Chilkoti A (2002) Characterization of a genetically engineered elastin-like polypeptide for cartilaginous tissue repair. Biomacromolecules 3(5):910–916
65. Merceron C, Portron S, Masson M, Fellah BH, Gauthier O, Lesoeur J et al (2010) Cartilage tissue engineering: From hydrogel to mesenchymal stem cells. Biomed Mater Eng 20(3):159–166
66. Lin Y, Luo E, Chen X, Liu L, Qiao J, Yan Z et al (2005) Molecular and cellular characterization during chondrogenic differentiation of adipose tissue-derived stromal cells in vitro and cartilage formation in vivo. J Cell Mol Med 9(4):929–939
67. Masuoka K, Asazuma T, Hattori H, Yoshihara Y, Sato M, Matsumura K et al (2006) Tissue engineering of articular cartilage with autologous cultured adipose tissue-derived stromal cells using atelocollagen honeycomb-shaped scaffold with a membrane sealing in rabbits. J Biomed Mater Res B Appl Biomater 79(1):25–34
68. Thomson JA, Itskovitz-Eldor J, Shapiro SS, Waknitz MA, Swiergiel JJ, Marshall VS et al (1998) Embryonic stem cell lines derived from human blastocysts. Science 282(5391):1145–1147
69. Kawaguchi J (2006) Generation of osteoblasts and chondrocytes from embryonic stem cells. Methods Mol Biol 330:135–148
70. Koay EJ, Hoben GM, Athanasiou KA (2007) Tissue engineering with chondrogenically differentiated human embryonic stem cells. Stem Cells 25(9):2183–2190
71. zur Nieden NI, Kempka G, Rancourt DE, Ahr HJ (2005) Induction of chondro-, osteo- and adipogenesis in embryonic stem cells by bone morphogenetic protein-2: effect of cofactors on differentiating lineages. BMC Dev Biol 5:1
72. Kramer J, Hegert C, Guan K, Wobus AM, Muller PK, Rohwedel J (2000) Embryonic stem cell-derived chondrogenic differentiation in vitro: activation by BMP-2 and BMP-4. Mech Dev 92(2):193–205

73. Nakagawa T, Lee SY, Reddi AH (2009) Induction of chondrogenesis from human embryonic stem cells without embryoid body formation by bone morphogenetic protein 7 and transforming growth factor beta1. Arthritis Rheum 60(12):3686–3692
74. Yang Z, Sui L, Toh WS, Lee EH, Cao T (2009) Stage-dependent effect of TGF-beta1 on chondrogenic differentiation of human embryonic stem cells. Stem Cells Dev 18(6):929–940
75. Oldershaw RA, Baxter MA, Lowe ET, Bates N, Grady LM, Soncin F et al (2010) Directed differentiation of human embryonic stem cells toward chondrocytes. Nat Biotechnol 28(11):1187–1194
76. Toh WS, Yang Z, Liu H, Heng BC, Lee EH, Cao T (2007) Effects of culture conditions and bone morphogenetic protein 2 on extent of chondrogenesis from human embryonic stem cells. Stem Cells 25(4):950–960
77. Vats A, Bielby RC, Tolley N, Dickinson SC, Boccaccini AR, Hollander AP et al (2006) Chondrogenic differentiation of human embryonic stem cells: the effect of the microenvironment. Tissue Eng 12(6):1687–1697
78. Bigdeli N, Karlsson C, Strehl R, Concaro S, Hyllner J, Lindahl A (2009) Coculture of human embryonic stem cells and human articular chondrocytes results in significantly altered phenotype and improved chondrogenic differentiation. Stem Cells 27(8):1812–1821
79. Hwang NS, Varghese S, Zhang Z, Elisseeff J (2006) Chondrogenic differentiation of human embryonic stem cell-derived cells in arginine-glycine-aspartate-modified hydrogels. Tissue Eng 12(9):2695–2706
80. Adachi N, Sato K, Usas A, Fu FH, Ochi M, Han CW et al (2002) Muscle derived, cell based ex vivo gene therapy for treatment of full thickness articular cartilage defects. J Rheumatol 29(9):1920–1930
81. Nawata M, Wakitani S, Nakaya H, Tanigami A, Seki T, Nakamura Y et al (2005) Use of bone morphogenetic protein 2 and diffusion chambers to engineer cartilage tissue for the repair of defects in articular cartilage. Arthritis Rheum 52(1):155–163
82. Kuroda R, Usas A, Kubo S, Corsi K, Peng H, Rose T et al (2006) Cartilage repair using bone morphogenetic protein 4 and muscle-derived stem cells. Arthritis Rheum 54(2):433–442
83. Ito Y, Fitzsimmons JS, Sanyal A, Mello MA, Mukherjee N, O'Driscoll SW (2001) Localization of chondrocyte precursors in periosteum. Osteoarthr Cartil 9(3):215–223
84. Choi YS, Noh SE, Lim SM, Lee CW, Kim CS, Im MW et al (2008) Multipotency and growth characteristic of periosteum-derived progenitor cells for chondrogenic, osteogenic, and adipogenic differentiation. Biotechnol Lett 30(4):593–601
85. Miura Y, Fitzsimmons JS, Commisso CN, Gallay SH, O'Driscoll SW (1994) Enhancement of periosteal chondrogenesis in vitro. Dose-response for transforming growth factor-beta 1 (TGF-beta 1). Clin Orthop Relat Res 301:271–280
86. Stevens MM, Marini RP, Martin I, Langer R, Prasad Shastri V (2004) FGF-2 enhances TGF-beta1-induced periosteal chondrogenesis. J Orthop Res 22(5):1114–1119
87. Fukumoto T, Sperling JW, Sanyal A, Fitzsimmons JS, Reinholz GG, Conover CA et al (2003) Combined effects of insulin-like growth factor-1 and transforming growth factor-beta1 on periosteal mesenchymal cells during chondrogenesis in vitro. Osteoarthr Cartil 11(1):55–64
88. Stevens MM, Qanadilo HF, Langer R, Shastri VP (2004) A rapid-curing alginate gel system: utility in periosteum-derived cartilage tissue engineering. Biomaterials 25(5):887–894
89. Haniffa MA, Collin MP, Buckley CD, Dazzi F (2009) Mesenchymal stem cells: the fibroblasts' new clothes? Haematologica 94(2):258–263
90. Mizuno S, Glowacki J (1996) Chondroinduction of human dermal fibroblasts by demineralized bone in three-dimensional culture. Exp Cell Res 227(1):89–97
91. French MM, Rose S, Canseco J, Athanasiou KA (2004) Chondrogenic differentiation of adult dermal fibroblasts. Ann Biomed Eng 32(1):50–56

92. Singh M, Pierpoint M, Mikos AG, Kasper FK (2011) Chondrogenic differentiation of neonatal human dermal fibroblasts encapsulated in alginate beads with hydrostatic compression under hypoxic conditions in the presence of bone morphogenetic protein-2. J Biomed Mater Res A 98:412–424
93. Deng Y, Hu JC, Athanasiou KA (2007) Isolation and chondroinduction of a dermis-isolated, aggrecan-sensitive subpopulation with high chondrogenic potential. Arthritis Rheum 56(1):168–176
94. Nicoll SB, Wedrychowska A, Smith NR, Bhatnagar RS (2001) Modulation of proteoglycan and collagen profiles in human dermal fibroblasts by high density micromass culture and treatment with lactic acid suggests change to a chondrogenic phenotype. Connect Tissue Res 42(1):59–69
95. Liu X, Zhou G, Liu W, Zhang W, Cui L, Cao Y (2007) In vitro formation of lacuna structure by human dermal fibroblasts co-cultured with porcine chondrocytes on a 3D biodegradable scaffold. Biotechnol Lett 29(11):1685–1690
96. Rutherford RB, Moalli M, Franceschi RT, Wang D, Gu K, Krebsbach PH (2002) Bone morphogenetic protein-transduced human fibroblasts convert to osteoblasts and form bone in vivo. Tissue Eng 8(3):441–452
97. Yin S, Cen L, Wang C, Zhao G, Sun J, Liu W et al (2010) Chondrogenic transdifferentiation of human dermal fibroblasts stimulated with cartilage-derived morphogenetic protein 1. Tissue Eng Part A 16(5):1633–1643
98. Hiramatsu K, Sasagawa S, Outani H, Nakagawa K, Yoshikawa H, Tsumaki N (2011) Generation of hyaline cartilaginous tissue from mouse adult dermal fibroblast culture by defined factors. J Clin Invest 121(2):640–657
99. Lee CH, Cook JL, Mendelson A, Moioli EK, Yao H, Mao JJ (2010) Regeneration of the articular surface of the rabbit synovial joint by cell homing: a proof of concept study. Lancet 376(9739):440–448
100. Karp JM, Leng Teo GS (2009) Mesenchymal stem cell homing: the devil is in the details. Cell Stem Cell 4(3):206–216
101. Fong EL, Chan CK, Goodman SB (2011) Stem cell homing in musculoskeletal injury. Biomaterials 32(2):395–409
102. Shapiro F, Koide S, Glimcher MJ (1993) Cell origin and differentiation in the repair of full-thickness defects of articular cartilage. J Bone Joint Surg 75(4):532–553
103. Bryant SJ, Anseth KS (2002) Hydrogel properties influence ECM production by chondrocytes photoencapsulated in poly(ethylene glycol) hydrogels. J Biomed Mater Res 59(1):63–72
104. Fedorovich NE, Alblas J, de Wijn JR, Hennink WE, Verbout AJ, Dhert WJ (2007) Hydrogels as extracellular matrices for skeletal tissue engineering: state-of-the-art and novel application in organ printing. Tissue Eng 13(8):1905–1925
105. Nicodemus GD, Bryant SJ (2008) Cell encapsulation in biodegradable hydrogels for tissue engineering applications. Tissue Eng Part B Rev 14(2):149–165
106. Herlofsen SR, Kuchler AM, Melvik JE, Brinchmann JE (2011) Chondrogenic differentiation of human bone marrow-derived mesenchymal stem cells in self-gelling alginate discs reveals novel chondrogenic signature gene clusters. Tissue Eng Part A 17:1003–1013
107. Wang W, Li B, Yang J, Xin L, Li Y, Yin H et al (2010) The restoration of full-thickness cartilage defects with BMSCs and TGF-beta 1 loaded PLGA/fibrin gel constructs. Biomaterials 31(34):8964–8973
108. Tigli RS, Cannizaro C, Gumusderelioglu M, Kaplan DL (2011) Chondrogenesis in perfusion bioreactors using porous silk scaffolds and hESC-derived MSCs. J Biomed Mater Res A 96(1):21–28
109. Wang Y, Bella E, Lee CS, Migliaresi C, Pelcastre L, Schwartz Z et al (2010) The synergistic effects of 3-D porous silk fibroin matrix scaffold properties and hydrodynamic environment in cartilage tissue regeneration. Biomaterials 31(17):4672–4681

110. Neves SC, Teixcira LSM, Moroni L, Reis RL, Van Blitterswijk CA, Alves NM et al (2011) Chitosan/poly(epsilon-caprolactone) blend scaffolds for cartilage repair. Biomaterials 32(4):1068–1079
111. Suh JK, Matthew HW (2000) Application of chitosan-based polysaccharide biomaterials in cartilage tissue engineering: a review. Biomaterials 21(24):2589–2598
112. Chang CH, Liu HC, Lin CC, Chou CH, Lin FH (2003) Gelatin-chondroitin-hyaluronan tricopolymer scaffold for cartilage tissue engineering. Biomaterials 24(26):4853–4858
113. Tan H, Chu CR, Payne KA, Marra KG (2009) Injectable in situ forming biodegradable chitosan-hyaluronic acid based hydrogels for cartilage tissue engineering. Biomaterials 30(13):2499–2506
114. Baier Leach J, Bivens KA, Patrick CW Jr, Schmidt CE (2003) Photocrosslinked hyaluronic acid hydrogels: natural, biodegradable tissue engineering scaffolds. Biotechnol Bioeng 82(5):578–589
115. Varghese S, Hwang NS, Canver AC, Theprungsirikul P, Lin DW, Elisseeff J (2008) Chondroitin sulfate based niches for chondrogenic differentiation of mesenchymal stem cells. Matrix Biol 27(1):12–21
116. Varghese S, Elisseeff JH (2006) Hydrogels for musculoskeletal tissue engineering. Adv Polym Sci 203:95–144
117. Ma Z, Mao Z, Gao C (2007) Surface modification and property analysis of biomedical polymers used for tissue engineering. Colloids Surf B Biointerfaces 60(2):137–157
118. Nguyen LH, Kudva AK, Guckert NL, Linse KD, Roy K (2011) Unique biomaterial compositions direct bone marrow stem cells into specific chondrocytic phenotypes corresponding to the various zones of articular cartilage. Biomaterials 32(5):1327–1338
119. Wang C, Varshney RR, Wang DA (2010) Therapeutic cell delivery and fate control in hydrogels and hydrogel hybrids. Adv Drug Deliv Rev 62(7–8):699–710
120. Shin H, Jo S, Mikos AG (2003) Biomimetic materials for tissue engineering. Biomaterials 24(24):4353–4364
121. Kuo YC, Wang CC (2011) Surface modification with peptide for enhancing chondrocyte adhesion and cartilage regeneration in porous scaffolds. Colloids Surf B Biointerfaces 84(1):63–70
122. Kaufmann D, Fiedler A, Junger A, Auernheimer J, Kessler H, Weberskirch R (2008) Chemical conjugation of linear and cyclic RGD moieties to a recombinant elastin-mimetic polypeptide—a versatile approach towards bioactive protein hydrogels. Macromol Biosci 8(6):577–588
123. Sawyer AA, Hennessy KM, Bellis SL (2007) The effect of adsorbed serum proteins, RGD and proteoglycan-binding peptides on the adhesion of mesenchymal stem cells to hydroxyapatite. Biomaterials 28(3):383–392
124. Shin H, Zygourakis K, Farach-Carson MC, Yaszemski MJ, Mikos AG (2004) Attachment, proliferation, and migration of marrow stromal osteoblasts cultured on biomimetic hydrogels modified with an osteopontin-derived peptide. Biomaterials 25(5):895–906
125. Bahney CS, Hsu CW, Yoo JU, West JL, Johnstone B (2011) A bioresponsive hydrogel tuned to chondrogenesis of human mesenchymal stem cells. FASEB J 25:1486–1496
126. Dong Y, Li P, Chen CB, Wang ZH, Ma P, Chen GQ (2010) The improvement of fibroblast growth on hydrophobic biopolyesters by coating with polyhydroxyalkanoate granule binding protein PhaP fused with cell adhesion motif RGD. Biomaterials 31(34):8921–8930
127. You M, Peng G, Li J, Ma P, Wang Z, Shu W et al (2010) Chondrogenic differentiation of human bone marrow mesenchymal stem cells on polyhydroxyalkanoate (PHA) scaffolds coated with PHA granule binding protein PhaP fused with RGD peptide. Biomaterials 32(9):2305–2313
128. Re'em T, Tsur-Gang O, Cohen S (2010) The effect of immobilized RGD peptide in macroporous alginate scaffolds on TGFbeta1-induced chondrogenesis of human mesenchymal stem cells. Biomaterials 31(26):6746–6755

129. Vonwil D, Schuler M, Barbero A, Strobel S, Wendt D, Textor M et al (2010) An RGD-restricted substrate interface is sufficient for the adhesion, growth and cartilage forming capacity of human chondrocytes. Eur Cell Mater 20:316–328
130. Salinas CN, Anseth KS (2008) The enhancement of chondrogenic differentiation of human mesenchymal stem cells by enzymatically regulated RGD functionalities. Biomaterials 29(15):2370–2377
131. Villanueva I, Weigel CA, Bryant SJ (2009) Cell–matrix interactions and dynamic mechanical loading influence chondrocyte gene expression and bioactivity in PEG-RGD hydrogels. Acta Biomater 5(8):2832–2846
132. Chang JC, Hsu SH, Chen DC (2009) The promotion of chondrogenesis in adipose-derived adult stem cells by an RGD-chimeric protein in 3D alginate culture. Biomaterials 30(31):6265–6275
133. Hwang NS, Varghese S, Lee HJ, Zhang Z, Ye Z, Bae J et al (2008) In vivo commitment and functional tissue regeneration using human embryonic stem cell-derived mesenchymal cells. Proc Natl Acad Sci USA 105(52):20641–20646
134. Connelly JT, Garcia AJ, Levenston ME (2007) Inhibition of in vitro chondrogenesis in RGD-modified three-dimensional alginate gels. Biomaterials 28(6):1071–1083
135. Matsuki K, Sasho T, Nakagawa K, Tahara M, Sugioka K, Ochiai N et al (2008) RGD peptide-induced cell death of chondrocytes and synovial cells. J Orthop Sci 13(6):524–532
136. Davidson ENB, van der Kraan PM, van den Berg WB (2007) TGF-beta and osteoarthritis. Osteoarthr Cartil 15(6):597–604
137. Park JS, Woo DG, Yang HN, Lim HJ, Park KM, Na K et al (2009) Chondrogenesis of human mesenchymal stem cells encapsulated in a hydrogel construct: neocartilage formation in animal models as both mice and rabbits. J Biomed Mater Res A 92(3):988–996
138. Park JS, Woo DG, Yang HN, Na K, Park KH (2009) Transforming growth factor beta-3 bound with sulfate polysaccharide in synthetic extracellular matrix enhanced the biological activities for neocartilage formation in vivo. J Biomed Mater Res A 91(2):408–415
139. van der Kraan PM, Blaney Davidson EN, Blom A, van den Berg WB (2009) TGF-beta signaling in chondrocyte terminal differentiation and osteoarthritis: modulation and integration of signaling pathways through receptor-Smads. Osteoarthr Cartil 17(12):1539–1545
140. Zheng L, Fan HS, Sun J, Chen XN, Wang G, Zhang L et al (2010) Chondrogenic differentiation of mesenchymal stem cells induced by collagen-based hydrogel: an in vivo study. J Biomed Mater Res A 93(2):783–792
141. Mackay AM, Beck SC, Murphy JM, Barry FP, Chichester CO, Pittenger MF (1998) Chondrogenic differentiation of cultured human mesenchymal stem cells from marrow. Tissue Eng 4(4):415–428
142. Elder BD, Athanasiou KA (2009) Systematic assessment of growth factor treatment on biochemical and biomechanical properties of engineered articular cartilage constructs. Osteoarthr Cartil 17(1):114–123
143. Liu XW, Hu J, Man C, Zhang B, Ma YQ, Zhu SS (2011) Insulin-like growth factor-1 suspended in hyaluronan improves cartilage and subchondral cancellous bone repair in osteoarthritis of temporomandibular joint. Int J Oral Maxillofac Surg 40:184–190
144. Longobardi L, O'Rear L, Aakula S, Johnstone B, Shimer K, Chytil A et al (2006) Effect of IGF-I in the chondrogenesis of bone marrow mesenchymal stem cells in the presence or absence of TGF-beta signaling. J Bone Miner Res 21(4):626–636
145. Park H, Temenoff JS, Tabata Y, Caplan AI, Raphael RM, Jansen JA et al (2009) Effect of dual growth factor delivery on chondrogenic differentiation of rabbit marrow mesenchymal stem cells encapsulated in injectable hydrogel composites. J Biomed Mater Res A 88(4):889–897
146. Schmidt MB, Chen EH, Lynch SE (2006) A review of the effects of insulin-like growth factor and platelet derived growth factor on in vivo cartilage healing and repair. Osteoarthr Cartil 14(5):403–412

147. Jiang Y, Chen LK, Zhu DC, Zhang GR, Guo C, Qi YY et al (2010) The inductive effect of bone morphogenetic protein-4 on chondral-lineage differentiation and in situ cartilage repair. Tissue Eng Part A 16(5):1621–1632
148. Miljkovic ND, Cooper GM, Marra KG (2008) Chondrogenesis, bone morphogenetic protein-4 and mesenchymal stem cells. Osteoarthr Cartil 16(10):1121–1130
149. Tamai N, Myoui A, Hirao M, Kaito T, Ochi T, Tanaka J et al (2005) A new biotechnology for articular cartilage repair: subchondral implantation of a composite of interconnected porous hydroxyapatite, synthetic polymer (PLA-PEG), and bone morphogenetic protein-2 (rhBMP-2). Osteoarthr Cartil 13(5):405–417
150. Majumdar MK, Wang E, Morris EA (2001) BMP-2 and BMP-9 promotes chondrogenic differentiation of human multipotential mesenchymal cells and overcomes the inhibitory effect of IL-1. J Cell Physiol 189(3):275–284
151. Sekiya I, Larson BL, Vuoristo JT, Reger RL, Prockop DJ (2005) Comparison of effect of BMP-2, -4, and -6 on in vitro cartilage formation of human adult stem cells from bone marrow stroma. Cell Tissue Res 320(2):269–276
152. Fukuda A, Kato K, Hasegawa M, Hirata H, Sudo A, Okazaki K et al (2005) Enhanced repair of large osteochondral defects using a combination of artificial cartilage and basic fibroblast growth factor. Biomaterials 26(20):4301–4308
153. Huang X, Yang D, Yan W, Shi Z, Feng J, Gao Y et al (2007) Osteochondral repair using the combination of fibroblast growth factor and amorphous calcium phosphate/poly(L-lactic acid) hybrid materials. Biomaterials 28(20):3091–3100
154. Miyakoshi N, Kobayashi M, Nozaka K, Okada K, Shimada Y, Itoi E (2005) Effects of intraarticular administration of basic fibroblast growth factor with hyaluronic acid on osteochondral defects of the knee in rabbits. Arch Orthop Trauma Surg 125(10):683–692
155. Nakayama J, Fujioka H, Nagura I, Kokubu T, Makino T, Kuroda R et al (2009) The effect of fibroblast growth factor-2 on autologous osteochondral transplantation. Int Orthop 33(1):275–280
156. Siebert CH, Schneider U, Sopka S, Wahner T, Miltner O, Niedhart C (2006) Ingrowth of osteochondral grafts under the influence of growth factors: 6-month results of an animal study. Arch Orthop Trauma Surg 126(4):247–252
157. Tanaka H, Mizokami H, Shiigi E, Murata H, Ogasa H, Mine T et al (2004) Effects of basic fibroblast growth factor on the repair of large osteochondral defects of articular cartilage in rabbits: dose-response effects and long-term outcomes. Tissue Eng 10(3–4):633–641
158. Weiss S, Hennig T, Bock R, Steck E, Richter W (2010) Impact of growth factors and PTHrP on early and late chondrogenic differentiation of human mesenchymal stem cells. J Cell Physiol 223(1):84–93
159. Holland TA, Bodde EW, Cuijpers VM, Baggett LS, Tabata Y, Mikos AG et al (2007) Degradable hydrogel scaffolds for in vivo delivery of single and dual growth factors in cartilage repair. Osteoarthr Cartil 15(2):187–197
160. Holland TA, Tabata Y, Mikos AG (2005) Dual growth factor delivery from degradable oligo(poly(ethylene glycol) fumarate) hydrogel scaffolds for cartilage tissue engineering. J Control Release 101(1–3):111–125
161. Tabata Y, Nagano A, Ikada Y (1999) Biodegradation of hydrogel carrier incorporating fibroblast growth factor. Tissue Eng 5(2):127–138
162. Fan H, Zhang C, Li J, Bi L, Qin L, Wu H et al (2008) Gelatin microspheres containing TGF-beta3 enhance the chondrogenesis of mesenchymal stem cells in modified pellet culture. Biomacromolecules 9(3):927–934
163. Jo S, Shin H, Shung AK, Fisher JP, Mikos AG (2001) Synthesis and characterization of oligo(poly(ethylene glycol) fumarate) macromer. Macromolecules 34(9):2839–2844
164. Tabata Y, Hijikata S, Muniruzzaman M, Ikada Y (1999) Neovascularization effect of biodegradable gelatin microspheres incorporating basic fibroblast growth factor. J Biomater Sci Polym Ed 10(1):79–94

165. Holland TA, Tabata Y, Mikos AG (2003) In vitro release of transforming growth factor-beta 1 from gelatin microparticles encapsulated in biodegradable, injectable oligo(poly(ethylene glycol) fumarate) hydrogels. J Control Release 91(3):299–313
166. Holland TA, Tessmar JK, Tabata Y, Mikos AG (2004) Transforming growth factor-beta 1 release from oligo(poly(ethylene glycol) fumarate) hydrogels in conditions that model the cartilage wound healing environment. J Control Release 94(1):101–114
167. Park H, Temenoff JS, Holland TA, Tabata Y, Mikos AG (2005) Delivery of TGF-beta1 and chondrocytes via injectable, biodegradable hydrogels for cartilage tissue engineering applications. Biomaterials 26(34):7095–7103
168. Park H, Temenoff JS, Tabata Y, Caplan AI, Mikos AG (2007) Injectable biodegradable hydrogel composites for rabbit marrow mesenchymal stem cell and growth factor delivery for cartilage tissue engineering. Biomaterials 28(21):3217–3227
169. Park H, Guo X, Temenoff JS, Tabata Y, Caplan AI, Kasper FK et al (2009) Effect of swelling ratio of injectable hydrogel composites on chondrogenic differentiation of encapsulated rabbit marrow mesenchymal stem cells in vitro. Biomacromolecules 10(3):541–546
170. Jaklenec A, Hinckfuss A, Bilgen B, Ciombor DM, Aaron R, Mathiowitz E (2008) Sequential release of bioactive IGF-I and TGF-beta 1 from PLGA microsphere-based scaffolds. Biomaterials 29(10):1518–1525
171. Bouffi C, Thomas O, Bony C, Giteau A, Venier-Julienne MC, Jorgensen C et al (2010) The role of pharmacologically active microcarriers releasing TGF-beta3 in cartilage formation in vivo by mesenchymal stem cells. Biomaterials 31(25):6485–6493
172. Bae SE, Choi DH, Han DK, Park K (2010) Effect of temporally controlled release of dexamethasone on in vivo chondrogenic differentiation of mesenchymal stromal cells. J Control Release 143(1):23–30
173. Park JS, Na K, Woo DG, Yang HN, Park KH (2009) Determination of dual delivery for stem cell differentiation using dexamethasone and TGF-beta3 in/on polymeric microspheres. Biomaterials 30(27):4796–4805
174. Park JS, Yang HN, Woo DG, Chung HM, Park KH (2009) In vitro and in vivo chondrogenesis of rabbit bone marrow-derived stromal cells in fibrin matrix mixed with growth factor loaded in nanoparticles. Tissue Eng Part A 15(8):2163–2175
175. Ivkovic A, Pascher A, Hudetz D, Maticic D, Jelic M, Dickinson S et al (2010) Articular cartilage repair by genetically modified bone marrow aspirate in sheep. Gene Ther 17(6):779–789
176. Madry H, Kaul G, Cucchiarini M, Stein U, Zurakowski D, Remberger K et al (2005) Enhanced repair of articular cartilage defects in vivo by transplanted chondrocytes overexpressing insulin-like growth factor I (IGF-I). Gene Ther 12(15):1171–1179
177. Madry H, Orth P, Kaul G, Zurakowski D, Menger MD, Kohn D et al (2010) Acceleration of articular cartilage repair by combined gene transfer of human insulin-like growth factor I and fibroblast growth factor-2 in vivo. Arch Orthop Trauma Surg 130(10):1311–1322
178. Pagnotto MR, Wang Z, Karpie JC, Ferretti M, Xiao X, Chu CR (2007) Adeno-associated viral gene transfer of transforming growth factor-beta1 to human mesenchymal stem cells improves cartilage repair. Gene Ther 14(10):804–813
179. Cucchiarini M, Madry H, Ma C, Thurn T, Zurakowski D, Menger MD et al (2005) Improved tissue repair in articular cartilage defects in vivo by rAAV-mediated overexpression of human fibroblast growth factor 2. Mol Ther 12(2):229–238
180. Cucchiarini M, Schetting S, Terwilliger EF, Kohn D, Madry H (2009) rAAV-mediated overexpression of FGF-2 promotes cell proliferation, survival, and alpha-SMA expression in human meniscal lesions. Gene Ther 16(11):1363–1372
181. Ayala R, Zhang C, Yang D, Hwang Y, Aung A, Shroff SS et al (2011) Engineering the cell-material interface for controlling stem cell adhesion, migration, and differentiation. Biomaterials 32(15):3700–3711

182. Park JS, Woo DG, Sun BK, Chung HM, Im SJ, Choi YM et al (2007) In vitro and in vivo test of PEG/PCL-based hydrogel scaffold for cell delivery application. J Control Release 124(1–2):51–59
183. Chen JP, Su CH (2011) Surface modification of electrospun PLLA nanofibers by plasma treatment and cationized gelatin immobilization for cartilage tissue engineering. Acta Biomater 7(1):234–243
184. Martins A, Pinho ED, Faria S, Pashkuleva I, Marques AP, Reis RL et al (2009) Surface modification of electrospun polycaprolactone nanofiber meshes by plasma treatment to enhance biological performance. Small 5(10):1195–1206
185. Kim K, Dean D, Lu A, Mikos AG, Fisher JP (2011) Early osteogenic signal expression of rat bone marrow stromal cells is influenced by both hydroxyapatite nanoparticle content and initial cell seeding density in biodegradable nanocomposite scaffolds. Acta Biomater 7(3):1249–1264
186. Vlacic-Zischke J, Hamlet SM, Friis T, Tonetti MS, Ivanovski S (2011) The influence of surface microroughness and hydrophilicity of titanium on the up-regulation of TGFbeta/BMP signalling in osteoblasts. Biomaterials 32(3):665–671
187. Lee JW, Kim YH, Park KD, Jee KS, Shin JW, Hahn SB (2004) Importance of integrin beta1-mediated cell adhesion on biodegradable polymers under serum depletion in mesenchymal stem cells and chondrocytes. Biomaterials 25(10):1901–1909
188. Hamilton DW, Riehle MO, Monaghan W, Curtis AS (2006) Chondrocyte aggregation on micrometric surface topography: a time-lapse study. Tissue Eng 12(1):189–199
189. Jeong CG, Hollister SJ (2010) A comparison of the influence of material on in vitro cartilage tissue engineering with PCL, PGS, and POC 3D scaffold architecture seeded with chondrocytes. Biomaterials 31(15):4304–4312
190. Kim K, Dean D, Wallace J, Breithaupt R, Mikos AG, Fisher JP (2011) The influence of stereolithographic scaffold architecture and composition on osteogenic signal expression with rat bone marrow stromal cells. Biomaterials 32(15):3750–3763
191. Kim K, Yeatts A, Dean D, Fisher JP (2010) Stereolithographic bone scaffold design parameters: osteogenic differentiation and signal expression. Tissue Eng Part B Rev 16(5):523–539
192. Nuernberger S, Cyran N, Albrecht C, Redl H, Vecsei V, Marlovits S (2011) The influence of scaffold architecture on chondrocyte distribution and behavior in matrix-associated chondrocyte transplantation grafts. Biomaterials 32(4):1032–1040
193. Li WJ, Mauck RL, Cooper JA, Yuan X, Tuan RS (2007) Engineering controllable anisotropy in electrospun biodegradable nanofibrous scaffolds for musculoskeletal tissue engineering. J Biomech 40(8):1686–1693
194. Martins A, Araujo JV, Reis RL, Neves NM (2007) Electrospun nanostructured scaffolds for tissue engineering applications. Nanomedicine (Lond) 2(6):929–942
195. Li WJ, Jiang YJ, Tuan RS (2006) Chondrocyte phenotype in engineered fibrous matrix is regulated by fiber size. Tissue Eng 12(7):1775–1785
196. Li WJ, Tuli R, Okafor C, Derfoul A, Danielson KG, Hall DJ et al (2005) A three-dimensional nanofibrous scaffold for cartilage tissue engineering using human mesenchymal stem cells. Biomaterials 26(6):599–609
197. Spadaccio C, Rainer A, Trombetta M, Vadala G, Chello M, Covino E et al (2009) Poly-L-lactic acid/hydroxyapatite electrospun nanocomposites induce chondrogenic differentiation of human MSC. Ann Biomed Eng 37(7):1376–1389
198. Wise JK, Yarin AL, Megaridis CM, Cho M (2009) Chondrogenic differentiation of human mesenchymal stem cells on oriented nanofibrous scaffolds: engineering the superficial zone of articular cartilage. Tissue Eng Part A 15(4):913–921
199. Binulal NS, Deepthy M, Selvamurugan N, Shalumon KT, Suja S, Mony U et al (2010) Role of nanofibrous poly(caprolactone) scaffolds in human mesenchymal stem cell attachment and spreading for in vitro bone tissue engineering–response to osteogenic regulators. Tissue Eng Part A 16(2):393–404

200. Hu J, Feng K, Liu X, Ma PX (2009) Chondrogenic and osteogenic differentiations of human bone marrow-derived mesenchymal stem cells on a nanofibrous scaffold with designed pore network. Biomaterials 30(28):5061–5067
201. Li WJ, Tuli R, Huang X, Laquerriere P, Tuan RS (2005) Multilineage differentiation of human mesenchymal stem cells in a three-dimensional nanofibrous scaffold. Biomaterials 26(25):5158–5166
202. Casper ME, Fitzsimmons JS, Stone JJ, Meza AO, Huang Y, Ruesink TJ et al (2010) Tissue engineering of cartilage using poly-epsilon-caprolactone nanofiber scaffolds seeded in vivo with periosteal cells. Osteoarthr Cartil 18(7):981–991
203. Li WJ, Chiang H, Kuo TF, Lee HS, Jiang CC, Tuan RS (2009) Evaluation of articular cartilage repair using biodegradable nanofibrous scaffolds in a swine model: a pilot study. J Tissue Eng Regen Med 3(1):1–10
204. Sharma B, Williams CG, Kim TK, Sun D, Malik A, Khan M et al (2007) Designing zonal organization into tissue-engineered cartilage. Tissue Eng 13(2):405–414
205. Ng KW, Ateshian GA, Hung CT (2009) Zonal chondrocytes seeded in a layered agarose hydrogel create engineered cartilage with depth-dependent cellular and mechanical inhomogeneity. Tissue Eng Part A 15(9):2315–2324
206. Gao J, Dennis JE, Solchaga LA, Goldberg VM, Caplan AI (2002) Repair of osteochondral defect with tissue-engineered two-phase composite material of injectable calcium phosphate and hyaluronan sponge. Tissue Eng 8(5):827–837
207. Schek RM, Taboas JM, Segvich SJ, Hollister SJ, Krebsbach PH (2004) Engineered osteochondral grafts using biphasic composite solid free-form fabricated scaffolds. Tissue Eng 10(9–10):1376–1385
208. Oliveira JM, Rodrigues MT, Silva SS, Malafaya PB, Gomes ME, Viegas CA et al (2006) Novel hydroxyapatite/chitosan bilayered scaffold for osteochondral tissue-engineering applications: Scaffold design and its performance when seeded with goat bone marrow stromal cells. Biomaterials 27(36):6123–6137
209. Guo X, Liao J, Park H, Saraf A, Raphael RM, Tabata Y et al (2010) Effects of TGF-beta3 and preculture period of osteogenic cells on the chondrogenic differentiation of rabbit marrow mesenchymal stem cells encapsulated in a bilayered hydrogel composite. Acta Biomater 6(8):2920–2931
210. Guo X, Park H, Liu G, Liu W, Cao Y, Tabata Y et al (2009) In vitro generation of an osteochondral construct using injectable hydrogel composites encapsulating rabbit marrow mesenchymal stem cells. Biomaterials 30(14):2741–2752
211. Holland TA, Bodde EW, Baggett LS, Tabata Y, Mikos AG, Jansen JA (2005) Osteochondral repair in the rabbit model utilizing bilayered, degradable oligo(poly(ethylene glycol) fumarate) hydrogel scaffolds. J Biomed Mater Res A 75(1):156–167
212. Guo X, Park H, Young S, Kretlow JD, van den Beucken JJ, Baggett LS et al (2010) Repair of osteochondral defects with biodegradable hydrogel composites encapsulating marrow mesenchymal stem cells in a rabbit model. Acta Biomater 6(1):39–47
213. Miller RE, Grodzinsky AJ, Vanderploeg EJ, Lee C, Ferris DJ, Barrett MF et al (2010) Effect of self-assembling peptide, chondrogenic factors, and bone marrow-derived stromal cells on osteochondral repair. Osteoarthr Cartil 18(12):1608–1619
214. Fan H, Hu Y, Qin L, Li X, Wu H, Lv R (2006) Porous gelatin-chondroitin-hyaluronate tri-copolymer scaffold containing microspheres loaded with TGF-beta1 induces differentiation of mesenchymal stem cells in vivo for enhancing cartilage repair. J Biomed Mater Res A 77(4):785–794
215. Fan H, Hu Y, Zhang C, Li X, Lv R, Qin L et al (2006) Cartilage regeneration using mesenchymal stem cells and a PLGA-gelatin/chondroitin/hyaluronate hybrid scaffold. Biomaterials 27(26):4573–4580

Interaction of Cells with Decellularized Biological Materials

Mathias Wilhelmi, Bettina Giere and Michael Harder

Abstract The idea to create the concept of cardiovascular "tissue engineering" is based on the recognition that until then all known allogeneic/xenogeneic biological or alloplastic implant materials were associated with shortcomings, which led to graft deterioration, degradation and finally destruction. Thus, it aims to develop viable cardiovascular structures, e.g. heart valves, myocardium or blood vessels, which ideally demonstrate mechanisms of remodeling and self-repair, a high microbiological resistance, complete immunological integrity and a functional endothelial cell layer to guarantee physiological hemostasis. In our current review we aim to identify basic limitations of previous concepts, explain why the use of decellularized matrices was a logical consequence and which limitations still exist.

Keywords Bioartificial organs · Extracellular matrix · Tissue engineering

Contents

1 Introduction	106
2 Native Grafts	106
3 Decellularized Grafts	108
4 Reseeding Concepts	109
5 Spontaneous In Situ Autologization	110
6 Immunogenicity of Extracellular Matrices	112
7 Conclusion	113
References	114

M. Wilhelmi (✉)
Division of Cardiac, Thoracic, Transplantation and Vascular Surgery,
Medizinische Hochschule Hannover, Carl-Neuberg-Str. 1,
30625 Hannover, Germany
e-mail: wilhelmi.mathias@mh-hannover.de

B. Giere · M. Harder
corlife GbR, Feodor-Lynen-Str. 23,
30625 Hannover, Germany

1 Introduction

In the search for a method "designed and constructed to meet the needs of each individual patient" [1], the idea of tissue engineering was created as an interdisciplinary field that applies the principles and methods of engineering and the life sciences to the development of biological substitutes that restore, maintain, and/or improve tissue function [1, 2]. The concept of "cardiovascular tissue engineering" aims to develop viable cardiovascular structures, e.g., heart valves, myocardium, and blood vessels, which demonstrate mechanisms of remodeling and self-repair, a high microbiological resistance, complete immunological integrity, and a functional endothelial cell layer to guarantee physiological hemostasis. Thus, regardless of the desired tissue type, all tissue engineering approaches rely on four essential components: (1) tissue-specific cells which form and vitalize the tissue, (2) signals (chemical and mechanical) which modulate cellular gene expressions and, thus, extracellular matrix (ECM) production, (3) cellular and humoral components of the recipient's immune system which allow and facilitate tissue integration or graft deterioration and destruction, and (4) matrix scaffolds which maintain these cells in a definite three-dimensional architecture.

2 Native Grafts

As early as the beginning of the eighteenth century, autologous vessel grafts were used for substitution or reconstructive vascular interventions [3–5]. However, allogeneic vascular grafts were not used until the 1940s. The clinical application of these prostheses was mainly based on works by Hufnagel [6] and Gross at al. [7, 8]. Harvested mainly from autopsies, these grafts were sterilized by cobalt radiation and predominately used for aortic replacement. In the 1950s, the first tissue banks were established and allogeneic vascular grafts were used to replace nearly any diseased central or peripheral artery. In the 1960s, Barret-Boyes [9] and Ross [10] were the first to use biological, allogeneic human heart valve prostheses in the clinical setting. In contrast to mechanical prostheses, the lack of distracting "clicking" noises and oral anticoagulation following the implantation of these biological valves to avoid thrombembolic complications made these valves increasingly attractive and the clinical demand by far exceeded their availability [11, 12]. However, degenerative changes were observed 8–10 years following implantation of both these valvular and vascular grafts, showing up on X-rays as severe calcifications and indicating ongoing degenerative processes, which ultimately led to complete graft loss [13–15] and thus required redo operations.

Another milestone in the history of biological grafts was laid by Rosenberg and Henderson [16, 17], who tried to reduce the immunogenicity of bovine carotid artery grafts by impregnation with dialdehyde starch. It was anticipated that induced cross-linkings between ECM proteins of these vessels would weaken the immunological responses of the recipient, while preserving graft structure/shape and prolonging its storage time. However, it emerged that dialdehyde starch was a bad choice, as all grafts demonstrated severe calcification and degradation. Another chemical agent—glutaraldehyde—was discovered and was found to be a much better choice. Many different tissues were impregnated very successfully with this agent, which, as a result, became readily available off-the-shelf. In 1973, Dardik and Darkik [18] evaluated glutaraldehyde-preserved human umbilical veins as an alternative biological bypass material, i.e., for lower extremity bypasses. However, as already postulated by Kunlin [19] in 1949, these grafts failed long term and, again, the greater saphenous vein was found to be the most suitable graft in this anatomic area.

Reviewing the underlying mechanisms of graft deterioration, on find two factors in particular which seem to play a major role in this phenomenon: (1) immunological reactions in the sense of subliminal tissue rejection [20, 21], which seem to be induced by the antigeneity of resident allogeneic cells, and (2) the method of preservation/fixation of these tissues with glutaraldehyde. Initially, this latter agent was used to reduce the immunogenicity of tissues via the cross-linking of collagen fibres to prolong its durability. In the meantime, however, it had become clear that glutaraldehyde increases the risk of calcification, potentially amplifies immunological reactions, and inhibits processes of in vivo regeneration [22]. Today it is believed that the antigeneic properties of allogeneic prostheses in the sense of histocompatibility differences are responsible for immunological responses and their resulting tissue rejection. Although, at least in theory, it is possible to modify immunological differences between donor tissue and the recipient immune system, e.g., via various methods of tissue preservation or low-dose immunosuppressive therapy, current clinical application of allogeneic grafts is mostly limited to special cases such as infections or elderly patients. In contrast, autologous vessel grafts, e.g., the greater saphenous vein, are still the first choice for reconstructive and substitutional interventions, especially in small- and medium-caliber vessels.

The clinical restriction of alloplastic vascular grafts to mainly large-vessel areas is explained by the clinical observation that autologous vessel grafts such as the greater saphenous vein still show superior patency rates. However, contrary to the assumption that the greater saphenous vein may represent a universally applicable vessel graft, it should be noted that this vessel is not available in every patient because of prior surgical interventions, varicosis, or deep vein thrombosis. Furthermore, it belongs to the venous and, consequently, low-pressure part of the cardiovascular system, predisposing ectatic and degenerative deformation when exposed to arterial/higher blood pressure load. Other autologous venous grafts, e.g., the femoral vein, or those obtained from the upper extremity exhibit the same structural disadvantages and are reported to be even less suitable than the greater saphenous vein. Autologous arteries, too, are not available in every patient and are shorter in length, so bypass grafts, e.g., at the lower extremity, are difficult to perform.

3 Decellularized Grafts

The use of decellularized tissues as matrix scaffolds seems to be obvious. Decellularized tissue is composed mainly of the ECM, and cellular residual components that are not washed out, such as DNA, RNA, cell membrane, and debris from cell organelles (nucleus, mitochondria, etc.). The degree of contamination by cellular components depends on the method and tissue used.

The ECM is a secreted product composed of functional proteins, glycosaminoglycans, glycoproteins, and small molecules arranged in a unique, tissue-specific, three-dimensional architecture. The composition and ultrastructure of the ECM are determined by the resident cells, the mechanical stress, and physiological conditions. In addition, proteases (such as kallikrein) and modulators (growth factors, cytokines) are associated with the ECM. All cellular activities, such as settlement, migration, and three-dimensional growth, are strongly determined by the structural and functional roles of the ECM and the physiological and biomechanical setting of the ECM.

Although the components of the ECM are highly homologous across different species, xenogeneic ECM is often rejected by the recipient. Chemical cross-linking passivates the antigenic epitopes, resulting in a material which is less immunogenic but also less biocompatible.

Decellularized tissue has been developed as a biologic scaffold for tissue engineering applications in virtually every body system, such as bladder, tendons, and cardiovascular structures. Interaction of cells with the decellularized tissue in terms of immunogenicity and repopulation can best be studied in artery and heart-valve substitutes, as these systems are in direct contact with blood and because their initial functionality does not depend on living cells, avoiding any necrotic or other adverse effects.

A working group led by Wilson [23] established a multistep decellularization process for heart valves, which was based on the use of hypotonic and hypertonic solutions, detergents, and enzymes to remove all cellular (predominate antigenetic) components of allogeneic canine heart valve prostheses. After 1 month following implantation of these valves in the pulmonary position in dogs, the valves were macroscopically intact and gave no indication of inflammatory reactions or other immunological side effects. Other working groups who used similar in vitro decellularization protocols prior to implantation reported comparable good results. Thus, the first commercially available decellularized and cryopreserved heart valve prosthesis was created. However, despite evidence that decellularized valves exhibit reduced immunogenicity in comparison with native control valves [24], Simon et al. [25] warned that the application of these decellularized valves may lead to accelerated destruction, especially when used in infants. The presumed reason for this phenomenon was an elevated activity of the immune system of infants in combination with a physiologically increased calcium metabolism at this age.

A working group headed by Huynh [26] used small intestinal submucosa and bovine type-I collagen to generate a new kind of vascular prosthesis. After the removal of all cellular components via hypotonic solutions, Huynh et al. [26] tested autologous, allogeneic, and xenogeneic grafts in large-animal models and reported excellent patency rates. In the histological analysis they found that these primary decellularized implanted grafts were spontaneously reseeded in vivo and, therefore, obviously underwent regenerative processes. However, in experiments using smaller animals such as rats, these small-caliber vessel grafts were found to be occluded at a very early stage owing to thrombus formation [27].

4 Reseeding Concepts

Trying to avoid the immunological influence of residual cellular fragments, Gulbins et al. [28] reseeded cryopreserved human allografts with autologous endothelial cells in vitro without prior decellularization and implanted these prostheses in animals—with moderate success. Another concept which can be taken as a logical consequence of the previously described method was the autologous endothelial reseeding of decellularized allogeneic and xenogeneic heart valve prostheses. Following very promising data in animal models, initial results in human patients are now available (xenogeneic heart valves reseeded with autologous human endothelial cells) [29].

Another approach to autologous endothelial reseeding of decellularized heart valve prostheses was implemented in Hanover in close cooperation with the University of Chişinău (Moldava). On the basis of the consideration that decellularized, i.e., xenogeneic, matrix scaffolds may still induce immunological reactions because of interspecies differences, human allografts which had been decellularized using an elaborate protocol were reseeded with autologous endothelial cells obtained as mononuclear cells isolated from individual blood samples. Positive stains for von Willebrand factor, CD31 (platelet/endothelial cell adhesion molecule 1) and flk-1, as observed in monolayers of cells cultivated and differentiated on the luminal surface of the scaffolds in a dynamic bioreactor system, indicated the endothelial nature of these cells. Reseeded valves were implanted in a pulmonary position of two pediatric patients (aged 13 and 11 years) with congenital pulmonary valve failure. Postoperatively, a mild pulmonary regurgitation was documented in both children. On the basis of regular echocardiographic investigations, the hemodynamic parameters and cardiac morphology changed in 3.5 years as follows: increase of the pulmonary valve annulus diameter (18–22.5 and 22–26 mm, respectively), decrease of valve regurgitation (trivial/mild and trivial, respectively), one decrease (16–9 mmHg) and one increase (8–9.5 mmHg) of the mean transvalvular gradient, and one remaining (26 mm) and one decreasing (32–28 mm) right ventricular end-diastolic diameter. The body surface area increased (1.07–1.42 and 1.07–1.46 m^2, respectively) and no signs of valve degeneration were observed in either of the patients 7 years after the procedure. Thus, it could be shown that the tissue engineering of heart valves using autologous endothelial progenitor cells is a feasible and safe method—at least for pulmonary valve replacement. Tissue-engineered valves have the potential to remodel and grow according to the somatic growth of a child [30, 31].

The first vascular prosthesis completely generated on the basis of biological materials goes back to the work of Weinberg and Bell [32]. They seeded smooth muscle cells in vitro on a collagen gel, which mimicked the lamina media, added fibroblasts to the outer surface of this construct, and, thus generated a bioartificial adventitia. Following 2 weeks of in vitro culture, endothelial cells were subsequently added to the luminal surface to serve as an artificial lamina interna. Scanning electron microscopy evaluations revealed closed endothelial monolayers on the luminal surface of these grafts, which stained positive for von Willebrand factor. However, although histological data were very promising, biomechanical tests revealed that this construct had only very limited structural stability, so Dacron nets had to be wrapped around the grafts to allow at least physiological pressure loads [32]. L'Heureux et al. [33] adopted up this method, but changed some of the culture conditions. They noticed that it was possible to positively influence graft stability and pressure tolerance/burst strength [34]. In an attempt to further evaluate the in vivo behavior of decellularized and autologous reseeded vascular grafts, Koenneker et al. [35] implanted xenogeneic bovine internal arteries as arteriovenous shunts into sheep. Following periods of 3 and 6 months, most endothelial reseeded ($n = 6/6$; 6/7) and decellularzed ($n = 5/6$; 5/7) grafts were patent and not significantly stenosed. However, histological analyses revealed complete endothelial surface coverage of reseeded grafts but not of decellularized grafts and, most importantly, only decellularized grafts exhibited pronounced tissue calcification. Thus, these data further support the hypothesis that autologous endothelial graft reseeding induces immunotolerance to some extent and protects even xenogeneic decellularized matrices against immunological processes, deterioration, and destruction [35].

Another approach was described by Campbell et al. [36], who implanted silastic tubings into the peritoneal cavity of rats. After 2 weeks, they observed that fibroblasts and mesothelial cells seeded on the outer surface of these tubings. The resulting tubular tissue sheet was then dissected and everted so that the previously outer mesothelial cells subsequently built up the inner surface and thus, mimicked the lamina interna. In animal models these bioartificial vessel grafts showed physiological reactivity towards vasoactive agents and were patent for up to 4 months.

5 Spontaneous In Situ Autologization

Spontaneous in situ autologization of decellularized tissue is a successful procedure which has been shown by numerous groups to be effective for heart valves. Spontaneous in situ autologization fulfills a major prerequisite to maintain a durable valve replacement [37, 38]. Numerous animal experiments have been conducted to demonstrate the in vivo re-endothelialization and cellular infiltration of deeper tissue

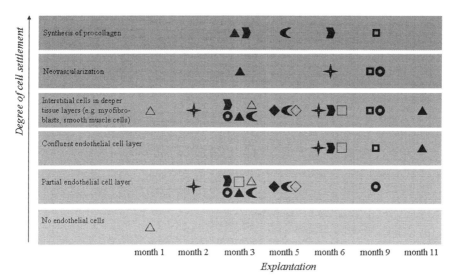

Fig. 1 Spontaneous in situ autologization of decellularized tissues in various animal models. ◆ Akhyari et al. [64], ○ Baraki et al. [40], ✦ Ota et al. [39], ☐ Dohmen et al. [42], ◗ Dohmen et al. [44], ◖ Elkins et al. [43], ◻ da Costa et al. [65], △ Leyh et al. [45], ▲ Lichtenberg et al. [41], ▲ Erdbrügger et al. [38]. The division of the time axis is not equidistant

layers. Regardless of the animal model (xenogeneic, allogeneic), the method of decellularization, and the type of valve replacement (aortic or pulmonary valve), the results are extremely homogeneous. Figure 1 shows the degree of cell-population settlement as a function of time. The element which all experiments have in common is that decellularized tissues were implanted in an orthotopic position (with the exception of [39]). An early-stage endothelialization is mainly observed after 3 months. In some experiments, it was possible to observe the invasion of interstitial cells and the production of procollagen [38, 40–43]. After 6 months, a confluent endothelium and a colonization of the deeper layers of tissue was observed in most cases [39, 44, 45]. An exception to these findings is the study of Baraki et al. [40], where, even after 9 months, no confluent endothelium was determined. In these experiments, however, aortic valves were implanted. The average pressure in the aorta is about 100 mmHg [46] and in the pulmonary artery it is about 14 mmHg [47]. It is conceivable that the increased blood flow in the aorta impedes or delays a cellular settlement. Furthermore, increased mechanical stress at the coaptation sites of the valve leaflets results in a very thin thrombotic layer that prevents the adhesion of cells. The ingrowth of interstitial cells, such as myofibroblasts and smooth muscle cells, neovascularization, and synthesis of procollagen between months 3 and 11 are evidence for tissue remodelling of the adventitial side and the subsequent reconstruction of deeper layers of tissue.

There are only few clinical spontaneous in situ autologization data available. The reasons for this are the lack of in vivo imaging systems and the low number of explanted heart valve substitutes based on decellularized tissue. In 2006,

Erdbrügger et al. [38] published data from a clinical trial involving patients who had received a decellularized porcine valve replacement. The substitute was explanted after 10 months because of an aneurysm at the ligature. Histological analysis on the explant demonstrated endothelial cells on the surface and recolonization of deeper tissue layers with recipient cells. Similar results were reported by Dohmen et al. [48] in 2007; however, the patient received a decellularized heart valve, which had been reseeded in vitro before implantation. The patient was reoperated on after 3 months. A monolayer of endothelial cells and interstitial cells was found in deeper tissue layers of the heart valve biopsy.

6 Immunogenicity of Extracellular Matrices

The failure of allogeneic heart valve substitutes may have immunological causes [49–54]. Basically, the genetic differences of the cellular surface antigens, the human leukocyte antigen (HLA) molecules, are responsible for immunological compatibility between donor and recipient. Therefore, incompatibility between recipient and donor in terms of blood groups (A, B, O) and HLAs may trigger immunological reactions [53]. The resulting graft rejection is a host-versus-graft disease. Use of decellularized tissue may overcome this disadvantage. The cellular components of the starting material are removed, so the resulting decellularized tissue mainly consists of an ECM. The elimination of cellular components results in a reduction of alloreactivity.

In 2005, da Costa et al. [55] published a study involving 20 patients who underwent heart valve replacement: 11 patients received decellularized grafts and nine patients received conventional cryopreserved homografts. The immune status with regard to the formation of HLA antibodies was collected after 5, 10, 30, 90, and 180 days. The basal levels in homograft recipients increased 1 month after surgery and were still elevated 6 months after the surgery. In contrast, no increase in HLA class I and HLA class II titer was determined in seven patients who received decellularized grafts. One patient showed a slight increase of HLA class I levels and two additional patients exhibited an abnormal increase (Fig. 2).

The human, pulmonary, decellularized, and cryopreserved CryoValve® SG heart valve from CryoLife® has been available on the market sine 2000. CryoValve® SG valves are already in clinical use and publications have report a lower alloreactivity compared with conventional vital homografts. Several groups have shown that HLA antibody levels after CryoValve® SG implantation were significantly reduced. Working groups based around Zehr [56] and Elkins [57] reported that the CryoValve® SG valve was tolerated by most patients (91%): after 1 month 86% were HLA negative, 88% were HLA negative after 3 months, and 95% were HLA negative after 1 year. By contrast, HLA immunogenicity of cryopreserved homografts is much higher. In studies involving pediatric patients ($n = 9$), Shaddy et al. [52] demonstrated an average increase in HLA class I level of 3.2% before surgery to 63% 25 days after surgery and to 99.7% 3 months after the operation. Further analysis showed that these antibody levels persisted at a level of about 87% for up to 1 year after surgery.

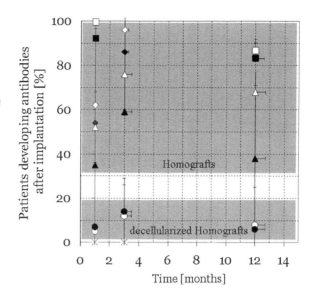

Fig. 2 Patients developing antibodies after implantation of homografts or decellularized homografts. Homografts: *open diamonds* class I, *filled diamonds* class II (Hawkins et al. [59]); *open triangles* class I, *filled triangles* class II (Hawkins et al. [63]); *open squares* panel reactive antibodies (Shaddy et al. [52]); *filled squares* panel reactive antibodies with DDT (Shaddy et al. [52]). Decellularized homografts: *open circles* class I, *filled circles* class II (Hawkins et al. [59]); *asterisks* class I (Elkins et al. [43])

Bechtel et al. [58] compared the immunological data from 24 CryoValve® SG recipients with the data from 22 patients who had received a standard cryopreserved homograft. No anti-HLA antibodies were detectable in the CryoValve® SG group between months 1 and 6 after surgery. But 66% of the homograft recipient group patients developed positive HLA titers [58]. The data are consistent with the findings of Hawkins et al. [59].

Improper decellularization seems to be one major factor for the fatal outcomes of decellularized porcine heart valves [25]. Significant cellular components were still detectable [60]. Other decellularized porcine heart valves have produced unsatisfactory clinical results, as they are repelled relatively quickly [61].

Numerous publications of clinical data have shown a reduced immunogenicity of decellularized human tissue [55–59, 62, 63]. The depletion of the immunogenic cellular components by the decellularization process in vivo leads to a significant decrease of HLA class I and HLA class II antibody stimulation compared with conventional homografts (and compared with xenogeneic matrices) and thus to a better acceptance by the immune system. This is an important basis for long-term durability of heart valve replacements.

7 Conclusion

The shortcomings of all known alloplastic implant materials, namely, infectious and thrombembolic complications, blazed a trail for biological and bioartificial tissue substitutes. However, although these grafts found their way into clinical practice and improved methods for tissue decellularization were established over the last few years, residual immunological differences between donor tissue and the recipient immune system still restrict current clinical applications of allogeneic and xenogeneic grafts. Thus, prospective scientific efforts will be focused on the development of strategies for in vitro or improved/accelerated in situ autologization of these tissues to overcome this last major barrier of bioartificial tissue implantation.

References

1. Vacanti JP, Langer R (1999) Tissue engineering: the design and fabrication of living replacement devices for surgical reconstruction and transplantation. Lancet 354(Suppl 1):32–34
2. Skalak R, Fox C (1988) Tissue engineering. In: Skalak R, Fox C (eds) Workshop on tissue engineering 26–29.02.1988, Granlibakken, Lake Tahoe, CA, USA. Liss, New York
3. Murphy JB (1897) Resection of arteries and veins injured in continuity—end-to-end suture: experimental results and clinical research. Med Res 31:73–88
4. Carrel A (1902) La technique operatoire des anastomoses vascularies at le transplantation des visceres. Lyon Med 89
5. Goyanes J (1906) Nuevos trabajos de cirurgia vascular. Siglo Med 53:446–561
6. Hufnagel CA (1947) Preserved homologous arterial transplants. Bull Am Coll Surg 32:231
7. Gross RE, Bill AH (1948) Preliminary observations on the use of the human arterial grafts in the treatment of certain cardiovascular defects. N Engl J Med 239:578–591
8. Gross RE, Bill AH, Preice EC (1949) Methods for preservation and transplantation of arterial grafts: observations on arterial grafts in dogs; report on transplantation of preserved arterial grafts in nine human cases. Surg Gynecol Obstet 88:68–71
9. Barratt-Boyes BG (1965) A method for preparing and inserting a homograft aortic valve. Br J Surg 52:847–856
10. Ross D (1967) Homograft replacement of the aortic valve. Br J Surg 54:842–843
11. Szilagyi DE, McDonald RT, Smith RF (1957) Biologic fate of human arterial homografts. Arch Surg 75:506–529
12. Crawford ES et al (1960) Evaluation of late failures after reconstructive operations for occlusive lesions of the aorta and iliac, femoral, and popliteal arteries. Surgery 47:79–104
13. Outdot J (1951) La greffe vasculaire dans les thromboses du Carrefour aortique. Presse Med 59:234
14. Outdot J, Beaconsfield P (1953) Thromboses of the aortic bifurcation treated by resection and homograft replacement. Arch Surg 66:365–374
15. Dubost C, Allary M, Oeconomos N (1952) Resection of an aneurysm of the abdominal aorta: reestablishment of the continuity by a preserved human arterial graft, with result after five months. Arch Surg 64:405–408
16. Rosenberg NG, Henderson J (1956) The use of segmental arterial implants prepared by enzymatic modification of heterologous blood vessels. Surg Forum 6:242
17. Rosenberg N (1976) The bovine arterial graft and its several applications. Surg Gynecol Obstet 142(1):104–108
18. Dardik I, Darkik H (1973) Vascular heterograft: human umbilical cord vein as an aortic substitute in baboon. A preliminary report. J Med Primatol 2(5):296–301
19. Kunlin J (1949) Le traitement de l'ischemie obliterante par la greffe veineuse longue. Arch Mal Couer 42: 371–372
20. Wilhelmi MH et al (2003) Role of inflammation and ischemia after implantation of xenogeneic pulmonary valve conduits: histological evaluation after 6 to 12 months in sheep. Int J Artif Organs 26(5):411–420
21. Wilhelmi MH et al (2003) Role of inflammation in allogeneic and xenogeneic heart valve degeneration: immunohistochemical evaluation of inflammatory endothelial cell activation. J Heart Valve Dis 12(4): 520–526

22. O'Brien MF et al (1999) The Synergraft valve: a new acellular (nonglutaraldehyde-fixed) tissue heart valve for autologous recellularization first experimental studies before clinical implantation. Semin Thorac Cardiovasc Surg 11:194–200
23. Wilson GJ et al (1995) Acellular matrix: a biomaterials approach for coronary artery bypass and heart valve replacement. Ann Thorac Surg 60:S353–S358
24. Bechtel JFM et al (2003) Evaluation of a decellularized homograft valve for reconstruction of the right ventricular outflow tract in the Ross-procedure. In: Second biennial meeting of the Society for Heart Valve Disease; 28 June–01 July 2003, Palais des Congres—Porte Maillot, Paris, p 347
25. Simon P et al (2003) Early failure of the tissue engineered porcine heart valve SYNERGRAFT in pediatric patients. Eur J Cardiothorac Surg 23:1002–1006; discussion 1006
26. Huynh T et al (1999) Remodeling of an acellular collagen graft into a physiologically responsive neovessel. Nat Biotechnol 17(11):1083–1086
27. Schmidt CE, Baier JM (2000) Acellular vascular tissues: natural biomaterials for tissue repair and tissue engineering. Biomaterials 21:2215–2231
28. Gulbins H et al (2003) Implantation of an autologously endothelialized homograft. J Thorac Cardiovasc Surg 126:890–891
29. Dohmen PM et al (2002) Tissue engineering of an auto-xenograft pulmonary heart valve. Asian Cardiovasc Thoracic Ann 10:25–30
30. Cebotari S et al (2006) Clinical application of tissue engineered human heart valves using autologous progenitor cells. Circulation 114(1 Suppl):I132–I137
31. Cebotari S et al (2002) Construction of autologous human heart valves based on an acellular allograft matrix. Circulation 106(12 Suppl 1):I63–I68
32. Weinberg CB, Bell E (1986) A blood vessel model constructed from collagen and cultured vascular cells. Science 231(4736):397–400
33. L'Heureux N et al (1998) A completely biological tissue-engineered human blood vessel. FASEB J 12(1):47–56
34. Edelman ER (1999) Vascular tissue engineering: designer arteries. Circ Res 85(12):1115–1117
35. Koenneker S, Teebken OE, Bonehie M, Pflaum M, Jockenhoevel S, Haverich A, Wilhelmi MH (2010) A biological alternative to alloplastic grafts in dialysis therapy: evaluation of an autologised bioartificial hemodialysis shunt vessel in a sheep model. Eur J Vasc Endovasc Surg 40(6):810–816
36. Campbell JH, Efendy JL, Campbell GR (1999) Novel vascular graft grown within recipient's own peritoneal cavity. Circ Res 85(12):1173–1178
37. Mertsching H, Leyh R, Haverich A (2001) Tissue engineering of autologous heart valves. Results of 3, 6, and 9 months implantation in a growing sheep model. Paper presented at the EACTS/ESTS joint meeting, Lisbon
38. Erdbrügger W, Konertz W, Dohmen PM et al (2006) Decellularized xenogenic heart valves reveal remodeling and growth potential in vivo. Tissue Eng 12(8):2059–2068
39. Ota T, Taketani S, Iwai S, Miyagawa S et al (2007) Novel method of decellularization of porcine valves using polyethylene glycol and gamma irradiation. Ann Thorac Surg 83(4):1501–1507
40. Baraki H, Tudorache I, Braun M et al (2009) Orthotopic replacement of the aortic valve with decellularized allograft in a sheep model. Biomaterials 6240–6246
41. Lichtenberg A, Tudorache I, Cebotari S et al (2006) Preclinical testing of tissue-engineered heart valves re-endothelialized under simulated physiological conditions. Circulation 114(1 Suppl): I559–I565
42. Dohmen PM, da Costa F, Holinski S et al (2006) Is there a possibility for a glutaraldehyde-free porcine heart valve to grow? Eur Surg Res 38(1):54–61
43. Elkins RC, Goldstein S, Hewitt CW et al (2001) Recellularization of heart valve grafts by a process of adaptive remodeling. Semin Thorac Cardiovasc Surg 13(4 Suppl 1):87–92
44. Dohmen PM, Costa F, Lopes SV et al (2005) Results of a decellularized porcine heart valve implanted into the juvenile sheep model. Heart Surg Forum 8(2):100–104; discussion E104
45. Leyh RG, Wilhelmi M, Rebe P et al (2003) In vivo repopulation of xenogeneic and allogeneic acellular valve matrix conduits in the pulmonary circulation. Ann Thorac Surg. 75(5):1457–1463; discussion 1463

46. Speckmann EJ, Hescheler J, Köhling R (2008) Physiologie, 5th edn. Urban & Fischer, Munich
47. Schmidt RF, Lang F (2007) Physiologie des Menschen, 30th edn. Springer, Berlin
48. Dohmen PM, Hauptmann S, Terytze A, Konertz WF (2007) In vivo repopularization of a tissue-engineered heart valve in a human subject. J Heart Valve Dis 16(4):447–449
49. Rajani B, Mee RB, Ratliff NB (1998) Evidence for rejection of homograft cardiac valves in infants. J Thorac Cardiovasc Surg 115(1):111–117
50. Breinholt JP 3rd, Hawkins JA, Lambert LM et al (2000) A prospective analysis of the immunogenicity of cryopreserved nonvalved allografts used in pediatric heart surgery. Circulation 102:179–182
51. Bechtel JF, Bartels C, Schmidtke C et al (2001) Does histocompatibility affect homograft valve function after the Ross procedure? Circulation 104:I25–I28
52. Shaddy RE, Thompson DD, Osborne KA et al (1997) Persistence of human leukocyte antigen (HLA) antibodies after one year in children receiving cryopreserved valved allografts. Am J Cardiol 80(3):358–359
53. Shaddy RE, Hawkins JA (2002) Immunology and failure of valved allografts in children. Ann Thorac Surg 74(4):1271–1275
54. Vogt PR, Stallmach T, Niederhäuser U et al (1999) Explanted cryopreserved allografts: a morphological and immunohistochemical comparison between arterial allografts and allograft heart valves from infants and adults. Eur J Cardiothorac Surg 15(5):639–644; discussion 644–645
55. da Costa FD, Dohmen PM, Duarte D et al (2005) Immunological and echocardiographic evaluation of decellularized versus cryopreserved allografts during the Ross operation. Eur J Cardiothorac Surg 27(4):572–578
56. Zehr KJ, Yagubyan M, Connolly HM et al (2005) Aortic root replacement with a novel decellularized cryopreserved aortic homograft: postoperative immunoreactivity and early results. J Thorac Cardiovasc Surg 130(4):1010–1015
57. Elkins RC, Lane MM, Capps SB et al (2001) Humoral immune response to allograft valve tissue pretreated with an antigen reduction process. Semin Thorac Cardiovasc Surg 13(4 Suppl 1):82–86
58. Bechtel JF, Müller-Steinhardt M, Schmidtke C et al (2003) Evaluation of the decellularized pulmonary valve homograft (Synergraft). J Heart Valve Dis 734–739; discussion 739–740
59. Hawkins JA, Hillman ND, Lambert LM et al (2003) Immunogenicity of decellularized cryopreserved allografts in pediatric cardiac surgery: comparison with standard cryopreserved allografts. J Thorac Cardiovasc Surg 126(1):247–252; discussion 252–253
60. Kasimir MT, Rieder E, Seebacher G et al (2006) Decellularization does not eliminate thrombogenicity and inflammatory stimulation in tissue-engineered porcine heart valves. J Heart Valve Dis 15(2):278–86; discussion 286
61. Rüffer A, Purbojo A, Cicha I et al (2010) Early failure of xenogenous de-cellularised pulmonary valve conduits–a word of caution, Eur J Cardiothorac Surg 38(1):78–85
62. Elkins RC, Dawson PE, Goldstein S et al (2001) Decellularized human valve allografts. Ann Thorac Surg 71:S428–S432
63. Hawkins JA, Breinholt JP, Lambert LM et al (2000) Class I and class II anti-HLA antibodies after implantation of cryopreserved allograft. J Thorac Cardiovasc Surg 119(2):324–330
64. Akhyari P, Kamiya H, Gwanmesia P, Aubin H, Tschierschke R, Hoffmann S, Karck M, Lichtenberg A (2010) In vivo functional performance and structural maturation of decellularized allogenic aortic valves in the subcoronary position. Eur J Cardiothorac Surg 38(5):539–546
65. Affonso da Costa FD, Dohmen PM, Lopes SV, Lacerda G, Pohl F, Vilani R, Affonso da Costa MB, Vieira ED, Yoschi S, Konertz W, Affonso da Costa I (2004) Comparison of cryopreserved homografts and decellularized porcine heterografts implanted in sheep. Artif Organs 28(4):366–370

Evaluation of Biocompatibility Using In Vitro Methods: Interpretation and Limitations

Arie Bruinink and Reto Luginbuehl

Abstract The in vitro biocompatibility of novel materials has to be proven before a material can be used as component of a medical device. This must be done in cell culture tests according to internationally recognized standard protocols. Subsequently, preclinical and clinical tests must be performed to verify the safety of the new material and device. The present chapter focuses on the first step, the in vitro testing according to ISO 10993-5, and critically discusses its limited significance. Alternative strategies and a brief overview of activities to improve the current in vitro tests are presented in the concluding section.

Keywords Biocompatibility · Limitations · Culture · In vivo · In vitro · Test battery · ISO10993 · ASTM F748 · Testing strategy · Cytotoxicity · Bioactivity

Contents

1	Introduction	118
	1.1 Standards for Biocompatibility Testing of Biomaterials	119
2	Cytotoxicity	121
	2.1 Factors Influencing Toxicity	121
	2.2 What is Measured in the ISO 10993-5 (2009) Standard Test for Cytotoxicity?	123
	2.3 Limitations of Parameters for Cytotoxicity	128
	2.4 Cytotoxicity Parameters and Prognostic Value for In Vivo Situations	132
3	Bioactivity	133
	3.1 General Limitations of In Vitro Bioactivity Measurements	136
4	Future Perspectives: Advanced In Vitro Systems Mimicking the In Vivo Niche	138
	4.1 Proposed Testing Strategy	141
References		144

A. Bruinink (✉)
Laboratory for Materials – Biology Interactions, Empa – Materials Science and Technology, Lerchenfeldstasse 5, CH-9014 St. Gallen, Switzerland
e-mail: Arie.Bruinink@empa.ch

R. Luginbuehl
Chemistry and Biology, RMS Foundation, Bischmattstrasse 12, CH-2544 Bettlach, Switzerland
e-mail: reto.luginbuehl@rms-foundation.ch

1 Introduction

The clinical success of a medical device, such as a catheter, stent, artificial knee implant, or simple forceps, is determined on the one hand by its functional properties, e.g. the mechanical characteristics of the materials and the design of the device. On the other hand, it is also defined by the way biological tissues will react that are in contact the implant. In that context, the term biocompatibility is defined as "the ability of a biomaterial to perform its desired function with respect to a medical therapy, without eliciting any undesirable local or systemic effects in the recipient or beneficiary of that therapy, but meanwhile generating the most appropriate beneficial cellular or tissue response in that specific situation, and optimizing the clinically relevant performance of that therapy" (Definition [5] based on [155]). Thus, biocompatibility testing is the fundamental requirement when developing new materials and their surfaces for medical devices and tissue engineered medical products (TEMPs). The cellular and tissue responses towards a material may be very diverse and they can be classified in many ways, but they can roughly be divided into (i) strong effects, i.e. modification of cell viability (cytotoxicity, genotoxicity), (ii) moderate to nearly negligible effects, i.e. irreversible to transient changes in cell functionality (e.g. complement activation, pharmacological effects), or (iii) the absence of measurable effects. The inclusion of "appropriate response with respect to its function" to in the definition emphasizes that biocompatibility is not a general characteristic but is defined by the location of implantation and envisioned function of the device material. This implies that the end-use application should already be known when evaluating biocompatibility and that the evaluation has to be adapted accordingly. Each end-use application and therapy may have different requirements. A biomaterial may fulfil all criteria for a specific therapy to be biocompatible, but may fail and cause an unwanted tissue reaction in another application and as a result must be defined in that case as being not biocompatible. For example, in the case of non-absorbable hernia nets, soft tissue integration is crucial, while in the case of intraocular lenses, absence of a tissue reaction and cell on-growth are important criteria for biocompatibility. The situation gets further complicated by the fact that the materials will elicit variable and different degrees of reactions in each host, i.e. each patient. For instance, the stainless steel alloy, UNS S31675, is known to be biocompatible and can be used as implant material. However, sometimes the host's immune system may react towards the material immediately or after a while, requiring the removal of the device, as the steel alloy contains 9–11% nickel (mass/mass). Although in the majority of cases the absence of the release of toxic compounds represent a key issue being evaluated, also in this regard the absence of cytotoxicity does not represent in general a key issue for being biocompatible that must always be fulfilled. Therefore, interpreting biocompatibility of investigated biomaterials with regard to the final use of the material is crucial. In the following, the different test methods and acceptance criteria for biocompatibility are critically discussed.

1.1 Standards for Biocompatibility Testing of Biomaterials

The market for medical devices and therapies is large and steadily growing. The global market for medical products and hospital supplies is over $220 billion [55]. Within this, the market for tissue engineering and cell therapy products is set to grow from a respectable $8.3 billion in 2010 to nearly $32 billion by 2018 [52]. The global market for minimally invasive devices and instruments was worth $14.8 billion in 2008, and could reach $23.0 billion in 2014 [14]. Orthopaedics is one of the largest segments of the medical device sector. The global orthopaedic instrumentation market is projected to surpass US $47 billion by the year 2015, driven by an aging global population, the rising incidence of age-related conditions such as osteoarthritis and sports-related injuries, and improving orthopaedic surgical procedures [141]. As a result, industry is strongly motivated to take part in the global competition and to enter this market or increase their share with new products and solutions which are heavily advertised by companies as the best on the market. In order to prevent fraud and negative consequences for the patients, federal agencies and notified bodies (NBs) survey the medical devices market and assess the performance of each device.

One premise of these products is that they have to fulfil criteria such as biocompatibility, and its assessment according to standard test methods has been one objective approach by which the NBs evaluate these products. These standards test methods have been included as an instrument by federal agencies and NBs to be able to assess the performance of each device and to prevent fraud and negative consequences to the patients from new medical devices entering the market. In the early 1980s the standards organization American Standards and Test Methods international (ASTM international) developed the first standards for testing cytotoxicity and skin irritation based on industrial needs and the demand of NBs [6, 7, 9]. The catalogue of standard tests was slowly broadened and the umbrella document F748 defining the requirements of biocompatibility testing was issued [8]. These documents were adopted by the International Organization for Standards (ISO) which issued their first standard on biocompatibility series ISO 10993-1 [70]. Ever since, many new standards on biocompatibility have been issued and revised by several organizations (see Table 1). Today, each new material that is considered for use in medical devices has to pass a whole battery of standard tests before it can be used in a product and put onto the market. As discussed above, biocompatibility depends largely on the end-use application. Therefore, the standards differentiate and classify not the material itself but the end-use applications. Typically, the material–tissue interaction is addressed regarding duration of contact and end-use, i.e. contacting tissue type. In addition, the ratio of contact area to host size may also matter and might have to be considered. At this point it is important to point out that the test outcome should be an intrinsic property of the material. This means that if a material of a given quality is tested for biocompatibility, taking the same exposure time and with the same material contact area to host ratio, the same result should always be obtained—independent of the

Table 1 Standards for testing specific responses

Specific response	USP	ASTM	ISO
Cytotoxicity	<87>	F813-07	10993-05
		F895-84	
		F1027-06	
		F1903-98	
Sensitization	–	F720-81	10993-10
		F2147-01	
		F2148-07	
Irritation or intracutaneous reactivity; mucous membrane irritation	<88>	F719-81	10993-10
		F749-98	
Systemic toxicity (acute toxicity)	<88>	F750-87	10993-11
Subchronic toxicity (subacute toxicity)	–	–	10993-11
Genotoxicity; reproductive or developmental toxicity	–	E1202-87	10993-3
		E1262-88	
		E1263-97	
		E1280-97	
		E1397-91	
		E1398-91	
Blood biocompatibility/complement activation	–	F2382-04	–
		F1984-99	
		F2065-00	
Immune response	–	F1905-98	10993-20
		F1906-98	
Hemocompatibility	–	F 756-08	10993-4
Chronic toxicity	–	–	10993-11
Carcinogenicity	–	F 1439-03	10993-3
Biodegradation	–	F1983-99	10993-9
			10993-13
			10993-14
			10993-15
Implantation short term–long term	<88>	F1408-97	10993-6
		F763-04	
		F1904-98	
		F981-04	
		F1983-99	

Note The listing is based on standards as issued in 2011. Requirements may change and standards may be withdrawn or new standards issued over time. Therefore, it is important to consult with the standards organizations on the validity of specific standards

variability within the biological tests. In order to pick -up modifications due to processing of the raw material (e.g. contamination, chemical disruption), standards tests are required to evaluate the final product. Process-based contamination is a cleanliness issue which is not a material property but a process consequence, and therefore it should not be addressed using biocompatibility approaches but by chemical and physical analysis as described in ASTM F2847. The biocompatibility standards conform regarding material–tissue contact duration, which is differentiated into three time periods: (i) <24 h, intra-operative contact, (ii) 24 h to 30 days, defined as short term implantation, and (iii) >30 days, which is called permanent or chronic implantation.

The classification of the end-use of a medical device product requires a more distinct approach. The first class is called external devices, and includes wound dressings, monitors, and splints. Here, intact and injured (breached) body surfaces are distinguished. The ISO standard suggests the addition of a third subclass, i.e. for devices coming in contact with mucous membranes. The second class is termed externally communicating devices, which include any devices that are applied and inserted via natural channels or trans-cutaneously. Examples include contact lenses, tracheal tubes, all types of catheters and hypodermic needles. While ISO 10993-1 differentiates only between indirect blood path, tissue/bone/dentin, and circulating blood, ASTM F748 makes a more pronounced differentiation between devices communicating with intact natural channels, communicating with body tissues and fluids, indirect blood path, and direct blood path. The third and last class is implantable devices. Again, ASTM and ISO standards diverge in their subcategorization. ISO has tissue/bone and blood as subcategories while ASTM differentiates between their contacting tissues, especially "devices principally contacting bones, devices principally residing in the subcutaneous space, devices principally contacting soft tissue and tissue fluids, and devices principally contacting blood".

All the standards require specific tests to be performed, dependent on the classification, subcategories and contact duration. The cytotoxicity, sensibilization and irritation/intracutaneous reactivity tests are required and common to all products and materials. Additional requirements depend on the subcategories and contact duration. Details should be sought in the respective ISO 10993-1 or ASTM F748 standards since requirements may change over time. It is important to realize that standards are not irrevocable documents, but undergo revisions and changes that allowing for incorporation of new insights or novel techniques. The committee F04 on Medical and Surgical Materials and Devices of ASTM international actively follows the needs of industry for new and refined standard test methods and guides; in particular, the subcommittees F04.42/43/44/46 on different aspects of Tissue Engineered Medical Products and F04.16 on biocompatibility are very active and consolidate very recent findings (see for details see [5]). In all the discussion on biocompatibility one has to keep in mind that it is just a rated definition without hard limits.

2 Cytotoxicity

2.1 Factors Influencing Toxicity

The appearance of cellular responses (summarized in Fig. 1) depends on the kind of compound and its concentration, and can be classified as toxic, cytoeffective, and no visible effects. Each of the responses can be assessed in vitro by a specific test regimen and evaluating the corresponding key parameter(s). The varieties of

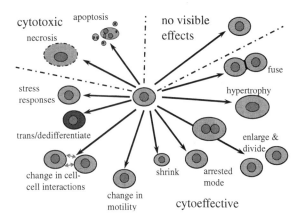

Fig. 1 Schematic representation of important cell responses versus biomaterials or drugs

pathways by which the response may be (primarily) induced are summarized in Fig. 2. The importance of these pathways in the appearance of the cell responses is strongly dependent on the exposure period and/or the period between start of the exposure and the time-point at which the cell response is assessed (which need not always to be the same). By prolonging the period between start of the exposure and time-point at which effects are measured, the possible mechanisms by which the cell responses are induced also increase, and thus the kind of appearances of the effects in the test concentration range. If the period between start of the exposure and measurement of the effects is very short, observed effects are mainly based on acute responses, i.e. the induction of cell necrosis. Apoptosis may also become manifest if the time period is prolonged, e.g. after 2 h when treating cardiomyocytes with 0.1 mM H_2O_2 [134], or in the case of treating SCC-9 cells with 50 μM quercetin after 72 h [64]. These two toxins reveal one of the problematic points with regard to standard testing methods: the appearance of the cell response is not a fixed event, but may vary tremendously and is highly specific as to the function of the compound and the cell phenotype; in this case the difference is a factor of 36. While necrosis and apoptosis are early effects, the consequences for cell proliferation can be measured only if the treatment period is further increased corresponding to the normal proliferation rate of the cells used. The consequences becomes more visible by progressing the treatment period. Effects on cell number manifest only after many days in cell culture, e.g. six days for 300 μM and at least ten days for 30 μM $AlCl_3$ on osteoprogenitor cells [15].

Compounds may interact in two different ways [27]. (1) By a direct interaction with the target cell protein inducing the effects. For this kind of response, a single moment of a critical concentration is needed to induce the effect. Cell necrosis is an effect which is probably induced by direct interaction. (2) By accumulation of the toxicant in a sensitive cell compartment. Effects are induced if a critical concentration is reached. For this accumulation, the compound has to be taken up by the cell and transported into the sensitive or target cell compartment.

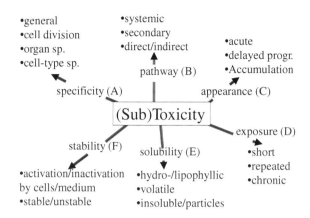

Fig. 2 Schematic overview of the different factors influencing cytotoxicity. (A–F) Examples of the different factors: A: [26, 29, 31, 32, 34, 100, 152]; B: [36]; C: [28, 39]; D: [39]; E: [11, 153]; F: [30, 36, 108, 27]

This uptake and transport may be rate-limited and will take some time, resulting in a delay of the time-point of effect induction and thus of effect appearance. Several factors may affect toxicity and its appearance: (i) The critical effective compound may be a constituent released by an implant itself. (ii) It may be a constituent altered by metabolization. (iii) It may be a compound whose formation is induced or catalysed by the released constituent—for example, reactive oxygen species (ROS) may be formed by an indirect reaction. (iv) The effects found in vitro may be strongly modified by other components that are present, which by themselves within a wide concentration range affect the cells differently or not at all [33, 37]. (v) In cell cultures containing several cell types, the toxicant may be produced by one type but affect the other cell type(s). Examples are the bioactivation of cyclophosphamide and isophenphos by liver cells resulting in a toxic product for nerve cells [36]. (vi) The effects are not necessarily observed during treatment but may appear several days after exposition (delayed or progressive effects). The existence of delayed or progressive effects can be proven by removing the toxicant after a short treatment period, replacing the culture medium with a fresh one without the toxicant. A strong indication for the presence of delayed effects is present if the effect of the initial treatment prevails several days later. Examples of compounds that induce such delayed effects are cisplatin [28] and single wall carbon nanotubes [39].

2.2 What is Measured in the ISO 10993-5 (2009) Standard Test for Cytotoxicity?

The ISO 10993-5 in vitro cytotoxicity test guideline does not define one single standard test method but it describes testing schemes that require decisions at given time-points. Three categories of tests are proposed for assessing the cytotoxicity of potentially released materials: (1) extract tests, (2) direct-contact tests and (3) indirect-contact tests (agar diffusion test, filter diffusion test, see Fig. 3 for schemes). One or more tests may be used for the evaluation depending on the nature of the sample material.

Cytotoxicity is assessed using different parameters based on cell and culture morphology (qualitatively), quantitative measurement of cell impairment, such as effects on cell growth (proliferation), and specific aspects of cell metabolism. Qualitative and quantitative measurements are reported to correlate very well [20]. Extract tests are normally based on a so-called extract obtained by exposing cell culture medium to the test material or compound of interest for 24 h at 37 °C. Subconfluent cell cultures are treated by measuring effects on cell functionality typically after 24 h, or low-density cultures are revealed by measuring effects after a prolonged time period of six days. In the latter case, effects are assessed by counting the number of colonies consisting of, as proposed, at least 50 cells. In the direct contact test, a test sample covering about 10% of the subconfluent cell layer is placed on top of that layer, while in the agar diffusion test an agar layer covers the cells instead of cell culture medium and the test samples are placed on top of the agar layer. In both tests, the sample is removed after 24–72 h exposure time and the cells are qualitatively and quantitatively assessed below and adjacent to the test samples. Another suggested test method is the filter test. For this cells are cultured until confluency on one side of the filter, which is then placed with the cell side on top of an agar layer. Subsequently, the test material is placed on the other side of the filter. Effects on cells are qualitatively assessed after 2 h exposure time.

The most striking limitation of all suggested tests in ISO 10993-5 is the short test period, i.e. 2 h (filter diffusion), 24 h (extract acute cytotoxicity), or 24–72 h (agar diffusion). By defining a reduction of 30% as the threshold for an extract to be toxic, only a cell lysing compound, a compound inducing apoptosis in a very short term, or very strong inhibitor of cell proliferation will be able to give rise to such a reduction after a treatment period of 24 h. For instance lidocaine (1 mM), known to reduce cell proliferation of the suggested cell line (NIH 3T3), yields a 65% reduction in cell count after 120 h, but only 30% reduction is observed after the required culture period of 72 h. No clear-cut effects on cell numbers are seen even 48 h after treatment [49]. Clear results are only obtained if the colony-forming test is chosen, having the prolonged incubation period of six days. Therefore, the short test duration is very limiting regarding the informative value and the kind of effects that can be assessed. In particular, effects based on accumulation and delayed/progressive effects will not be detected.

A second general limitation is the use of cell lines, in particular of cell lines that may not be relevant for the proposed use of the biomaterial. The standard asks indirectly for a justification of the cell sources used in tests, but limits the use of primary cells to applications that require specific sensitivity. The result is that industry will hardly ever test with primary cells or organ cultures, in order to reduce the risk of a cytotoxic outcome. Therefore, fibroblast L-929 or NIH 3T3 cells are used for most if not all medical devices that are in contact with muscle and skeletal tissues.

A1: Extract test: Acute cytotoxicity

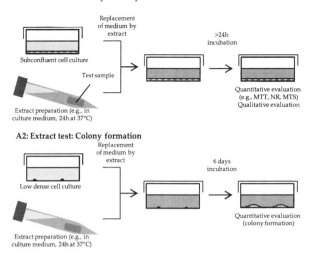

A2: Extract test: Colony formation

B: Direct contact test

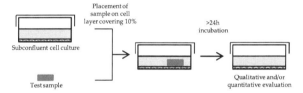

C1: Indirect contact test: Agar diffusion test

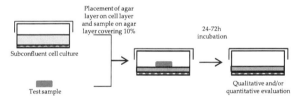

C2: Indirect contact test: Filter diffusion test

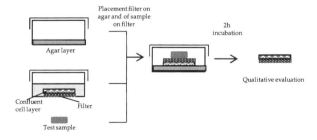

Fig. 3 The ISO 10993-5 tests. Schematic overview of the extract tests [acute cytotoxicity test (**A1**) and colony-forming test (**A2**)], direct-contact test (**B**), and indirect-contact tests [agar-diffusion test (**C1**) and filter-diffusion test (**C2**)]

2.2.1 Limitations of Extraction Procedure

Extraction of compounds from materials is not as simple as it sounds. The standard describes different extraction fluids for preparing extracts, and each one has its own limitation. Generally speaking, the proposed fluids are not designed to extract compounds from a material! It is a very mild procedure which detaches some compounds from the surface and may extract very soluble ones from the near surface bulk material, but the procedure does not reflect in vivo conditions in any way, i.e. prolonged contact between cells/tissue/tissue fluids and the material. The extraction yield can be easily <10%, depending on the extractable compounds, as compared to an exhaustive extraction. In particular, the approach is not suited for polymeric systems that may contain extractable hydrophobic constituents, as, for example, is well-known from total organic carbon (TOC) measurements [87].

The standard suggests three extraction fluids: (a) culture medium with serum, (b) physiological saline buffer, or (c) other "suitable extraction vehicles" including pure water or dimethyl sulfoxide (DMSO) <0.5%. The use of culture medium containing serum has the disadvantage that the material surface is immediately covered by a protein layer of albumin and fibronectin. This layer passivates the material interface, limiting the release of possible toxic components. Here it must be noted, however, that this kind of barrier is also formed after implantation of the material and thus in principle mimics the in vivo situation to a certain extent. The use of physiological saline buffer is only suited for the extraction of hydrophilic compounds including soluble salts. Other extraction fluids such as pure water or fluids containing minute amounts of DMSO allow for comprehensive extraction of neither hydrophobic and hydrophilic components. Furthermore, DMSO (1–2%) is known to alter the cell phenotype [51, 131]. However, the goal and at the same time also the problem of the extraction procedure is that it should capture all compounds that might be released from the surface or that might diffuse out of the material during the lifetime of the device—which might be many years. Obviously, hydrophilic soluble substances are not seen as a problem for collection in the extract, but most organic hydrophobic substances will escape the collection since their saturation in an aqueous medium is extremely low. Furthermore, diffusion processes in polymers are usually very slow and apolar solvents are required to speed the process up.

The ISO 10993-5 standard also dictates the extraction temperature and time. Four different standard conditions are given: 24 h at 37°C, 72 h at 50°C, 24 h at 70°C, and 1 h at 121°C. Among the different extraction temperatures suggested, 37°C is certainly a physiological temperature and matches cell culture and serum extraction media. The advantage of using higher temperatures (50/70/121°C) is that the release of components may be induced that would not be released under normal conditions. However, this may occur with the disadvantage that the released components may be modified due to the higher temperature. Therefore, by using higher extraction temperatures the absence of these modifications needs to be ensured to prevent the occurrence of false positives and negatives.

A further variable portrays the surface to extract fluid volume ratio. It is suggested that the surface to volume ratio should be 1.25–6 cm^2/ml depending on the thickness, or 0.1–0.2 g/ml or 6 cm^2/ml in the case of materials with indeterminate surface irregularities like foams [1]. Starting from the 100% stock extract, dilution series may be prepared with culture medium. The standard is, however, very inaccurate since it does not define the minimum surface to volume ratio.

2.2.2 Limitations of Direct Contact Test

As discussed briefly above, the sample material is placed on top of a subconfluent cell layer. That tight contact between material surface and cells is thought to reflect the material–cell interaction. Such tests have to be carried out with great care to avoid artefacts and misinterpretation of the results. The first problems may be encountered if samples are placed on the cell layer. Most often, the sample weight is either too high or too low. For instance, e.g. a sample made of a CoCrMo alloy has a density of 8.3 g/cm^3 and might crush easily the cells, while a dense polyurethane sample is slightly below 1 g/cm^3 and will float on the cell culture medium. The contact regime will be completely different for these two samples. Another problem that might occur is that the cells adhere to the material surface instead of to the tissue culture plastic. Rupture of the underlying cell layer due to sample removal or apparent "reduction" of the number of cells below the sample may be the result. In this case, it will be difficult to judge the contact test, although preferential adherence as compared to tissue culture plastic is a good sign.

Another drawback of contact tests is that released toxic components may be diluted in the culture media to such an extent that they no longer affect the cells. In this situation the released components are extracted during the tests instead of before testing as is the case in the extract tests. The difference between the extract test and direct contact test is the ratio between the material surface and medium and is well illustrated by the following situation. A circular film sample with a thickness of less than 0.5 mm is evaluated with extract and direct contact tests both using 3.5-cm diameter culture dishes with 3 ml of culture medium. In order to obtain an extract according to the ISO 10993-12, the surface to volume ratio must be 6 cm^2/ml. In the present example, this is the case if the diameter of the sample is 3.4 cm. However, that contradicts the requirement of the direct contact test where the sample should cover only 10% of the culture dish surface. The latter corresponds to 0.96 cm^2 and thus the circular sample should not exceed 1.1 cm in diameter. Furthermore, in the case of the extract test, cells are already treated from the beginning with the concentration of the constituents that is released over a period of 24 h. In the direct contact test this maximal concentration occurs only at the end of the treatment period, normally 24 h.

2.2.3 Limitations of Indirect Contact Test

(i) Agar diffusion test. This tests exhibits the same limitation regarding dilution of the toxicant as the direct contact test, since again only 10% of the cell surface should be covered. This test is even less sensitive because only one sample side comes into contact with the agar layer separating the cells and sample. In addition, the toxicants have to be able to diffuse freely through the agar and must not react with the agar. In cases where the chemical properties of the released compounds are not known, it may be very challenging to prove the latter. Furthermore, a good contact between the sample and the agar must be ensured. The advantage of the present test is that an incubation period of 24–72 h is suggested, of which 72 h is certainly preferred, enabling the measurement of cytotoxic effects via other pathways in addition to acute cytotoxicity. Another advantage of the agar diffusion test is that the samples can be heavier, since the agar layer protects the cell to a certain extent and the cell layer cannot be disrupted due to a sample removal.

(ii) Filter diffusion test. The advantage over the agar diffusion test is that the distance between cells and material is much smaller and mimics the direct contact test in this regard. The only limitation of the test is that a good contact between sample and filter must be ensured, i.e. the fluid layer is in contact with the cells and the sample surface over the whole filter surface. Furthermore, it must be demonstrated that the released constituents do not bind to the filter. The advantage of testing extracts instead of pure samples is that with extracts exposure to maximal constituent concentration is ensured from the moment the extract is applied. However, the exposure period is extremely short (2 h $\pm$ 10 min).

2.3 Limitations of Parameters for Cytotoxicity

Typically, only final values of the cytotoxicity parameters are considered after a specified exposure period in cell cultures according to ISO 10993-5. The history, i.e. what occurred in between, is not taken into account. For instance, the presence of fewer cells in an exposed well relative to control may be evoked by cell death, reduction of cell proliferation or a combination of both. The absence of an effect may be the result of a real absence of adverse effects or of a too short exposure period (for instance, inhibition of cell proliferation can certainly not be detected in the filter diffusion test). Furthermore, the cells used for these tests will affect the outcome. This is due to the variable sensitivity of cells and the activity of the cells (e.g. the MTT conversion activity varies between different cell types and may also differ between cell line and primary cells of the same cell type). As a rule of thumb, it can presumed that the relevance of the chosen cell types and therefore the test outcome are directly connected with the proposed application of the biomaterial. Furthermore, it must be noted that false positive results may be obtained if the biochemical assay is influenced by a cross-reaction between test sample and/or the compound itself and the constituents of the biochemical assay.

In the following examples, the use and limitations of some common tests will be discussed.

2.3.1 Example: MTT-Test for Measuring Viability or Cell Number

One of the most common tests used in cytotoxicity evaluation is the test based on the conversion of 3-(4,5-dimethylthiazol-2-yl)-2,5-diphenyltetrazoliumbromide (MTT) to its formazan product. MTT is membrane-impermeable and the cellular uptake of MTT is endocytosis-dependent [83]. The reduction is associated with the cytoplasm and non-mitochondrial membranes including the endosome/lysosome compartment [83]. How far mitochondria participate in formazan formation, as first suggested [127], remains unclear [18, 83, 86]. As reducing enzymes, NAD(P)H-oxidoreductases [19, 83] and glutathione S-transferase [161] are proposed, among others. The final formazan product is transported to the cell surface through exocytosis [83]. Some research teams use the test to enumerate cells [86]. This might be correct under normal conditions but it has been shown that in several situations where adverse effects of compounds and particles are assessed this is not the case [30, 72, 148, 152]. Another application is to evaluate cell viability. Here also cases have been reported where results obtained with the MTT test do not reflect viability measures using other kinds of assays [123, 146, 148]. Since both aspects, cell number and cell activity, will affect the level of MTT conversion, this test may be taken as an index of cell activity of the culture as a whole. Special attention has to be paid to the interaction between test sample and assay. For instance, evidence was found that the MTT formazan product is able to bind to the carbon nanotubes (CNT), reducing the dissolved MTT formazan yield. A pseudo-reduction of MTT formazan levels is therefore observed [16, 98]. In addition, CNT interferes with the assay by a second pathway. In a control experiment, under cell- and serum-free conditions it could be shown that CNT by themselves are able to reduce MTT [16]. Similar effects are observed in systems containing high salt levels, such as in combination with absorbable calcium phosphate scaffolds (Luginbuehl et al., unpublished data). One other disadvantage of the test is that the MTT by itself may affect cell functionality, as shown for astrocytes. As a result, some discrepancies of cell functionality data using other assays may occur [71]. This test should therefore be used with care; i.e. its limitations and interference potentials must be addressed before use.

2.3.2 Example: Neutral Red Uptake Test

Neutral red (3-amino-7-dimethylamino-2-methylphenazine hydrochloride, Basic Red 5, Toluylene red, NR) is a lipophilic free base used for staining in histology, as a pH indicator in the range of 6.8–8.0 and for measuring cell viability [10, 162, 26]. It can be measured by light absorbance (at 540 nm) but also by fluorescence (488 nm extinction, 590–600 nm emission). After addition, NR is taken up by living cells. It accumulates linearly in the lysosomes of living cells as a function of time [162]. There is one limitation to this test, due to the lipophilic characteristics of NR. It may bind to certain test samples or adsorb on components such as carbon black. The resulting high background levels depend on sample surface or concentration of components and have to be subtracted in analysis [98].

2.3.3 Example: MTS

The MTS [3-(4,5-dimethylthiazol-2-yl)-5-(3-carboxymethoxyphenyl)-2-(4-sulfophenyl)-2H-tetrazolium, inner salt] assay can be used as an alternative to the MTT test. As in the MTT test, the conversion of MTS to its formazan product is measured. The advantage over the MTT test is that the MTS formazan product is water-soluble and by that does not need to be solubilized. In analogy to the MTT, the MTS assay cannot be used for measuring antiproliferative activity since it also underestimates the antiproliferative effect [146, 148]. It may be assumed that the MTS conversion cannot be taken as an index of cell viability, like the MTT conversion. In toxicological studies comparing the results of MTS and neutral red assay, MTS systematically underestimated toxicity [11, 110]. Since cell number and cell activity both play a role in total culture MTT conversion activity level, this test may be taken as an index for total culture cell activity, like the MTT test.

2.3.4 Example: DCF Test for Measuring ROS

Reactive oxygen species (ROS) are important in several metabolic pathways, act as intracellular signalling molecules and are important in homeostasis. ROS are formed as metabolic products and exhibit a very high reactivity due to their unpaired oxygen electron. Below a certain concentration range, their adverse effects are reduced by numerous scavenger molecules that are produced by the cell for this purpose. If cells are stressed, e.g. due to drugs, particles, UV light or endotoxins, the production of ROS is increased, reaching concentrations which may be harmful for the cells. One of the common tests for evaluating ROS formation is the $H_2DCF-DA$ test [58, 103, 118, 139, 158]. In cell cultures, the $H_2DCF-DA$ test is primarily used for the detection of a variety of ROS species including peroxyl and hydroxyl radicals, the peroxynitrite anion and nitric oxide, as well as hydrogen peroxide [44, 104]. The $2',7'$-dichlorodihydrofluorescein diacetate ($H_2DCF-DA$) test is based on the assumption that this non-fluorescent dye is taken up by cells and deacetylated to its non-fluorescent congener $2',7'$-dichlorodihydrofluorescein (H_2DCF), which is thereafter entrapped within cells. ROS react with H_2DCF, converting it to the highly fluorescent $2',7'$-dichlorofluorescein (DCF) which is not membrane-permeable [13, 80, 137]. Recently, some doubts have arisen as to how far the DCF test is adequate for measuring ROS [22]. Additionally against current opinion, H_2DCF as well as DCF is found not to be

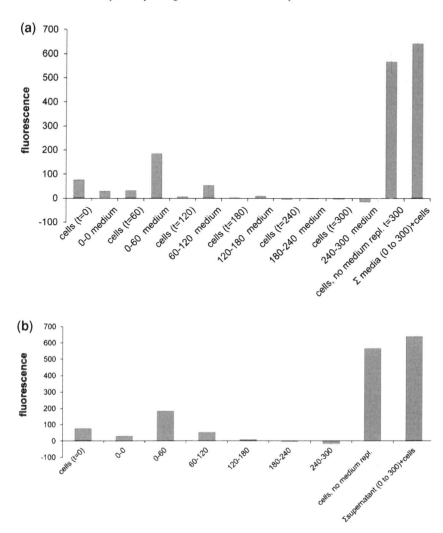

Fig. 4 Release of H₂DCF and DCF after H₂DCF-DA loading of A549 cells. A549 cells were loaded with H₂DCF-DA loading for 60 min. After two washes with HBSS medium, the supernatant of some of the cultures was collected every 60 min and new HBSS was added. Fluorescence of supernatants and cell cultures with medium replacement was measured every 60 min until 300 min after H₂DCF-DA loading (485 nm extinction; 528 nm emission). DCF fluorescence was seen in the supernatant and, at the start, also in the cells. The sum of the fluorescence of all supernatants was similar to that of the cultures in which the medium was not replaced. Thus it could be shown that DCF was formed in the cells and released in the supernatant. After 300 min all media and cultures received H₂O₂ (**b**). Fluorescence was measured just before (t = 0) and 15 min and 18 h afterwards. Under the influence of H₂O₂, additional DCF was formed in the supernatants, proving that H₂DCF was also released by the cells. H₂DCF-DA was found to be insensitive to H₂O₂. Data are presented as mean ±SEM over three independent experiments (A. Bruinink and U. Tobler, unpublished results)

entrapped within the cells (Fig. 4) and H$_2$DCF may be converted to a significant extent into DCF in a cell-free environment under the influence of various drugs like apocynin (100 μM) and chlorpromazine (10 μM) (A. Bruinink and U. Tobler, unpublished results).

2.4 Cytotoxicity Parameters and Prognostic Value for In Vivo Situations

Based on current data and tests, it must be stated that the prognostic value of in vitro cytotoxicity tests for animals and humans is limited and depends on the kind of released compounds and the in vitro model. In some cases a correlation is found [57, 160], but unfortunately in most cases not [17, 62, 136, 149]. The general limitations of current in vitro tests are that the in vivo situation is not mimicked. In vitro, tests are performed under static conditions in terms of fluid flow, while in vivo dynamic situations prevail resulting in a homeostatic condition in stead of an equilibrium regarding compound release (including particles). The homeostatic concentration is defined by the rate of compound release and its clearance. One key issue defining the homeostatic concentration at the implant surface is the microcirculation in the vicinity of the implant defining not only the clearance [59] but also the component release [67]. A clearance can only be achieved in vitro either by compound metabolism/degradation, by inactivation through binding to proteins, or via medium exchange (e.g. medium replacement, presence of an artificial fluid flow) [27, 59]. However, a medium exchange is usually not performed (unless the consumption of nutrients are makes it necessary) during the short cell culture period for cytotoxicity evaluation.

The lack of correlation between in vitro tests and clinical experience is probably not only related to the clearance but also to a variety of other factors:

1. *Exposure period*. One important factor is certainly the short exposure period in the ISO 10993-5 tests, as discussed above. The biological reactions in vivo may be induced by different pathways and continue beyond the evaluated exposure period of 2–72 h.
2. *Release kinetics*. The release of toxic implant constituents may occur only initially with a strong decline with time (burst release), or be nearly constant or even increase [65]. The outcome of the in vitro ISO 10993-5 is in both cases the same and does not distinguish between the different reactions. However, the response of the body to these two types of materials may differ strongly. For instance, Rosengren and co-workers inserted polyurethane discs with 0.5 or 1% zinc diethyldithiocarbamate and nontoxic discs in the abdominal wall of rats for 1 day up to 6 weeks and investigated the foreign body response [114]. They could show that the foreign body reaction disappeared with the reduction of toxic compound release (in vivo, tissues may react with an inflammatory response while in vitro cells may respond to toxic compounds with signs of cytotoxicity). Thus an initial release of toxic constituent need not be dramatic

as is generally seen, as long as it is limited to this initial period. (It even may be argued that a rapid initial toxic release of constituents at levels that are able to kill these bacteria without systemic effects is desirable for most types of implants, since the introduction of bacteria into the wound during implantation is unavoidable). As a result, materials which in vivo could perform well may be seen as unacceptable based on the ISO 10993-5 test.

3. *Cell-type specificity*. A suboptimal selection of cell type with which to perform the tests may result in false positive and false negative results. In the standard ISO 10993-5 guideline, cell lines are used whose sensitivity may differ greatly. Such variations are not only seen between primary cells and cell lines but also between cell lines (e.g. after a treatment period of six days 7.5 ppm of Fe_2O_3 nanoparticles was lethal for MSTO-211H pleural cells, whereas 30 ppm of these particles was almost ineffective for 3T3-NIH fibroblast cells [40].

In order to improve the prognostic value of in vitro tests, on the one hand several specific in vitro parameters must be analysed and this data should be combined with the known chemical information and biokinetic data (e.g. [88, 115]), and on the other hand the cytotoxicity of untreated and washed samples (e.g. after 24 h in medium or even longer) should be compared. In addition, the in vitro models should be improved to mimic more precisely the in vivo situation at the site of implantation.

3 Bioactivity

Early tissue reaction to a biomaterial that does not release cytotoxic components ranges between normal inflammatory reaction and subsequent wound healing, to strong foreign body reaction, expressed in prolonged inflammatory reaction, accumulation of multinucleated macrophages and formation of a thick fibrous capsule at the tissue implant interface. In normal wound healing, different sequential phases can be discriminated, starting with (a) stopping bleeding by producing a temporary blood cloth, followed by (b) an inflammatory phase characterized by swelling, debridement (removal of debris, foreign bodies and bacteria mainly by phagocytosis by inflammatory cells), (c) closing the wound with new tissue (proliferative phase) and finally (d) a remodelling phase (Fig. 5). If a foreign body reaction develops against the biomaterial, the inflammatory phase of the healing process is greatly prolonged. Since the foreign body, i.e. the biomaterial, cannot be removed, the acute inflammation turns into a chronic type of inflammation characterized by the formation of foreign body giant cells and a thick fibrotic, generally avascular, capsule around the biomaterial, isolating it from the host. The capsule is typically not tight around the biomaterial and the void is filled with wound liquid.

The influences of a biomaterial on cell functionality can be divided into the general groups of (i) chemical and spatial influences, (ii) mechanical influences,

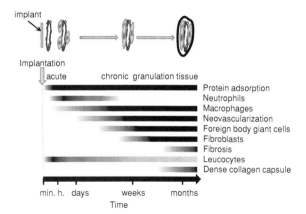

Fig. 5 Schematic representation of the response time sequence after implantation of a medical device. Adapted from Anderson [3]

and (iii) topography and porosity influences. The chemical properties include aspects of chemical functionalities of the cells/tissue, biomacromolecule coating, chemical patterning of surfaces, or absorption and degradation behaviour including secondary metabolic compounds (e.g. [74]). The mechanical properties of a biomaterial, i.e. its elastic modulus, pressure, and flow, influence to a large extent cell behaviour and tissue formation adjacent to the biomaterial [54]. Most cells are mechano-sensitive and have at their surfaces elements that respond to mechanical stimulations. Good examples of that responsiveness are muscle cells, chondrocytes, bone cells, nerve cells and dendritic cells. Last but not least, cells are influenced by the surrounding topography. One of the best investigated examples is the so-called SLA surface-grid blast and acid-etched surfaces of dental implants. That treatment introduces macro-, micro- and nanometer 3D surface structures on titanium metal surfaces favouring the functionality of osteoblast cells and allowing enhanced and fast osseointegration of the implants [42, 129]. Separating the reaction towards each characteristic is, however, only of theoretical interest. Cells always respond to all of these characteristics of a biomaterial [84, 125].

The extent of these influences varies from general to cell-type specific. Cell-type-specific effects may results in a shift in the original tissue composition of the cells in the environment of the implant and thus result in a (pathological) change in its functional characteristics. For instance, in the bone environment, bone tissue may alter locally in soft tissue after placing a new implant. Surface nanostructures and chemistry, including surface charges and the bioenvironment, i.e. local tissues and fluids, will in their combined action influence which set of proteins will adsorb to the surface, in which orientation they will adhere and in how far after adhesion these proteins will change their conformation, with denaturation as the most extreme situation [92, 113, 135, 150, 156]. Protein adsorption occurs immediately upon contact with the biomaterial and fibronectin and albumin adsorb predominantly, due to their serum concentrations. Depending on the material, equilibrium in protein adhesion regarding protein quantity, composition and z-potential is reported to be reached after 1–24 h after immersion in protein solutions, during which time small proteins and proteins with low affinity are rearranged and replaced by those with higher affinity [2, 43, 68, 76, 128]. Total protein adsorption and the final adsorbed protein composition is affected by surface charge and surface energy (surface wetting capability) [4, 93, 107, 120, 157].

Protein denaturation may induce cascades of unwanted events, resulting in a foreign body reaction. Cells adhere to the implant surface through accessible integrin binding motifs of adsorbed proteins. Integrins are cell membrane proteins that mediate the contact of the cell with these proteins. Cells express several types of integrins. The relative expression of these integrins is defined by the type of surface [126] and is probably defined by the type of accessible integrin binding motifs of the adsorbed proteins as specific ligands. Different proteins adsorbed to the surface result in different cell adherence profiles [143]. Furthermore, the integrin expression pattern is related to the functional differentiation of the cells [82, 101]. The cell be able to form a stable connection between cells and implant surface only if integrins can cluster to focal adhesions. The ability to form focal adhesions is defined by the intermolecular spacing size of the surface-bound integrin adhesion ligands [48]. A spacing of above 90 nm has been reported to inhibit focal adhesion formation [121]. This might also be one reason why cells have difficulty in adhering to certain (sub-) micrometer structured surfaces [21]. Upon maturation of these focal adhesions, actin filaments network are strengthened and the cytoskeleton reorganizes accordingly. Finally, with sufficient cellular forces (cytoskeletal contractility and/or globular actin motion), adhesions mature into long-lasting entities. This last step is essential for cell contractility and decisive underpinning of mechanosensing and cellular physical integrity [84].

Whereas the effects of nanostructuring are probably mainly protein-based, the effects of microstructure are assumed to be based on surface-induced modification of cell shape [23, 38, 75]. The cell shape is thought to directly affect cell functionality [23, 90, 124]. Although much research has been carried out, so far the crucial cues that steer and define cell functionality cannot be exactly identified. Furthermore, cell stiffness adapts depending on the elastic modulus of the biomaterial surface. It is known that cells can adapt to a variation in the stiffness of their environment within 0.1 s [94] and a modification of the substrate's elastic modulus is known to affect cell morphology, cytoskeleton structure and adhesion [159]. The elasticity of the material surface defines the strain in the cytoskeleton of the cells, which affects cell physiology in vitro and in vivo [35, 138].

Since tissue formation adjacent to a material is directly affected and defined by the cell–surface interactions, it may be assumed that material surfaces which best mimic the targeted tissue environment perform best if all influences such as chemistry, mechanical moduli and topography are adapted and considered [130, 154].

In vitro, bioactivity can be measured on a molecular level by assessing specific gene expressions as a function of time and/or synthesis of certain proteins. On the cellular level, it is evaluated by measuring cell adherence, proliferation and/or differentiation.

3.1 General Limitations of In Vitro Bioactivity Measurements

The advantage of in vitro systems is that they are simple and that all parameters can be defined. Furthermore they are cheaper and less time-consuming than in vivo tests. However, current in vitro systems also have their drawbacks:

1. *Cell source.* As for cytotoxicity tests, cell line cells used for these studies are often of animal, not human, origin (e.g. mouse MC3T3 or human MG63 cells for osteoblasts [46, 74, 105, 119]).
2. *Assumptions.* The assessments are based on certain assumptions which are often not correct, since the prognostic values of in vitro bioactivity parameters versus in vivo performance have not been systematically evaluated. For example, the assumed correlations between in vivo performance and the degree of cell adherence, cell proliferation rate and cell differentiation are not valid (e.g. a parameter of early osteoblast differentiation, alkaline phosphatase (ALP) activity, is increased [74]).
3. *Heat-inactivated serum.* Heat inactivation alters the conformational state of heat-labile proteins. It is known that heat inactivation of these proteins may affect cell functionality and cell–surface interactions [25, 37]. Furthermore, culture medium containing a dilution of heat-inactivated fetal calf serum may give a result for a protein layer not comparable to one present after implantation in an environment of 100% interstitial fluid or serum, both of which are not heat-inactivated [97]. As a result, not only cell spreading but also proliferation and differentiation may be affected. The use of media with denatured proteins thus certainly represents one important drawback of current in vitro systems for selecting biomaterials, investigating cell material interactions and elucidating the underlying mechanisms.

In most studies in which the prognostic value of in vitro studies is a theme, a comparison of these studies with animal studies is made. However, since the prognostic value of animal studies for the human situation is also limited, the outcome of such a comparison is questionable and it might be discussed how far such a comparison should be made at all.

3.1.1 Example: Cell Adherence

Generally, it can be assumed that cell adherence is directly related to the cell's ability to interact with the surface. Experimentally, cells are seeded on top of the samples and cell adherence is measured by quantifying cell number at defined times thereafter. The incubation period used varies depending on the research group: 1 h [105], 2 h [99], 3 h [74], 4 h [81, 117], 5 h [106], 0.5–6 h [140] or 24 h [85]. However, comparing in vitro with in vivo data, such a correlation could not always been found (a correlation is seen in, e.g. [81, 99], but not in, e.g. [105, 117]). One explanation could be that cell adherence is measured before protein adsorption equilibrium is reached. A solution would be to immerse the sample in the culture medium for 24 h, at which time-point it may assumed that protein adsorption equilibrium is reached, before seeding cells on top of it. Furthermore, depending on the type of proteins that are adsorbed onto the surface, the period for obtaining maximal cell adherence has been reported to differ from 30 min to 6 h, suggesting that 6 h might be the optimal seeding period in a cell adherence test [143].

3.1.2 Example: Cell Proliferation

It can be assumed that tissue integration of a material correlates with optimal cell proliferation. If cell lines are used, they are based in most cases on tumour cells of specific organs, e.g. from bone or connective tissue. Thus, the question has to be addressed why cells that are selected to readily proliferate and whose genetic background is not identical to the original cell type should be a good model for primary cells. Therefore, it is not surprising that contradictionary reports have been published regarding their prognostic value for in vivo performance. For instance, the optimum for titanium surface roughness for maximal MC3T3-E1 osteoblastic cell proliferation (Ra: 0.0125–6.3 µm) [46] is similar to that found to be optimal for in vivo bone response and implant fixation (Ra: 0.5–8.5 µm) [122]. However, in another study comparing smooth titanium surfaces (Ra 0.1–0.2 µm) with rough surfaces (Ra 3 µm), an inverse relationship between osteoblastic MG63 cell proliferation and bone–implant contact (screws, in sheep vertebrae) was found [119]. The use of primary cells instead of cell lines, however, improves the prognostic value only marginally. For instance, the team of Ravanetti investigated different titanium surfaces and found an inverse relationship between human primary osteoblast cell number as measured after 4 weeks and in vivo performance (bone contact) after 2 weeks in rabbits [112]. In contrast, Brama and co-workers described a direct correlation of primary human osteoblast proliferation with in vivo performance when comparing titanium and titanium carbide-coated titanium materials [24].

It is important to point out that not one but various cell types are present at the location where the implants are placed and that if multiple cell types adhere to the surface all of these will proliferate. The cell reaction on the surface might differ depending on surface and cell types [78]. Furthermore, these cells will interact with each other resulting in a modification of the proliferation rates [41, 151]. It can be hypothesized that in this case it is probably not important to have the highest proliferation rate of osteoblast progenitor cells, but to have an advantage relative to other cell types with which they have to compete. This would imply that proliferation must always been seen in the context of other cells and might be a reason why this parameter, as it is assessed now, is limited in its prognostic value.

3.1.3 Example: Cell Differentiation

An important measure of the cell–material interaction is the maintenance of cell phenotype or the capability of progenitor cells to differentiate into the tissue-respective phenotype, e.g. for an applications in bone, that mesenchymal progenitor cells differentiate into osteoblasts. Therefore, phenotype-specific gene markers or proteins must be assessed. These assessments have to be performed at different time-points to allow a statement regarding the expression kinetics. The time-point of measurement is critical here, since the up-regulation of mRNA markers and protein concentrations have their own optima [73, 164]. The expression kinetics are influence by the material. The time-points of maximal and also of overall expression can be shifted. Thus by investigating one time-point only, it is not possible to make any useful statement. This may be another reason for the contradictionary in vitro and in vivo results [62].

So far, unfortunately, a systematic evaluation comparing in vitro and clinical data is still missing. The types of influences that a material evoke are broad and limitations of in vitro tests may vary largely between the type of influence (for instance surface topography or drug) and parameter that is measured in vitro. However, it must be noted that, especially for such complex processes like differentiation, the current in vitro systems might be far too simple to expect a good correlationship between in vitro performance and clinical success.

4 Future Perspectives: Advanced In Vitro Systems Mimicking the In Vivo Niche

In vitro and in vivo systems are hard to compare due to the differences in their complexity. In vitro systems are designed to be as simple as possible, i.e. there is only one cell type which is cultivated under two-dimensional conditions, on tissue culture treated polystryrene (TCPS) and with an optimized culture medium containing partly overdosed but mostly very underdosed factors. In contrast, in vivo there is always more than one cell type involved and cells act in a three-dimensional environment in a dynamic fashion surrounded by interstitial fluids and extracellular matrix (ECM). It is this artificial, for the cells, pathological environment to which cells adapt and respond. For cell lines this pathological adaption may even be more striking. It is in this context that the biocompatibility of materials is currently evaluated in vitro. The fact that cells are in a pathological situation which is related in hardly any aspect to the in vivo counterpart probably represents the predominant reasons for the limitations in translating in vitro findings into preclinical applications. So far, no final solutions have been presented and the bridging of the gap between in vitro and in vivo is still a matter of further research. However, various efforts have been undertaken which are very promising.

- *From single to multiple cell types.* Recently, there have been many attempts to address this issue. On the one hand, different mature cells types are cultivated in the same culture, either completely mixed or separated via membranes. On the other hand, obtained knowledge from stem cell research leads to the awareness that an in vitro mixture of progenitor cells in different states of differentiation and sometimes dedicated to different phenotypes yields more significant results than single cell types (e.g. [132]). If a material is placed in a multi-cell-type environment, each cell interacts with another in a synergistic way which is crucial for tissue formation in vivo and in particular for tissue repair and tissue homeostasis. The various cell types mutually affect cell proliferation, state of differentiation, and functionality of other cells, either directly via membrane-to-membrane contact or indirectly via released factors [41, 109, 116, 147]. If a material is placed in such a multi-cell- type environment, different cell types will adhere to the surface and will thereafter compete for the space. The selected cell-type composition should depend on the target application of the implant material. In the case of materials used in a bone environment, key cell types may include mesenchymal progenitor cells, osteoblasts, osteoclasts, fibroblasts and endothelial cells, whereas for a topical application of a material, fibroblasts, keratinocytes and endothelial cells are probably the key players. The cell type with the strongest competitive force will finally prevail at the material surface and will probably determine which kind of tissue is formed at the surface. The material surface characteristics strongly affect the competitive force based on cell proliferation and cell migration, which may occur cell-type-dependently as shown by Vrana and co-workers [142]. The competitive force can be determined by seeding different cell types on the material of interest and subsequently evaluating the change in cell number of each cell type as a function of the presence of the other cell type(s). Furthermore, the effect on the state of differentiation of each of the different cell types might also be measured by assessing cell-type-specific marker proteins [89, 151].
- *From 2D to 3D.* In current cell culture biomaterial evaluations, single cells are seeded on top of the material to evaluate its bioactivity. Under in vivo conditions, an implant is placed within a tissue and thus tissues and less single cells will contact the implant. It is generally accepted that cells in a 3D environment behave differently from those in a 2D environment [12, 56, 66, 102]. Nearly 25 years ago Sutherland started to use multicellular spheroids as an experimental model to elucidate processes taking place in a tumour [133]. Since the multicellular spheroids made of tumour cell line cells closely resemble solid tumours, it has recently been suggested to use this model as a high-throughput test system for antitumour drug development [77]. Similarly we have proposed the use of cell reaggregates of primary cells to evaluate biomaterials [95, 96]. For this, cell reaggregates are prepared of a defined cell number of one of the key cell types of the tissue that after implantation will contact the implant. The latter reaggregate, which can be seen as a kind of organoid, is placed on the test material and outgrowth is assessed. Cells in this situation have the choice to grow out on the material or to stay within their own context. One driving force to grow out of the

reaggregate is the stiffness gradient [63] which is probably formed when soft reaggregate comes into contact with the stiffer biomaterial. Stiff material generally attracts cells. We suggested to take the surface that is covered by the outgrown cells as an index of bioactivity. In addition, the cell morphology of the outgrown cells can be assessed as an additional parameter. The point which might affect the outcome is the difference in the oxygen tension gradient in vivo, with probably the lowest tension at the implant surface, and in this in vitro system where the lowest oxygen tension is not at the reaggregate–test material surface but in the centre of the reaggregate. Oxygen tension gradients may greatly affect cell migration activity and cell proliferation rate [47, 111].

- *From culture medium to body fluid mimicry*. Culture medium contains a mixture of salts, amino acids and dilute quantities of serum proteins (mostly heat-inactivated fetal calf serum, 10% concentration), and specific growth factors. Under in vivo conditions, the cell environment is crowded with proteins resulting in a reduction of diffusion and local high concentrations of released cell products. In this environment, nonspecific reactions play an important role. Unlike the in vivo environment, procollagen produced by the cells, for instance, only slowly coverts to collagen under common in vitro conditions. One solution for this nonphysiological situation is given by the team of Raghunath. They could show that, by the addition to the medium of charged macromolecules with a large hydrodynamic radius, the situation is changed and collagen is formed as in vivo [79]. Additionally, differentiation of various cell types is promoted in such an environment [45]. These effects show the biological relevance of the excluded volume effect by macromolecules and suggest that by adding macromolecules the original microenvironment can be better mimicked [45].
- *From static to dynamic*. The in vivo situation is characterized by the presence of microcirculation in the vicinity of the implanted material [59]. Dolder and co-workers reported that proliferation and differentiation towards osteoblasts (ALP activity, calcium deposition) of rat bone marrow stromal cells is (or at least tends to be) increased in the presence of a fluid flow [50]. This effect might be due to locally increased nutrients and oxygen tension, or to fluid-flow-induced mechanical loading [12]. It has been shown that fluid flow affects the outcome of material evaluation (e.g. scaffold mesh size [69]). By developing an in vitro set-up, the question arises which situation should be mimicked, since the microcirculation is strongly dependent on the site of implantation. Besides fluid flow variation, fluctuations in hydrodynamic compression and strain are normally occurring within the body, affecting cell performance [144].
- *Oxygen tension*. Control and adaptation of oxygen tension in cell culture is hardly ever considered even though it is a very important factor in how cells react [53]. While hypoxia is a pathological condition for most cell types, it is a key factor for chondrocyte development and behaviour. In native tissue, chondrocytes are exposed to oxygen concentrations ranging from 1–5%, since the distance to the vessels supplying the synovial membrane is exceptionally large compared to other tissues [163]. Cells react to a hypoxic environment by switching their metabolism from aerobic to anaerobic, up-regulating stress

proteins to withstand the toxic insults of hypoxia, and releasing vasoactive factors to re-establish proper perfusion and thus oxygen supply. Under these conditions, chondrocytes up-regulate collagen II production and aggrecan production [53]. An elegant approach to mimicking the in vivo oxygen tension gradient as well as an ECM environment is to place a collagen or agar gel layer with embedded cells on top of the test material. With this set-up, another important in vivo aspect would be taken into consideration—the dimensionality (3D instead of 2D). Collagen gels with embedded cells are currently used for various investigations including gene transfer [61], effects of matrix stiffness [63], mechanical load and tissue engineering [145]).

All these are single individual attempts to improve the test system by taking at least one key issue of the in vivo niche in the direct vicinity of the implant into account. However, a combination of several of these modifications of the test system would certainly help to better mimic this niche and as a result to further improve the test system.

4.1 Proposed Testing Strategy

By assessing in vitro biocompatibility of new materials in principle, answers to ten important questions are sought. In Fig. 6 these questions are listed together with which kind of tests we think could be used to find an answer. Since current testing strategies for biocompatibility have strong limitations regarding their prognostic value, we propose a multi-level sequential approach in which cytotoxicity is evaluated in the first of two levels. The first two levels are similar but not identical to the ISO 10993-5 test. In subsequent tests, bioactivity is investigated. For this, no in vitro test standards are currently present. The proposed multi-level test sequence is shown in Fig. 7. The complexity of the tests and their specificity increases from one level to the next. Each level focuses on another aspect relevant for biocompatibility and the biological performance of a material. In total, this test sequence should give an answer to all the questions mentioned in Fig. 6. The set-up presented assumes that the material is implanted for at least five days. The cells used for the first two levels are based on cell lines that are representative for the site of implantation of the material. Subsequent levels are based on primary cells (optimally human cells), again specific for the site of implantation. The various levels can be described as follows:

- Level 1—direct contact test (induction of cell necrosis): The first level is identical to the ISO 10993-5 direct contact test. With this test, a statement on the induction of acute cell necrosis can be made (within 24 h) (=>Answer to question 1).
- Level 2—extract test (effects on cells mass): For the second level, slight modifications are suggested to the ISO 10993-5 extract test. Instead of subconfluent cultures to which the extract is given, a seeding density is chosen

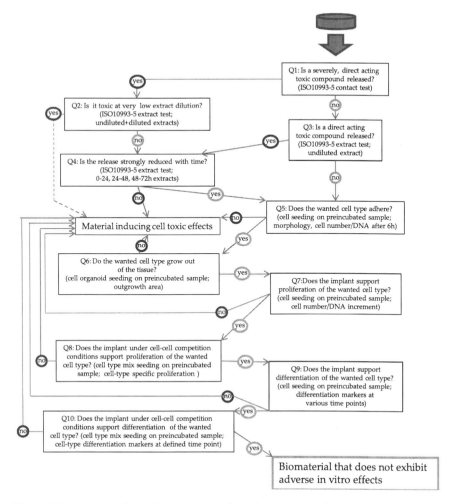

Fig. 6 Schematic overview of the sequence of ten important questions (Q1–10) which arise when assessing for in vitro biocompatibility

resulting in subconfluent cultures after 6 days. Furthermore, three culture medium extracts are prepared instead of one: one according to ISO 10993-5 (24-h extract at 37 C). For the second extract, the sample used to obtain the first 24-h extract is transferred to the same quantity of fresh medium and incubated for another 24 h at 37 C (24–48-h extract). A 48–72 h extract is prepared similarly. With these extracts of the sample, compounds are extracted that are not released during the first 24 h or 48 h of incubation. Cells are allowed to adsorb and adhere for 24 h before addition of the extract media at different concentrations. As diluent, a medium is used that was treated similarly but without the test sample. Total culture cell mass (measured by parameters such as total culture cell protein or DNA) is assessed five days after addition of the

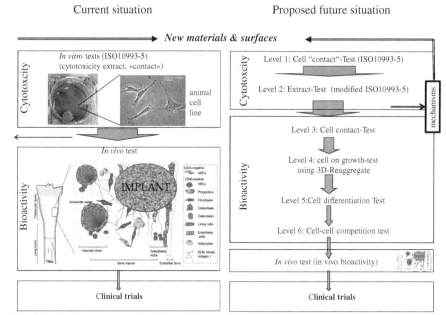

Fig. 7 Schematic representation of the current and proposed in vitro test battery for assessing biocompatibility. Schematic drawing of teh bone nicht: modified from 60

extracts. During the five days incubation period, the cell mass of the controls should increase at least five-fold relative to the initial mass. A deviation of observed cell mass would indicate that the material compromises the cell proliferation rate. The difference between the three types of extracts enables a statement of how far an initial toxicity may be of relevance (=>Answer to questions 2–4).

All samples for the next four levels are incubated in culture medium for 24 h before cells are seeded on them, ensuring that the protein layer composition is in equilibrium. Furthermore, at the same time toxic constituents are excluded that are released in the first 24 h in an initial burst type of release. As the time-point of measurement, five days after seeding is taken as a reference point but may be increased depending on the application, target tissue and cells. The subsequent tests should be feasible for most cell types but for some cell types the proposed test needs to be adapted.

- Level 3—cell adherence and spreading test: Cells are seeded on top of the test material at a density such that they reach 50% confluence after five days. Cells are stained for nuclei, actin and eventually for a typical differentiation marker. Cell spreading is qualitatively assessed. This test evaluates how far cells are able to attach to the test material surface, a process which should occur before apoptosis is induced [91], and how far the test material supports "normal" cell spreading and proliferation. This parameter of course can also be negatively influenced by test material constituents that are (sub-)toxic for the seeded cell type (=>Answer to questions 5 and 7).

- Level 4—cell on-growth test: In this level, material integration in tissues is tested. Normally tissues are in direct contact with an implant material. Cell aggregates containing around 15,000 cells of one or more typical cell types of the targeted site of the implantation niche are prepared and placed on top of the samples. The cells have the choice to stay within the tissue or to leave the tissue-like environment and contact and migrate onto the test material. Cell outgrowth is analysed after five days in culture. The area covered by cells is determined and compared with reference surfaces. A reduced coverage gives an indication of the cells intrinsic competitive strength to cover the test material surface and, if more cell types are included, also relative to each other (=>Answer to question 6).
- Level 5—cell differentiation test: In this test level, cells are seeded on top of the samples at a density ensuring sub-confluency after five days in culture. Thereafter, mRNA is isolated and the relative concentration of mRNA of specific genes determined. The outcome yields information on how far the material supports cell differentiation and allows statements on how far differentiation is influenced by the material. Statements regarding the extent and time evolution of cell differentiation can be made if more than one time-point is investigated (=>Answer to question 9).
- Level 6—cell–cell competition test: The final level would allow for assessment of cell affinity to the substratum in a cell–cell competitive manner. Labelled cells of different cell types in a defined ratio and amount are seeded on top of the material. After five days in culture, the cell number of each cell type is assessed as well as the expression of cell-type-specific differentiation markers. The latter and final test has the potential to predict which cell type finally covers the implant surface (=>Answer to questions 8 and 9).

The idea behind this suggested set-up is that each test is more stringent, reducing the number of promising materials from one level to the next. Materials that perform adequately in all test levels are considered to be optimal and ready for subsequent in vivo testing. The proposed test battery has the potential to reduce the time to market and cost per new developed implant that will positively pass the human trials.

Acknowledgments The present work was supported by the European Community's 7th Framework Programme under grant agreement nos. NMP3-LA-2008-214685 (project Magister) and NMP3-LA-2008-213939 (project POCO).

References

1. 10993–12/TC194 I (2007) ISO 10993–12:2007
2. Alves CM, Reis RL, Hunt JA (2010) The competitive adsorption of human proteins onto natural-based biomaterials. J R Soc Interface 7:1367–1377

3. Anderson JM (2001) Biological responses to materials. Annu Rev Mater Res 31:81–110
4. Arima Y, Iwata H (2007) Effect of wettability and surface functional groups on protein adsorption and cell adhesion using well-defined mixed self-assembled monolayers. Biomaterials 28:3074–3082
5. ASTM F720 (1981) Standard Practice for Testing Guinea Pigs for Contact Allergens: Guinea Pig Maximization Test. West Conshohocken, PA 19428-2959, USA, ASTM international
6. ASTM F895 (1984) Standard Test Method for Agar Diffusion Cell Culture Screening for Cytotoxicity. West Conshohocken, PA 19428-2959, USA, ASTM international
7. ASTM F748 (2006) Standard Practice for Selecting Generic Biological Test Methods for Materials and Devices. West Conshohocken, PA 19428-2959, USA, ASTM international
8. ASTM (2011) "F04 on medical and surgical materials and devices." Retrieved 28.6.2011, 2011. http://www.astm.org/COMMIT/SUBCOMMIT/F04.htm
9. ASTM F719 (1981) Standard Practice for Testing Biomaterials in Rabbits for Primary Skin Irritation ASTM international West Conshohocken, PA 19428-2959 USA
10. Babich H, Borenfreund E (1987) Structure-activity relationship (SAR) model established in vitro with the neutral red cytotoxicity assay. Toxicol In Vitro 1:3–9
11. Bakand S, Winder C, Khalil C et al (2006) A novel in vitro exposure technique for toxicity testing of selected volatile organic compounds. J Environ Monit 8:100–105
12. Baron MJ, Tsai C-J, Donahue SW (2010) Mechanical stimulation mediates gene expression in MC3T3 osteoblastic cells differently in 2D and 3D environments. J Biomech Eng 132:e041005
13. Bass DA, Parce JW, Dechatelet LR et al (1983) Flow cytometric studies of oxidative product formation by neutrophils: a graded response to membrane stimulation. J Immunol 130:1910–1917
14. bcc-Research (Market-Forecasting). (2011). Healthcare. The Market for Minimally Invasive Medical Devices. Report Code: HLC051E Published: March 2009 Retrieved 28.6.2011, 2011. http://www.bccresearch.com/report/minimally-invasive-medical-devices-hlc051e.html
15. Bellows CG, Heersche JNM, Aubin JE (1999) Aluminum accelerates osteoblastic differentiation but is cytotoxic in long-term rat calvaria cell cultures. Calcif Tissue Int 65:59–65
16. Belyanskaya L, Manser P, Spohn P et al (2007) The reliability and limits of the MTT reduction assay for carbon nanotubes-cell interaction. Carbon 45(13):2643–2648
17. Benigni R, Bossa C, Guilliani A et al (2010) Exploring in vitro/in vivo correlation: Lessons learned from analyzing phase I results of the US EPA's ToxCast project. J Environ Sci Health C 28:272–286
18. Bernas T, Dobrucki J (2002) Mitochondrial and nonmitochondrial reduction of MTT: interaction of MTT with TMRE, JC-1, and NAO mitochondrial fluorescent probes. Cytometry 47:236–242
19. Berridge MV, Herst PM, Tan AS (2005) Tetrazolium dyes as tools in cell biology: new insights into their cellular reduction. Biotechnol Annu Rev 11:127–152
20. Bhatia SK, Yetter AB (2008) Correlation of visual in vitro cytotoxicity ratings of biomaterials with quantitative in vitro cell viability measurements. Cell Biol Toxicol 24:315–319
21. Bigerelle M, Giljean S, Anselme K (2011) Existence of a typical threshold in the response of human mesenchymal stem cells to a peak and valley topography. Acta Biomater (in press)
22. Bonini MG, Rota C, Tomasi A et al (2006) The oxidation of 2′, 7′-dichlorofluorescin to reactive oxygen species: a self-fulfilling prophesy? Free Radic Biol Med 40:968–975
23. Born A-K, Rottmar M, Lischer S et al (2009) Correlating cell architecture with osteogenesis: first steps towards live single cell monitoring. Eur Cell Mater 18:49–62
24. Brama M, Rhodes N, Hunt J et al (2007) Effect of titanium carbide coating on the osseointegration response in vitro and in vivo. Biomaterials 28:595–608
25. Brodbeck WG, Colton E, Anderson JM (2003) Effects of adsorbed heat labile serum proteins and fibrinogen on adhesion and apoptosis of monocytes/macrophages on biomaterials. J Mater Sci Mater Med 14:671–675

26. Bruinink A (1992) Serum-free monolayer cultures of fetal chick brain and retina: immunoassays of developmental markers, mathematical data analysis, and establishment of optimal culture conditions. In: Zbinden G (ed) The Brain in Bits and Pieces. M.T.C, Zollikon, pp 23–50
27. Bruinink A (2008) In vitro toxicokinetics and dynamics: modelling and interpretation of toxicity data. In: Gad SC (ed) Preclinical Development Handbook. John Wiley, New York, pp 509–550
28. Bruinink A, Birchler F (1993) Effects of cisplatin and ORG.2766 in chick embryonic brain cell cultures. Arch Toxicol 67:325–329
29. Bruinink A, Reiser P (1991). Ontogeny of MAP2 and GFAP antigens in primary cultures of embryonic chick brain: effect of substratum, oxygen tension, serum and Ara-C. Int J Dev Neuroscience 9(3):269–279
30. Bruinink A, Sidler C (1997) The neurotoxic effects of ochratoxin-A are reduced by protein binding but are not affected by L-phenylalanine. Toxicol Appl Pharmacol 146:173–179
31. Bruinink A, Zimmermann G, Riesen F (1991) Neurotoxic effects of chloroquine in vitro. Arch Toxicol 65:480–484
32. Bruinink A, Reiser P, Müller M et al (1992) Neurotoxic effects of bismuth in vitro. Toxicol in Vitro 6:285–293
33. Bruinink A, Sidler C, Birchler F (1996) Neurotrophic effects of transferrin on embryonic chick brain and neural retina cell cultures. Int J Dev Neuroscience 14:785–795
34. Bruinink A, Faller P, Sidler C et al (1998) Growth inhibitory factor and zinc affect neural cell cultures in a tissue specific manner. Chemico-Biol Interact 115:167–174
35. Bruinink A, Siragusano D, Ettel G et al (2001) The stiffness of bone marrow cell-knit composites is increased during mechanical load. Biomaterials 22:3169–3178
36. Bruinink A, Yu D, Maier P (2002) Short-term assay for the identification of neurotoxic compounds and their liver derived stable metabolites. Toxicol In Vitro 16(6):717–724
37. Bruinink A, Tobler U, Hälg M et al (2004) Effects of serum and serum heat-inactivation on human bone derived osteoblast progenitor cells. J Mater Sci: Mater Med 15:497–501
38. Bruinink A, Kaiser J-P, Meyer DC (2005) Effect of biomaterial surface morphologies on bone marrow cell performance. Adv Eng Mater 7:411–418
39. Bruinink A, Hasler S, Manser P (2009) In vitro effects of SWCNT: role of treatment duration. Phys Status Solidii B 246:2423–2427
40. Brunner TJ, Wick P, Manser P et al (2006) In vitro cytotoxicity of oxide nanoparticles: Comparison to asbestos, silica and the effect of solubility. Environ Sci Technol 40: 4374–4381
41. Burguera EF, Bitar M, Bruinink A (2010) Novel in vitro co-culture methodology to investigate heterotypic cell–cell interactions. Eur Cell Mater 19:166–179
42. Buser D, Broggini N, Wieland M et al (2004) Enhanced bone apposition to a chemically modified SLA titanium surface. J Dent Res 83:529–533
43. Casals E, Pfaller T, Duschl A et al (2010) Time evolution of the nanoparticle protein corona. ACS Nano 4:3623–3632
44. Chang T, Wang R, Wu L (2005) Methylglyoxal-induced nitric oxide and peroxynitrite production in vascular smooth muscle cells. Free Radic Biol Med 38:286–293
45. Chen C, Loe F, Blocki A et al (2011) Applying macromolecular crowding to enhance extracellular matrix deposition and its remodeling in vitro for tissue engineering and cell-based therapies. Adv Drug Delivery Rev 63:277–290
46. Davies JT, Lam J, Tomlins PE et al (2010). An in vitro multi-parametric approach to measuring the effect of implant surface characteristics on cell behaviour. Biomed Mater 5:015002
47. Decaris ML, Lee CI, Yoder MC et al (2009) Influence of the oxygen microenvironment on the proangiogenic potential of human endothelial colony forming cells. Angiogenesis 12:303–311
48. Deeg JA, Louban I, Aydin D et al (2011) Impact of local versus global ligand density on cellular adhesion. Nano Lett 11:1469–1476

49. Desai SP, Kojima K, Vacanti CA et al (2008) Lidocaine inhibits NIH-3T3 cell multiplication by increasing the expression of cyclin-dependent kinase inhibitor 1A (p21). Anesth Analg 107:1592–1597
50. Dolder Jvd, Bancroft GN, Sikavitsas VI et al (2003) Flow perfusion culture of marrow stromal osteoblasts in titanium fiber mesh. J Biomed Mater Res A 64:235–241
51. Draper JS, Pigott C, Thomson JA et al (2002) Surface antigens of human embryonic stem cells: changes upon differentiation in culture. J Anat 200:249–258
52. Driscoll P (2011). Tissue engineering and cell therapy market growth. HealthWorks collective Retrieved 28.6.2011, 2011, from http://healthworkscollective.com/patrickdriscoll/21222/tissue-engineering-and-cell-therapy-applications-technologies-and-global-marke
53. Egli RJ, Wernike E, Grad S et al (2011) Physiological cartilage tissue engineering: Effect of oxygen and biomechanics. Int Rev Cell Mol Biol 289:37–89
54. Engler AJ, Sen S, Sweeney HL et al (2006) Matrix elasticity directs stem cell lineage specification. Cell 126:677–689
55. Eureka-Medical (2011) Medical device market. Retrieved 28.6.2011, 2011, from http://www.eurekamed.com/medical-device-market.html
56. Fischbach C, Kong HJ, Hsiong SX et al (2009) Cancer cell angiogenic capability is regulated by 3D culture and integrin engagement. Proc Natl Acad Sci USA 106:399–404
57. Forsby A, Bal-Price AK, Camins A et al (2009) Neuronal in vitro models for the estimation of acute systemic toxicity. Toxicol In Vitro 23:1564–1569
58. Garza KM, Soto KF, Murr LE (2008) Cytotoxicity and reactive oxygen species generation from aggregated carbon and carbonaceous nanoparticulate materials. Int J Nanomed 3:83–94
59. Gatzka C, Schneider E, Knothe Tate ML et al (1999) A novel ex vivo model for investigation of fluid displacements in bone after endoprosthesis implantation. J Mater Sci Mater Med 10:801–806
60. Grassel S, Ahmed N (2007) Influence of cellular microenvironment and paracrine signals on chondrogenic differentiation. Front Biosci 12:4946–4956
61. Haberl S, Pavlin M (2010) Use of collagen gel as a three-dimensional in vitro model to study electropermeabilization and gene electrotransfer. J Membrane Biol 236:87–95
62. Habibovic P, Woodfield T, de GK et al (2007) Predictive value of in vitro and in vivo assays in bone and cartilage repair—what do they really tell us about the clinical performance? Adv Exp Med Biol 585:327–360
63. Hadjipanayi E, Mudera V, Brown RA (2009) Guiding cell migration in 3D: a collagen matrix graded directional stiffness. Cell Motil Cytoskeleton 66:121–128
64. Haghiac M, Walle T (2005) Quercetin induces necrosis and apoptosis in SCC-9 oral cancer cells. Nutr Cancer 53:220–231
65. Haider W, Munroe N, Tek V et al (2011) Cytotoxicity of metal ions released from nitinol alloys on endothelial cells. JMEPEG 20:816–818
66. Hillmann G, Gebert A, Geurtsen W (1999) Matrix expression and proliferation of primary gingival fibroblasts in a three dimensional cell culture model. J Cell Sci 112:2823–2832
67. Hoang Thi TH, Chai F, Leprêtre S et al (2010) Bone implants modified with cyclodextrin: Study of drug release in bulk fluid and into agarose gel. Int J Pharm 400:74–85
68. Holmberg M, Hou X (2009) Competitive protein adsorption of albumin and immunoglobulin G from human serum onto polymer surfaces. Langmuir 26:938–942
69. Holtorf HL, Datta N, Jansen JA et al (2005) Scaffold mesh size affects the osteoblastic differentiation of seeded marrow stromal cells cultured in a flow perfusion bioreactor. J Biomed Mater Res A 74(2):171–180
70. International Organization for Standards (2009) Biological evaluation of medical devices—Part 1: evaluation and testing within a risk management process. Geneva, Switzerland
71. Isobe I, Yanagisawa K, Michikawa M (2001) 3-(4, 5-Dimethylthiazol-2-yl)-2, 5-diphenyltetrazolium bromide (MTT) causes Akt phosphorylation and morphological changes in intracellular organellae in cultured rat astrocytes. J Neurochem 77:274–280

72. Jabbar SA, Twentyman PR, Watson JV (1989) The MTT assay underestimates the growth inhibitory effects of interferons. Br J Cancer 60:523–528
73. Jarrahy R, Huang W, Rudkin GH et al (2005) Osteogenic differentiation is inhibited and angiogenic expression is enhanced in MC3T3–E1 cells cultured on three-dimensional scaffolds. Am J Physiol J Cell Physiol 289:C408–C414
74. Jun S-H, Lee E-J, Yook SW et al (2010) A bioactive coating of a silica xerogel/chitosan hybrid on titanium by a room temperature sol–gel process. Acta Biomater 6:302–307
75. Kaiser JP, Reinmann A, Bruinink A (2006) The effect of topographic characteristics on cell migration velocity. Biomaterials 27:5230–5241
76. Kandori K, Mukai M, Yasukawa A et al (2000) Competitive and cooperative adsorptions of bovine serum albumin and lysozyme to synthetic calcium hydroxyapatites. Langmuir 16:2301–2305
77. Khaitan D, Dwarakanath BS (2006) Multicellular spheroids as an in vitro model in experimental oncology: applications in translational medicine. Expert Opin Drug Discov 1:663–675
78. Kunzler TP, Drobek T, Schuler M et al (2007) Systematic study of osteoblast and fibroblast response to roughness by means of surface-morphology gradients. Biomaterials 28: 2175–2182
79. Lareu RR, Subramhanya KH, Peng Y et al (2007) Collagen matrix deposition is dramatically enhanced in vitro when crowded with charged macromolecules: The biological relevance of the excluded volume effect. FEBS Lett 581:2709–2714
80. LeBel CP, Ischiropoulos H, Bondy SC (1992) Evaluation of the probe 2'7'-dichlorofluorescin as an indicator of reactive oxygen species formation and oxidative stress. Chem Res Toxicol 5:227–231
81. Lee J-Y, Choo J-E, Park H-J et al (2008) Synthetic peptide-coated bone mineral for enhanced osteoblastic activation in vitro and in vivo. J Biomed Mater Res A 87:688–697
82. Liu Y, Peterson DA, Kimura H et al (1997) Mechanism of cellular 3-(4, 5-dimethylthiazol-2-yl)-2, 5-diphenyltetrazolium bormide (MTT) reduction. J Neurochem 69:581–593
83. Liu H, Niu A, Chen S-E et al (2011) β3-Integrin mediates satellite cell differentiation in regenerating mouse muscle. FASEB J 25:1914–1921
84. Loosli Y, Luginbuehl R, Snedeker JG (2010) Cytoskeleton reorganization of spreading cells on micro-patterned islands: a functional model. Philos Transact A Math Phys Eng Sci 368:2629–2652
85. Rd Lordo Franco, Chiesa R, Beloti MM et al (2009) Human osteoblastic cell response to a Ca-and P-enriched titanium surface obtained by anodization. J Biomed Mater Res A 88:841–848
86. Loveland BE, Johns TG, Mackay IR et al (1992) Validation of the MTT daye assay for enumeration of cells inproliferrrative and antiproliferative assays. Biochem Int 27:501–510
87. Luginbuehl R, Gasser B, Frauchiger V (2006) Residue analysis on implants. J ASTM Int 3:JAI13390
88. Luttringer O, Theil FP, Poulin P et al (2003) Physiologically based pharmacokinetic (PBPK) modeling of disposition of epiroprim in humans. J Pharm Sci 92:1990–2007
89. McBane JE, Battiston KG, Wadhwani A et al (2011) The effect of degradable polymer surfaces on co-cultures of monocytes and smooth muscle cells. Biomaterials 32:3584–3595
90. Meredith JE, Fazeli B, Schwartz MA (1993) The extracellular matrix as a survival factor. Mol Biol Cell 4:953–961
91. Meredith DO, Eschbach L, Riehle MO et al (2007) Microtopography of metal surfaces influence fibroblast growth by modifying cell shape, cytoskeleton, and adhesion. J Orthop Res 25:1523–1533
92. Michel R, Pasche S, Textor M et al (2005) The influence of PEG architecture on protein adsorption and conformation. Langmuir 21:12327–12332
93. Michiardi A, Aparicio C, Buddy D, Ratner BD et al (2007) The influence of surface energy on competitive protein adsorption on oxidized NiTi surfaces. Biomaterials 28:586–594

94. Mitrossilis D, Fouchard J, Pereira D et al (2011) Real-time single-cell response to stiffness. Proc Natl Acad Sci USA107:16518–16523
95. Moczulska M, Bitar M, Swieskowski W et al (2010) Comparison of two cell culture set-up for identifications of optimal textile scaffold regarding cell response. Eur Cells Mat 20 Suppl 1:34
96. Moczulska M, Bitar M, Święszkowski W et al (2011) Biological characterization of woven fabric using 2-and 3-dimensional cell cultures. J Biomed Mater Res Part A (in press)
97. Monopoli MP, Walczyk D, Campbell A et al (2011) Physical-chemical aspects of protein corona: Relevance to in vitro and in vivo biological impacts of nanoparticles. J Am Chem Soc 133:2525–2534
98. Monteiro-Riviere NA, Inman AO, Zhang LW (2009) Limitations and relative utility of screening assays to assess engineered nanoparticle toxicity in a human cell line. Toxicol Appl Pharmacol 234:222–235
99. Morra M, Cassinelli C, Cascardo G et al (2006) Collagen I-coated titanium surfaces: mesenchymal cell adhesion and in vivo evaluation in trabecular bone implants. J Biomed Mater Res A 78:449–458
100. Müller JP, Bruinink A (1994) Neurotoxic effects of aluminium on embryonic chick brain cultures. Acta Neuropathol 88:359–366
101. Olivares-Navarrete R, Raz P, Zhao G et al (2008) Integrin alpha2beta1 plays a critical role in osteoblast response to micron-scale surface structure and surface energy of titanium substrates. Proc Natl Acad Sci USA 105:15767–15772
102. Ouyang A, Ng R, Yang ST (2007) Long-term culturing of undifferentiated embryonic stem cells in conditioned media and three-dimensional fibrous matrices without extracellular matrix coating. Stem Cells 25:447–454
103. Pacurari M, Yin XJ, Zhao J et al (2008) Raw single-wall carbon nanotubes induce oxidative stress and activate MAPKs, AP-1, NF-kappaB, and Akt in normal and malignant human mesothelial cells. Environ Health Perspect 116:1211–1217
104. Panduri V, Weitzman SA, Chandel NS et al (2004) Mitochondrial-derived free radicals mediate asbestos-induced alveolar epithelial cell apoptosis. Am J Physiol Lung Cell Mol Physiol 286:L1220–1227
105. Park KN, Kim HJ, Wee JH et al (2008) Human melanoma cells adhesion on the surface of polymeric chemoattractants. Tissue Engin Regen Med 5:932–938
106. Park J-W, Kim H-K, Kim Y-J et al (2010) Osteoblast response and osseointegration of a Ti–6Al–4V alloy implant incorporating strontium. Acta Biomater 6:2843–2851
107. Pasche S, Vörös J, Griesser HJ et al (2005) Effects of ionic strength and surface charge on protein adsorption at PEGylated surfaces. J Phys Chem B 109:17545–17552
108. Pelkonen O, Raunio H (1997) Metabolic activation of toxins: tissue-specific expression and metabolism in target organs. Environ Health Perspect 105:767–774
109. Potapova IA, Gaudette GR, Brink PR et al (2007) Mesenchymal stem cells support migration, extracellular matrix invasion, proliferation, and survival of endothelial cells in vitro. Stem Cells 25:1761–1768
110. Puerto M, Pichardo S, Jos A et al (2009) Comparison of the toxicity induced by microcystin-RR and microcystin-YR in differentiated and undifferentiated Caco-2 cells. Toxicon 54:161–169
111. Raheja LF, Genetos DC, Wong A et al (2011) Hypoxic regulation of mesenchymal stem cell migration: the role of RhoA and HIF-1alpha. Cell Biol. Int. May 16. :PMID: 21574962 (epub ahead of print)
112. Ravanetti F, Borghetti P, Angelis ED et al (2010) In vitro cellular response and in vivo primary osteointegration of electrochemically modified titanium. Acta Biomater 6: 1014–1024
113. Rezek B, Michalıkova L, Ukraintsev E et al (2009) Micro-pattern guided adhesion of osteoblasts on diamond surfaces. Sensors 9:3549–3562
114. Rosengren A, Faxius L, Tanaka N et al (2005) Comparison of implantation and cytotoxicity testing for initially toxic biomaterials. J Biomed Mater Res A 75:115–122

115. Rostami-Hodjegan A, Tucker GT (2007) Simulation and prediction of in vivo drug metabolism in human populations from in vitro data. Nature Rev 6:140–148
116. Sanchez C, Deberg MA, Piccardi N et al (2005) Subchondral bone osteoblasts induce phenotypic changes in human osteoarthritic chondrocytes. Osteoarthritis Cartilage 13: 988–997
117. Schade R, Sikiric MD, Lamolle S et al (2010) Biomimetic organic–inorganic nanocomposite coatings for titanium implants in vitro and in vivo biological testing. J Biomed Mater Res A 95:691–700
118. Schubert D, Dargusch R, Raitano J et al (2006) Cerium and yttrium oxide nanoparticles are neuroprotective. Biochem P Biophys Res Commun 342:86–91
119. Schwartz Z, Raz P, G Z et al (2008) Effect of micrometer-scale roughness of the surface of Ti6Al4V pedicle screws in vitro and in vivo. J Bone Joint Surg Am 90:2485–2498
120. Scotchford CA, Gilmore CP, Cooper E et al (2002) Protein adsorption and human osteoblast-like cell attachment and growth on alkylthiol on gold self-assembled monolayers. J Biomed Mater Res 59:84–99
121. Selhuber-Unkel C, Erdmann T, Lopez-Garcia M et al (2010) Cell adhesion strength Is controlled by intermolecular spacing of adhesion receptors. Biophys J 98:543–551
122. Shalabi MM, Gortemaker A, Van't Hof MA et al (2006) Implant surface roughness and bone healing: a systematic review. J Dent Res 85:496–500
123. Sims JT, Plattner R (2009) MTT assays cannot be utilized to study the effects of STI571/Gleevec on the viability of solid tumor cell lines. Cancer Chemother Pharmacol 64:629–633
124. Singhvi R, StephanopoulosS G, Wang DIC (1994) Review: effects of substratum morphology on cell physiology. Biotechnol Bioeng 43(8):764–771
125. Sinha RK, Tuan RS (1996) Regulation of human osteoblast integrin expression by orthopedic implant materials. Bone 5(18):451–457
126. Sinha RK, Morris F, Shah SA et al (1994) Surface composition of orthopaedic implant metals regulates cell attachment, spreading, and cytoskeletal organisation of primary human osteoblasts in vitro. Clin Orthop Relat Res 305:258–272
127. Slater TF, Sawyer B, Sträuli U (1963) Studies on succinatetetrazolium reductase systems. III. Points of coupling of four different tetrazolium salts. Biochem Biophys Acta 77:383–393
128. Söderling E, Herbst K, Yli-Urpo A (1996) Protein adsorption to a bioactive glass with special reference to precorrosion. J Biomedical Mater Res 31:525–531
129. Sommer B, Felix R, Sprecher C et al (2005) Wear particles and surface topographies are modulators of osteoclastogenesis in vitro. J Biomed Mater Res A 72:67–76
130. Stadlinger B, Pilling E, Huhle M et al (2007) Influence of extracellular matrix coatings on implant stability and osseointegration: an animal study. J Biomed Mater Res B Appl Biomater 83:222–231
131. Su T, Waxman DJ (2004) Impact of dimethyl sulfoxide on expression of nuclear receptors and drug-inducible cytochromes P450 in primary rat hepatocytes. Arch Biochem Biophys 424:226–234
132. Sukmana I, Vermette P (2010) The effects of co-culture with fibroblasts and angiogenic growth factors on microvascular maturation and multi-cellular lumen formation in HUVEC-oriented polymer fibre constructs. Biomaterials 31:5091–5099
133. Sutherland RM (1988) Cell and environment interactions in tumor microregions: the multicell spheroid model. Science 8:177–184
134. Suzuki K, Kostin S, Person V et al (2001) Time course of the apoptotic cascade and effects of caspase inhibitors in adult rat ventricular cardiomyocytes. J Mol Cell Cardiol 33:983–994
135. Thevenot P, Hu W, Tang L (2008) Surface chemistry influences implant biocompatibility. Curr Topics Med Chem 8:270–280
136. Timbrell JA, Delaney J, Waterfield CJ (1996) Correlation between in vivo and in vitro toxic effects of foreign compounds. Comp Haematol Int 6:232–236
137. Trayner ID, Rayner AP, Freeman GE et al (1995) Quantitative multiwell myeloid differentiation assay using dichlorodihydrofluorescein diacetate (H, DCF-DA) or dihydrorhodamine 123 (H, R123). J Immunol Methods 186:275–284

138. Tse JR, Engler AJ (2011) Stiffness gradients mimicking in vivo tissue variation regulate mesenchymal stem cell fate. Plos One 6:e15978
139. Vejrazka M, Micek R, Stipek S (2005) Apocynin inhibits NADPH oxidase in phagocytes but stimulates ROS production in non-phagocytic cells. Biochim Biophys Acta 1722: 143–147
140. Vigier S, Helary C, Fromigue O et al (2010) Collagen supramolecular and suprafibrillar organizations on osteoblasts long-term behavior: Benefits for bone healing materials. J Biomed Mater Res A 94:556–567
141. Vocus/PRWEB (2011) Global orthopedic instrumentation market to exceed US$47 billion by 2015, according to a new report by Global Industry Analysts. Retrieved 28.6.2011, 2011, from http://www.prweb.com/releases/orthopedic_products/instrumentation/prweb8072105.htm
142. Vrana NE, Dupret A, Coraux C et al (2011) Hybrid titanium/biodegradable polymer implants with an hierarchical pore structure as a means to control selective cell movement. Plos One 6:e20480
143. Wachem vPB, Vreriks CM, Beugeling T et al (1987) The influence of protein adsorption on interactions of cultured human endothelial cells with polymers. J Biomed Mater Res 21:701–718
144. Walboomers XF, Elder SE, Bumgardner JD et al (2006) Hydrodynamic compression of young and adult rat osteoblast-like cells on titanium fiber mesh. J Biomed Mater Res A 76:16–24
145. Wallace DG, Rosenblatt J (2003) Collagen gel systems for sustained delivery and tissue engineering. Adv Drug Deliv Rev 55:1631–1649
146. Wang L, Sun J, Horvat M et al (1996) Evaluation of MTS, XTT, MTT and 3HTdR incorporation for assessing hepatocyte density, viablility and proliferation. Methods Cell Sci 18:249–255
147. Wang IE, Shan J, Choi R et al (2007) Role of osteoblast-fibroblast interactions in the formation of the ligament-to-bone interface. J Orthop Res 25:1609–1620
148. Wang P, Henning SM, Heber D (2010) Limitations of MTT and MTS-based assays for measurement of antiproliferative activity of green tea polyphenols. Plos One 5:e10202
149. Warheit DB, Sayes CM, Reed KL (2009) Nanoscale and fine zinc oxide particles: Can in vitro assays accurately forecast lung hazards following inhalation exposures? Environ Sci Technol 43:7939–7945
150. Webster TJ, Schadler LS, Siegel RW et al (2001) Mechanisms of enhanced osteoblast adhesion on nanophase alumina involve vitronectin. Tissue Eng 7:291–301
151. Wein F, Tobler U, Brose C et al (2010) The development of a triple-celltype system (TCS) that mimics the bone environment to study cell–cell-cell interactions/competition and biocompatibility. Eur Cells Mater 20 Suppl 1:51
152. Wick P, Manser P, Spohn P et al (2006) In vitro evaluation of possible adverse effect of nanosized materials. Physica Status Solidi 243:3556–3560
153. Wick P, Manser P, Limbach LK et al (2007) The degree and kind of agglomeration affect carbon nanotube cytotoxicity. Toxicol Lett 168:121–131
154. Wilkinson A, Hewitt RN, McNamara LE et al (2011) Biomimetic microtopography to enhance osteogenesis in vitro. Acta Biomater 7:2919–2925
155. Williams DF (2008) On the mechanisms of biocompatibility. Biomaterials 29:2941–2953
156. Xu L-C, Siedlecki CA (2007) Effects of surface wettability and contact time on protein adhesion to biomaterial surfaces. Biomaterials 28:3273–3283
157. Yamazaki K, Ikeda T, Isono T et al (2011) Selective adsorption of protein molecules on phase-separated sapphire surfaces. J Colloid Interface Sci (epub ahead of print)
158. Yang H, Liu C, Yang D et al (2009) Comparative study of cytotoxicity, oxidative stress and genotoxicity induced by four typical nanomaterials: the role of particle size, shape and composition. Appl Toxicol 29:69–78
159. Yeung T, Georges PC, Flanagan LA et al (2005) Effects of substrate stiffness on cell morphology, cytoskeletal structure, and adhesion. Cell Motil Cytoskeleton 60:24–34

160. Ying Y, Xingfen Y, Wengai Z et al (2010) Combined in vitro tests as an alternative to in vivo eye irritation tests. Altern Lab Anim 38:303–314
161. York JL, Maddox LC, Zimniak P et al (1998) Reduction of MTT by glutathione S-transferase. Biotechniques 25:622–628
162. Zhang S-Z, Lipsky MM, Trump BF et al (1990) Neutral red (NR) assay for cell viability and xenobioticc-induced cytotoxicityin primary cultures of human and rat hepatocytes. Cell Biol Toxicol 6:219–234
163. Zhou S, Cui Z, Urban JP (2004) Factors influencing the oxygen concentration gradient from the synovial surface of articular cartilage to the cartilage-bone interface: a modeling study. Arthritis Rheum 50(12):3915–3924
164. Zhou H-D, Bu Y-H, Tang A-G et al (2006) Time course of the osteoprotegerin gene expression in human primary osteoblasts as well as MG-63 cell lines and the effect of 17β-estradiol on it. J Chin Clin Med 1:61–66

Artificial Scaffolds and Mesenchymal Stem Cells for Hard Tissues

Margit Schulze and Edda Tobiasch

Abstract Medicine was revolutionized in the last two centuries and its advances have more than doubled life expectancy. Nevertheless, some problems are as old as mankind and although the underlying causes might have changed, the problems themselves have not. Musculoskeletal disorders and tooth loss are such problems; they are the major reasons for the ever-growing need for bone replacement, which cannot always be realized by autologous material. New, multidisciplinary strategies are needed for the development of novel materials to meet the demand. Stem-cell-based approaches combined with newly designed scaffold materials seem to be promising tools for constructing tissue replacements. Human mesenchymal stem cells and their remarkable differentiation potential are an interesting cell source for the development of bio-engineered tissues. Scaffolds based on natural and synthetic materials with or without the use of bioactive molecules are constructed to mimic the natural environment. They can improve proliferation and differentiation of the scaffold-seeded cells. Combined, they can provide specific remedies for hard tissue replacement, which will be discussed in this chapter.

Keywords Bioactive factors · Biomaterials · Bone · Composites · Dental · Differentiation · Hard tissue · Mesenchymal stem cells · Polymers · Proliferation · Regenerative medicine · Scaffolds · Tissue engineering

Abbreviations

2D	Two-dimensional
3D	Three-dimensional
AAV	Adeno-associated virus
ALS	Amyotrophic lateral sclerosis
ATP	Adenosine-5′-triphosphate
ATSC	Adipose tissue derived stem cell
BMSC	Bone marrow stromal cell

M. Schulze · E. Tobiasch (✉)
Deptartment of Natural Sciences,
University of Applied Sciences Bonn-Rhine-Sieg,
von-Liebig-Str. 20, 53359 Rheinbach, Germany
e-mail: edda.tobiasch@h-bonn-rhein-sieg.de

BMP	Bone morphogenic protein
cAMP	Cyclic adenosine monophosphate
CD	Cluster of Differentiation
CSD	Critical size defect
CVD	Chemical vapour deposition
DFC	Dental follicle cell
DNA	Deoxyribonucleic acid
DPLSC	Dental periodontal ligament stem cell
DPSC	Dental pulp stem cell
ECM	Extracellular matrix
ESCs	Embryonic stem cells
FDM	Fused deposition modelling
HA	Hydroxyapatite
HLA-DR	Human leukocyte antigen-DR
hMSC	Human MSC
HSCs	Hematopoietic stem cells
IL	Interleukin
iPS	Induced pluripotent stem cells
ISCT	International Society for Cellular Therapy
Klf4	Krueppel-like factor 4
LB	Langmuir–Blodgett
LbL	Layer-by-layer
Lin28	(Cell) lineage abnormal 28
MSCs	Mesenchymal stem cells
Oct4	Octamer binding transcription factor 4
P	Purinergic
P2X	Purinergic receptors (ligand-gated ion channels)
P2Y	Purinergic receptors (G protein-coupled)
PCL	Poly(ε-caprolactone)
PCL/TCP	Poly(ε-caprolactone)/tri-calcium phosphate
PEO	Poly(ethylene oxide)
PEOT/PBT	Poly(poly(ethylene oxide)terephthalate-co-(butylene)terephtalate)
PGA	Poly(glycolic acid)
PHMGCL	Poly(hydroxymethyl glycolide-co-ε-caprolactone)
PLA	Poly(lactic acid)
PLGA	Poly(lactic-co-glycolide)
PVD	Physical vapour deposition
RA	Retinoic acid
RGD	(one letter code of amino acids)
rhBMP-7	Recombinant human bone morphogenic protein
RhoA	Ras homolog gene family, member A
ROCKII	Rho-associated protein kinase II
SATB2	Special AT-rich sequence-binding protein 2
SC	Stem cell

SCAP	Stem cells from the apical papilla
SCID	Severe combined immunodeficiency
SEM	Scanning electron microscope
SES	Screw extrusion system
SHED	Stem cells of human exfoliated deciduous teeth
SLA	Selective laser ablation
SLS	Selective laser sintering
SMCs	Smooth muscle cells
Sox2	Sex determining region Y-related High-Mobility Group box 2
TGF-β	Transforming growth factor β
TIP	Tension-induced proteins
TP	Tricalcium phosphate
TRP	Transient receptor potential
UTP	Uridine-5′-triphosphate
VEGF	Vascular endothelial growth factor
Wnt	Wingless integration (signaling pathway)

Contents

1	Introduction	155
2	SCs for Bone Regeneration	157
	2.1 SCs and Their Potential	157
3	Scaffolds for Bone Regeneration	163
	3.1 Conventional Scaffold Materials and Fabrication Methods	164
	3.2 Conventional Scaffold Fabrication Methods	166
	3.3 Nanomaterials and Novel Fabrication Methods	167
4	Scaffold–Cell Interaction	169
	4.1 Surface Modification Methods	169
	4.2 MSCs on Artificial Surfaces	176
5	Scaffold–Cell Interaction in Hard Tissue Engineering	181
	5.1 MSCs on Scaffolds for Bone Defects	181
	5.2 MSCs for Dental Tissues	182
6	Future Developments	184
References		185

1 Introduction

Modern medicine has provided several breakthrough successes in drug research, such as the discovery of antibiotics providing the ability to cure bacteria-caused infectious diseases, and in surgery, such as solid organ transplantation providing help for patients with diseased organs. The increased demand for organs caused by a variety of different reasons, for instance cancer, infectious and metabolic diseases, or degeneration due to old age, cannot be covered by donation. Regenerative medicine, a new research field dedicated to rebuilding damaged organs and tissues, may provide alternatives in the future.

An interesting and rather new approach in regenerative medicine is the use of stem cells (SC), promising with their capacity for self-renewal and organ repair an unlimited source of material. But cells in vitro do not grow in three dimensions, but in a monolayer, and an organ with its highly regulated structures and various cell types cannot easily be constructed. Even simple tissues such as bone still provide obstacles which have to be overcome before tissue or organ replacement with SCs can be successfully applied in patients.

Musculoskeletal disorders such as osteoporosis, osteoarthritis, and bone defects represent major health problems worldwide with increasing incidence due to extended life expectancy and new recreational behavior. Critical-size bone defects, such as conditions after limb-preserving tumor surgeries and trauma-based osseous defects, still remain a huge challenge for reconstructive orthopedic surgery. Due to lack of alternatives, autologous bone transplantation is still the gold standard for treatment of osseous defects but, on the other hand, the sources of supply for autografts are limited and associated with several problems including infection risks, severe additional pain and donor site morbidity. Guided tissue regeneration with undifferentiated or differentiated SCs might be an alternative.

The necessity for tooth replacement is another widespread problem in industrial nations. The combination of tooth decay, formerly known as dental caries, and periodontal diseases has led to ten million lost teeth per year in Germany alone. To produce a stable implant often requires bone reconstruction and to achieve regeneration of the periodontium, the formation of the soft and mineralized connective tissues (root cementum, connective tissue fibers, bone) is required. Several therapies, such as implantation of autografts, allografts, alloplastic materials, or guided tissue regeneration, have been evaluated. However, results vary widely and are largely unpredictable [1].

For both approaches, cells and scaffold materials with specific characteristics are indispensable. The cells should be available in abundance, preferably from the actual patient, to avoid undesired immune reactions. For this, SCs seem to present the best source. A functionalized scaffold with appropriate design to provide skeletal mechanotransduction for increased proliferation and/ or enhance site-specific differentiation for the SCs used, thus leading to osteogenesis and bone remodeling, is the aim which must be met for the three-dimensional (3D) cell support material. And finally the interaction mechanisms between the biological organism and the cell-loaded scaffold must be explored to produce a customized tissue replacement with good potential for clinical implementation. The status quo for SCs and scaffold materials is discussed in the following.

2 SCs for Bone Regeneration

2.1 SCs and Their Potential

SCs are the biological precursors of each unique cell type of all existing tissues and organ within a body. They are generally defined by two major properties: they can self-renew for extended time periods and they are to various degrees undifferentiated, with the ability to differentiate into specialized cells under the appropriate conditions [2, 3]. SCs are classified according to their differentiation potential and their source.

Totipotent (or omnipotent) SCs have the potential to differentiate into every cell type needed to structure a complete, viable organism. For this purpose, they form the embryonic and the extra-embryonic tissues. Pluripotent cells are capable of differentiating into all cells of the three germ layers composing an organism, namely the ectoderm, mesoderm and endoderm [3, 4]. Multipotent SCs are lineage-restricted to a number of closely related cell types, mainly from one of the germinal layers. Oligopotent SCs have the potential to develop into a few cell types from a specific family of cells and unipotent cells give rise to one cell type only and are therefore also termed precursor cells. Pluripotent SCs are derived from the inner cell mass of the blastocyst of an embryo. Depending on the tissue source, adult SCs have a multipotent, oligopotent or unipotent differentiation potential.

SCs can be used in basic research as a key tool for the investigation of early development of an organism or a specific cell type. For this, the identification of regulatory genes and signaling pathways is crucial. Huge efforts have been undertaken in recent years to define key marker genes for the individual steps of commitment in the lineage-specific differentiation of SCs, and a broad set of more or less specific markers is available for the various lineages. As well as basic research, the abilities of self-renewal and differentiation of SCs can provide new medical perspectives to re-establish cellular function for the regeneration of damaged tissues or organs. The oldest and best-established SC therapy is hematopoietic SC transplantation. Transplantation of bone marrow-derived or peripheral blood-derived SCs is used to reconstitute the hematopoietic system after chemotherapy treating blood cancers such as myeloma or leukemia [5, 6]. Taken from a patient, SCs can also be used as test material to better understand genetic diseases, which can then be followed by specific drug development [7, 8]. The treatment of patients with injuries from accidents or degenerative diseases with SCs also holds great promise for cell-based tissue reconstruction in the future, because mature, finally differentiated cells are often not available in sufficient amounts or quality or can no longer replicate in vitro. If the cells are obtained for the individual patient directly, severe side effects such as tissue rejection can be eliminated. A new approach is the creation of artificial SCs by reprogramming. These SCs are expected to combine the positive characteristics of embryonal SCs, their pluripotency, with the positive features of adult SCs, the prospect of an autologous source. As well as the regeneration of heart muscle [9, 10] which is a major focus at the moment, SCs are discussed as supplements for treating diabetes [11], muscular dystrophy [12], spinal injuries [13], retinal degeneration [14], liver damages [15, 16], as well as Parkinson's disease [17] and Alzheimer's disease [18, 19], among others. All in all, SCs provide an invaluable resource for patient-specific therapies in a variety of diseases in the future.

On the other hand, several obstacles have to be overcome before a safe and efficient treatment of patients is warranted. One hurdle is to force the cells with normal growth in cell culture in a monolayer to spread into the third dimension. A number of studies have investigated the use of SCs in combination with scaffold materials and growth factors to replace, for example, lost or damaged bone [20, 21]. This will be discussed below. Another problem is that the developmental fate of SCs seems to be guided, but not restricted, by the surrounding tissue and the application of undifferentiated SCs contains the risk of spontaneous differentiation into undesirable cell types. It has been demonstrated that transplanted bone marrow-derived SCs can spontaneously differentiate into the osteogenic lineage when applied to the heart [22]. To exclude tumor formation from undifferentiated SCs and differentiation into undesired cell types, it is absolutely fundamental to optimize the control of differentiation. An overview of the current status with further particulars is given below.

2.1.1 Embryonic SCs

Pluripotent embryonic stem cells (ESCs) are isolated from the inner cell mass of the blastocyst [23]. The high differentiation potential of ESCs gives rise to huge expectations for cell-based therapies in regenerative medicine (see Fig. 1).

However, the isolation and subsequent use of SCs derived from a human embryo brings about strong ethical concerns [24] and is strictly limited by law in most countries. These cells are therefore mainly used in approaches where other SCs have limitations, such as in differentiation towards the neuronal lineages [25] or towards tissue cells where no SCs have been found to date. The application of ESCs in bone replacement is rather limited and focuses on basic research for a better understanding of the neural regulation of bone, the marrow, and its specific microenvironment (for a review, see [26]).

2.1.2 Induced Pluripotent SCs

A new type of pluripotent SCs has been developed recently, which seem to be an alternative to ESCs without ethical issues, the so-called induced pluripotent stem cells (iPS). These iPS can be obtained from an adult individual by the genetic reprogramming of fully differentiated somatic cells using a set of four specific transcription factors, such as Oct4, Sox2, Klf4, and c-Myc [27–29] or Oct4, Sox2, Lin28, and Nanog [30] (see Fig. 2). Shortly after the first reports on iPS, the transcription factor set used was reduced and the methods have been altered and improved.

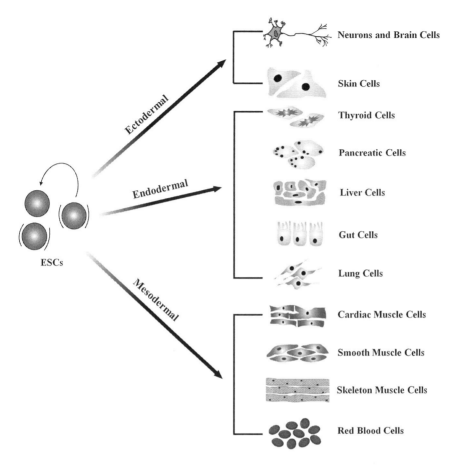

Fig. 1 The differentiation potential of ESCs. ESCs have the capacity for self-renewal. They can originate tissue-specific SCs from the three germinal layers, ectoderm, endoderm and mesoderm, and naturally have the ability to differentiate into all tissue-specific cell lineages

New reports have demonstrated that the lower the differentiation stage of the original cell type, the less the amount of transcription factors needed for the reprogramming, hinting to a ranking in the necessity of the four factors with Sox2 and Oct4 being the major ones. Consequently, iPS can be achieved with only one factor, Oct4, using precursor cells as source [31–33]. The first reprogramming strategies involved retroviral transfection or a multi-protein expression vector combined with the piggyBac transposon system for the delivery of the necessary transcription factor genes [34, 35]. However, both types of pluripotent cells, ESCs and iPS, are prone to cause cancer, as shown after transplantation into SCID mice, where they form teratomas [36]. This tumor risk is even higher for iPS reprogrammed with retroviruses, since these viruses integrate randomly, thus

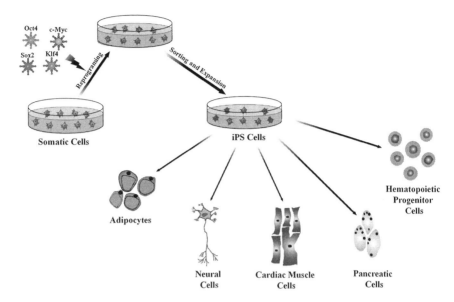

Fig. 2 Reprogramming somatic cells and the differentiation potential of iPS. iPS can be originated by viral transduction of a set of four transcription factors, e.g., Oct4, c-Myc, Sox2 and Klf4. These cells are pluripotent and have been successfully induced into endoderm, ectoderm and mesoderm cell lineages, such as adipocytes, hepatocytes, neural cells, cardiac muscle cells, pancreatic cells, and hematopoietic progenitors

giving rise to tumor formation on their own. Novel strategies use non-integrating viral vectors such as adenoviral or AAV-derived vectors to reduce the risk of insertion mutagenesis and subsequent cancer development, but these techniques diminish the already limited reprogramming efficiency [35, 37].

The obtained iPS display most characteristics of ESCs and can hardly be distinguished from them with respect to morphology, proliferation, gene expression, differentiation potential or surface antigens. However, differences can be found in the epigenetic status of the cells. Although the epigenetic status of iPS is very similar to that of ESCs, the iPS seem to keep at least part of the epigenetic features of the original cell [38]. In addition, recent data unexpectedly hint at rejection events after transplantation into inbred mouse strains [39]. If this report can be supported with further data, especially using cells differentiated from iPS, a major advantage of these cells—the use of autologous material with the potential of ESCs—is jeopardized.

The high risk of teratoma formation or simply the novelty of the cells might be the reason why relatively few data exist on the use of iPS for hard tissues. Nevertheless they also seem to have advantages when compared to adult SCs approaches, although iPS need rather a long time (12 weeks) to differentiate towards osteoblasts. When seeded in a gelfoam matrix and investigated in vitro and after transplantation in mice also in vivo, the authors report, as well as the expected data on osteodifferentiation, the recruitment of vasculature and microvascularization of the implant [40]. A very interesting and new approach is the use of the system also to create iPS for their differentiation. To enhance the osteogenic differentiation of iPS, Chen and colleagues transduced the iPS with a transcription factor, the nuclear matrix protein SATB2. They state that SATB2 facilitates the differentiation of iPS towards the osteoblast lineage [41].

2.1.3 Adult SCs

Some years ago most organs and tissues were thought to be regeneration-incompetent. Adult SCs were expected only in tissues with a high turnover, such as blood [42], skin [43, 44] or gastrointestinal (GI) tissues [45], since they are the repair system for the body and maintain the normal turnover of regenerative organs. However, since then, more and more stem and precursor cells have been found in adult tissues, including bone marrow [46–48], liver [49, 50], oral tissues [51, 52], brain [53], muscle [54, 55], fat tissue [56, 57] and recently also heart [58] and lung [59, 60]. These tissue-derived SCs greatly vary in their proliferative and differentiative potential.

A low proliferative potential of the SCs might be the reason why some organs seem to have a very low regenerative capacity, such as the heart, when compared to other organs like the liver, which have a far better repair system. Tissue-determined cells like the satellite cells in muscle or the precursor cells of the GI tract, skin and liver, on the other hand, have a limited lineage potential, whereas others SCs are multipotent, such as the fat-derived mesenchymal stem cells (MSCs), and can therefore differentiate in various tissue types. The MSCs with their broad differentiation potential are of particular interest for tissue or organ replacement strategies in regenerative medicine. Among the growing group of adult stem and precursor cells, MSCs are therefore a major focus of interest in research.

MSCs and Ectomesenchymal SCs

MSCs are defined according to minimal criteria by the International Society for Cellular Therapy (ISCT) [61]. One criterion is adherence to plastic. The SC character should be further confirmed by its multipotent in vitro differentiation potential into the adipogenic, chondrogenic and osteogenic lineages as demonstrated by specific stainings. In addition, MSCs should express the SC markers CD73, CD90 and CD105. Contamination with other cell types, such as macrophages, should be excluded by using the negative markers CD34, CD45, CD14 or CD11b, CD79α or CD19 and HLA-DR.

MSCs can be found in a variety of tissues, including umbilical cord blood, muscle, dermis, and bone marrow. and in marked abundance in adipose tissue. For years, bone marrow-derived MSCs have been investigated in detail, because bone marrow was already the most commonly used source for another adult SC type,

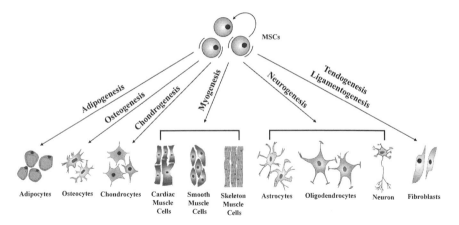

Fig. 3 The differentiation potential of MSCs. In accordance with HSCs, more committed progenitor cells are discussed, but they have not yet been defined. MSCs have the potential to differentiate into multiple mesenchymal-derived lineages. Adipogenesis leads to adipocytes and osteogenesis to osteocytes. Similar, chondrogenesis will end with chondrocytes. Myogenesis will generate cardiac, smooth, and skeletal muscle cells and neurogenesis will lead to astrocytes, oligodendrocytes, and neurons. After tendogenesis and ligamentogenesis, fibroblasts are produced

the hematopoietic stem cell (HSC). But the isolation procedure from bone marrow is painful and the yield of cells which can be obtained is low, and thus various other tissues have been investigated as sources of MSCs, with umbilical cord and fat being the most prominent [62–67].

Fat tissue-derived MSCs are especially attractive as a basic cell source for allografts, since they can be obtained from healthy and young donors with long telomers and thus an extended lifespan [68, 69]. They have eliminated some of the key challenges in potential tissue repair: the lack of sufficient base material and the risk of putative cell rejection by the immune defense system of the recipient, because these cells can be also isolated in abundance from the patient's own liposuction material and differentiated towards the desired cell type.

MSCs have the ability to differentiate into several different cell lineages, such as chondrocytes, astrocytes, myocytes, adipocytes, or osteoblasts (see Fig. 3), when cultivated under appropriate conditions using specific hormonal inducers and/or growth factors [3, 61, 66, 67, 70–74]. There are controversies about the limitation of the differentiation potential of MSCs. Several studies have shown differentiation of MSCs into cell types belonging to a germ layer other than mesoderm. Examples are differentiation into neuron-like cells [75], islet-like cells [76, 77] or hepatocytes [78]. They also posses the capacity to differentiate into cardiomyocytes [79] and endothelial cells [79, 80]. This might be due to the fact that the mesenchyme does not completely originate from the mesoderm, but partly stems from both of the other germ layers as well. A different hypothesis explains the potential of MSCs to differentiate into endodermal or exodermal cell types by transdifferentiation processes [8]. Due to this large differentiation capacity, using MSCs for autologous transplantations would be possible for several major diseases such as heart failure and diabetes or degenerative diseases such as bone defects. Furthermore, MSCs seem to be not only hypoimmunogenic and thus suitable for allogenic transplantation [66], but they are also able to induce immunosuppression upon transplantation [81].

Another group of adult SCs which has attracted attention is ectomesenchymal SCs, derived from oral tissues. This SC group includes dental pulp stem cells (DPSCs) and stem cells of human exfoliated deciduous teeth (SHEDs), both deriving from the pulpa, dental periodontal ligament stem cells (DPLSCs), dental follicle cells (DFCs) and stem cells from the apical papilla (SCAPs). These dental-derived progenitor cells or SCs have the potential to differentiate into dental cell types, such as ameloblasts, odontoblasts or cementoblasts. These properties make them valuable tools for dental regenerative medicine. In addition, it has been shown that some of these cells can also differentiate into osteoblasts or chondroblasts [82], which makes them valuable for additional, more general approaches in regenerative medicine. One major branch of research focuses on SC-based tissue engineering for the reconstruction of large bone defects and the osseointegration of tooth implants. This is also the topic of the following sections.

3 Scaffolds for Bone Regeneration

It is widely acknowledged that for the repair of musculoskeletal disorders such as bone defects and dental implants, tissue engineering approaches have to combine cells capable of osteogenic activity with an appropriate scaffolding material. Optimal bio-engineered scaffolds have to provide appropriate initial mechanical properties, promote the formation of new bone, and be gradually, evenly and homogeneously degraded without causing significant inflammatory responses or genetic alterations, in parallel allowing the new bone to remodel and assume the mechanical support function.

The biomaterials existing to date are not sufficiently optimized, in particular regarding the control of MSC differentiation. Consequently, there is an urgent need to design tissue-engineered scaffolds that offer an improved level of functionality over those currently available, adapted to be functionalized and to have direct influence on MSC growth and differentiation. For SC-directed bone repair to be clinically successful, a scaffold must be identified and optimized to support not only cellular adhesion and recruitment, but specifically also osteoinduction and osteoconduction. In the following, fundamental studies and recent results are summarized for scaffold material design, conventional and novel fabrication methods, and surface modification technologies. Particular focus is given to the influence of polymer scaffolds on adhesion and differentiation of hMSC.

3.1 Conventional Scaffold Materials and Fabrication Methods

According to Williams Dictionary, scaffolds for tissue engineering for regenerative medicine are defined as *porous substrate materials* (metals, ceramics, polymers) guiding cell adhesion, differentiation, proliferation, and growth [83]. Scaffolds are designed to promote new tissue formation by providing adequate 3D architecture for tailored porosity and appropriate surfaces for cellular adhesion and migration. Pioneer studies in scaffold development for tissue engineering were performed by Langer and Griffith [84–86]. An comprehensive overview of the variety of appropriate scaffold materials is given by Park [87] with specific focus on scaffold-based bone engineering by Hutmacher [88]. Concerning their chemical nature, scaffolds are prepared from metals, ceramics and natural or synthetic polymers. In addition to pure ceramic or polymer compounds, a broad variety of corresponding composites, i.e., ceramic–polymer composites, have been developed, in particular for bone tissue engineering. Those composites are designed to gain synergetic effects by combining required mechanical properties with osteoconductive characteristics.

Ceramics. Biocompatible ceramics include hydroxyapatite (HA), tricalcium phosphate, calcium phosphates (TP) and their composites, especially combined with natural or synthetic polymers. Pure ceramic materials are brittle and lack interconnected pores required for cell proliferation and angiogenesis, limiting the use of ceramic scaffolds to rather small defects. Thus, ceramic phosphate composites containing natural or synthetic polymers are more attractive for large bone tissue engineering, combining improved mechanics with osteoconductive properties [89]. Those ceramic–polymer composites are synthesized with ceramic particles embedded into the polymer matrix and exposed on the surface to improve osteoconductive effects. Mechanical properties are mainly influenced by calcium phosphate particle size and its distribution within the polymer matrix. The composite stiffness increases with decreasing particle size. HA bioceramics have been examined with mesenchymal stem and progenitor cells to study SC attachment, migration and differentiation into osteoblasts [90]. A recent review summarized the biological response, in particular the cell attachment influenced by ionic dissolution products from bioactive glasses and glass–ceramic composites [91].

Natural polymers. Naturally derived proteins or carbohydrate polymers are widely used as scaffolds for tissue engineering. Natural polymeric materials used for bone regeneration mainly include polysaccharides such as cellulose and corresponding derivatives [92], alginates (e.g., polyanionic co-polysaccharides [93], agar, and agarose derivatives), chitosan [94], hyaluronates [95], fibrin, fibronectin, collagen and gelatine and corresponding derivatives [96]. Since natural scaffold materials are often used in a gel-like phase, biological agents can be incorporated via gel formulation [88].

Collagens are the principal structural proteins in mammals, widely distributed in the body and a major component of the extracellular matrix (ECM): they occur in skin, bone, cartilage, tendons, ligaments, and blood vessels. Fibrillar collagens form a triple-helix containing three separate peptide strands twisted around one another. Collagens show high-tensile strength that makes them an essential structural factor providing mechanical strength to hard tissues. Collagens carry ligands supporting cell attachment and thus influence cell migration but also differentiation. In addition, these ligands (reactive functional groups such as hydroxy or amino) along the collagen backbone also enable interaction with bioactive molecules, i.e., growth factors, provided by drug releasing systems. Collagen composite materials developed in particular for bone regeneration include various combinations of fibrillar collagen with HA, polysaccharides, polyethylene glycol, cellulose derivatives or sodium hyaluronates. Chitosan fibrous scaffolds obtained by wet spinning were coated with different densities of type II collagen to evaluate the effect of this coating on MSC adhesion and chondrogenesis. The cell attachment and distribution after seeding correlated with the density of type II collagen. Cell number, matrix production, and expression of genes specific for chondrogenesis were improved after culture in collagen-coated chitosan constructs [97]. Mauney and colleagues performed in vitro and in vivo studies of differentially demineralized bone scaffolds using biologically-derived collagenous materials such as intestinal submucosa or demineralized bone matrix as substrate to facilitate the growth and differentiation of cells [98]. Besides collagen, its denatured derivative gelatine is used to prepare scaffold composites. Thus osteogenic differentiation of bone marrow-derived SCs could be demonstrated using mixed gelatine and chitosan-oligosaccharide scaffolds [99].

Synthetic polymers. Tissue-derived materials carry the risks of immune rejection, blood coagulation or tissue hypertrophy, and thus synthetic polymers are a very attractive alternative. Synthetic polymer scaffolds provide the opportunity to tailor physical properties such as molecular weight, molecular weight distribution, and correlated mechanical properties. Major challenges are the design of 3D architectures with defined porosity and a tailor-made surface adapted to specific requirements concerning cell adhesion.

Synthetic scaffold materials for bone tissue engineering mainly comprise polyesters, the most common being poly(lactic acid) (PLA), poly(glycolic acid) (PGA), and poly(caprolactone) (PCL). In addition to homo-polymers, a huge variety of synthetic co-polymers has been studied in the last two decades, and recently reviewed by Zippel and colleagues [100]. In general, the polymers themselves are biocompatible, and many of them are bioresorbable. Further requirements include injectability and biodegradability. For bone regeneration, scaffolds have to possess appropriate mechanical stability. Biodegradation rate and mechanical properties can be varied through variation of molecular weight and molecular weight distribution. Depending on the polymer synthesis methods, linear polymers, branched structures, and 3D networks can be prepared. One of the remaining problems is the formation and accumulation of a certain amount of degradation products in a short time period due to bulk degradation. Although the degradation products (e.g., lactic and glycolic acids) are also present in normal metabolic pathways, these amounts may result in local inflammation.

Moroni and colleagues synthesized novel polyether–ester co-polymers, such as poly(poly(ethylene oxide)terephthalate-co-(butylene) terephtalate) (PEOT/PBT), and studied the influence of porosity, molecular network mesh size and swelling on dynamic mechanical properties of the corresponding scaffold materials. PEOT/PBT co-polymers are characterized by high elasticity, robustness, and strength. In addition, they possess good processability due to a temperature-dependent cross-linking of hydrophilic poly(ethylene oxide) (PEO) and semi-crystalline poly(butylene terephthalate) segments. Mechanical properties and biodegradation rate has been controlled via variation of the molecular weight of the starting poly(ethylene glycol) segments and the weight ratio of PEO and PBT blocks. These polyether–ester co-polymers have been investigated for tissue regeneration and already tested for clinical applications (PolyActive^{TM}, IsoTis Orthopaedics), i.e., for bone filling materials [101]. Porous scaffold composite materials based on synthetic PLA mineralized with calcium phosphate have been designed by Kim and colleagues. Studies demonstrated a significant influence of scaffold composition on growth and differentiation of bone marrow-derived MSCs [102]. Recently, Li and colleagues reported in vitro mineralization and in vivo bone regeneration studies performed in a rat calvarial defect model using novel resveratrol-conjugated poly(ε-caprolactone) (PCL) composites. The incorporation of resveratrol results in increased alkaline phosphatase activity of rat bone marrow stromal cells (BMSCs) and enhanced mineralization of the cell–scaffold composites in vitro. The calvarial defects implanted with resveratrol-conjugated PCL showed a higher X-ray density than the defects implanted with control PCL. Bone-like structures, positively immunostained for bone sialoprotein, were shown to be more extensively formed in the resveratrol-conjugated PCL. Thus, incorporation of resveratrol into the acrylic acid-functionalized porous PCL scaffold led to a significant increase in osteogenesis [103].

3.2 Conventional Scaffold Fabrication Methods

Well-established polymer processing techniques include various moulding and casting processes, spinning, sintering, and extrusion techniques. The fabrication of 3D scaffolds includes the generation of pores via particle or selective leaching, phase separation and different gas forming methods, and various textile formation processes such as braiding, weaving, and knitting. Polymers have been investigated in form of foams, sponges, gels, and hydrogels as scaffold and release materials to deliver biologically active agents inducing tissue growth factors, as reviewed in detail by Sachloz and Moroni [96, 101].

Gels and Hydrogels are the most widely used scaffold materials providing the possibility of encapsulating cells, i.e., to generate engineered cartilage or to protect beta-cells against the immune system in type 1 diabetic patients. Hydrogels made from both non-resorbable polymers such as polyesters and polyamides and biodegradable polymers based on collagen, glycolic acid, lactic acid or hyaluronic acid are used in surgery. Gels and hydrogels used for tissue engineering applications are prepared starting from natural biopolymers (polysaccharides, e.g., hyaluronic acid derivative), synthetic polymers (e.g., poly(hydroxyethyl methacrylate)) or semi-synthetic derivatives (e.g., collagen–PLA composites). 3D network formation is performed via radical or photopolymerization induced by ultraviolet irradiation. It has been demonstrated that self-assembling peptide hydrogel structures support the differentiation and transdifferentiation of cells. Stem or progenitor cells are encapsulated within these self-assembling peptide hydrogel structures. The peptide hydrogel nanoscale environment renders the cells available for instruction by differentiation factors such as growth factors or ECM components, enabling the cells to differentiate or transdifferentiate within the structures. Due to limited mechanical and viscoelastic properties, hydrogels are mainly used for controlled drug release [104].

Polymeric foams can be used as both scaffolds and drug delivery matrices. One of the few synthetic polymers approved for human clinical use are porous foams made of a racemic poly(lactide-co-glycolide) copolymer. Microcellular foams are made from biodegradable or non-biodegradable polymers with pores throughout the material having a diameter of about 1–200 μm [105]. In addition, polymer surfaces may be textured as a result of foaming. This is of vital interest, since surface morphology and roughness have been demonstrated to influence the physiological response to an implant, including cell attachment, morphology, and differentiation [106].

3.3 Nanomaterials and Novel Fabrication Methods

The discovery of fullerenes and carbon nanotubes produced a tremendous development of novel nanomaterials and their investigation for use in many different applications [107]. Nanostructured biomaterials including nanoparticles, nanofibers, nanosurfaces, nanocomposites, and nanosphere-immobilized biomaterials have gained increasing interest in regenerative medicine, since these materials often mimic the ECM. Nanomaterials have thus been intensively studied in the last decade for utilization in tissue engineering and scaffold fabrication. Preparation, characterization, and invitro analysis of novel structured nanofibrous scaffolds for bone tissue engineering have recently been reviewed by Wang and co-workers [108].

Nanomaterials. Materials designed in nanoscale used for bone regeneration include nanospheres and nanoparticles [109, 110], nanotubes, in particular carbon nanotubes [111–113], and nanodendrimers based on carboxymethylchitosan/poly(amido amine) [114]. Applications are mainly focused on utilization of nanomaterials to improve mechanical properties of scaffold materials [115]. Lim studied micropatterning and characterization of electrospun PCL/gelatine nanofiber tissue scaffolds by femtosecond laser ablation for tissue engineering applications [116]. Nanofibers are prepared via electrospinning, phase separation or self-assembling techniques. Biodegradable polymer nanofibers mimic the nanofibrillar structure of ECM. The nanoscaled collagen fibrillar structure (50–500 nm in diameter) has been found to enhance cell–matrix interactions [117]. The most promising approaches are represented by nanocomposites, reviewed by Zhang [110]. It could be shown that the early osteogenic signal expression of rat BMSCs is influenced by both HA nanoparticle content and initial cell seeding density in biodegradable nanocomposites scaffolds [109]. Sitharaman and co-workers introduced a novel nanoparticle-enhanced biophysical stimulus based on the photoacoustic effect. Results showed that the photoacoustic effect influences differentiation of bone marrow-derived marrow stromal cells grown on poly(lactic-co-glycolic acid) polymer films into osteoblasts. Osteodifferentiation of MSCs due to photoacoustic stimulation is significantly enhanced by the presence of single-walled carbon nanotubes in the polymer [115]. Nanostructured mesoporous silicon can be used for discriminating in vitro calcification of electrospun scaffold composites [118]. Carbon nanotubes possess exceptional mechanical, thermal, and electrical properties, facilitating their use as reinforcements or additives in various biomaterials to improve their mechanical behavior in particular. Carbon nanotubes are synthesized and added to conventional polymer scaffolds to promote and guide bone tissue growth and regeneration [113]. Another approach recently reported combines controlled synthesis of colloidal nanoparticles with freeze-drying technique for bone tissue engineering applications. Porous nanocomposite scaffolds based on poly(vinylalcohol) and colloidal HA nanoparticles were prepared. In vitro experiments with osteoblast cells indicated an appropriate penetration of the cells into the scaffold's pores and cell growth support [119].

Novel Scaffold Fabrication Technologies have been developed in the past decade that open new opportunities for 3D scaffold design [117, 120, 121]. In particular, electrospinning and different rapid prototyping techniques including 3D printing, fused deposition modelling (FDM), stereolithography, selective laser ablation (SLA), and selective laser sintering (SLS) are considered to be the most promising techniques for smart scaffold fabrication [101, 122], resulting in new materials, nanostructured surfaces, and novel 3D architectures. Rapid prototyping technologies thus enable the production of scaffolds with a controllable interconnected pore network, allowing improved cell migration and nutrient exchange. Electrospinning provides fibrous scaffolds mimicking the dimensions and topology of ECM fibers. Filaments can be formed on the nanometer scale and used as medical membranes and scaffolds for tissue engineering.

A broad variety of materials was tested including natural compounds such as collagen and synthetic polymers, e.g., PLA, PGA, PCL, and corresponding co-polymers. The preparation and characterization of a 3D printed scaffold based on a functionalized polyester for bone tissue engineering applications was reported by Seyednejad and co-workers. Porous scaffolds were prepared based on a hydroxyl functionalized polymer, poly(hydroxymethylglycolide-co-ε-caprolactone) (PHMGCL). Scaffolds consisting of PHMGCL or PCL were produced via 3D plotting resulting in a high porosity and an interconnected pore structure. Human MSCs were seeded onto the scaffolds to evaluate the cell attachment properties and differentiation. Results demonstrated that cells filled the pores of the PHMGCL scaffold within one week, displayed increased metabolic activity, and supported osteogenic differentiation [123].

4 Scaffold–Cell Interaction

4.1 Surface Modification Methods

From a chemical point of view, surfaces characteristics include hydrophilicity versus hydrophobicity, polar versus non-polar functionalities, and neutral versus charged surfaces, all defined by nature and amount of functional chemical groups. Depending on electron negativity strength of the atoms, non-polar or polar structures can be created, and hydrophilicity can be changed in a controlled manner that in turn significantly influences cell adhesion processes. Tailor-made surfaces can be designed via chemical modification, including physical, mechanical, and chemical adhesion (chemisorption). A stronger connection is realized via chemical grafting resulting in covalent bonds. In Table 1 bond energies and corresponding adhesion forces for different chemical bonds are summarized.

According to Dupré, the adhesion power of any substance on a surface depends on the surface energies of both substance to be adhered to and substrate, and their interface energy [124]. A comprehensive review entitled *Biomolecular engineering at interfaces* including discussion of the most relevant aspects influencing cell adhesion on scaffold surfaces and corresponding analytical methods is given by He and colleagues [125]. A detailed analysis of physical and biochemical effects for a scaffold-based approach to directing SC neural and cardiovascular differentiation is given by Chew and Low [126].

4.1.1 Biomimetic Surfaces and Controlled Drug Release

Surface modification in biomedicine, in particular tissue engineering, is mainly realized via creation of a so-called *biomimetic surface*. Engineering cell and tissue behavior at device surfaces is focused on modifying the material surface to interact selectively with a specific cell type through biomolecular recognition processes. The cell surface has a variety of receptors that bind with other cells or specific proteins, which compose the environment surrounding the cells known as the ECM. A promising approach is the biomimetic modification of the material in which peptides containing the adhesion domains of the ECM proteins are attached to the base material. The central hypothesis of biomimetic surface engineering is that peptides which mimic part of the ECM affect cell attachment to the material, and surfaces modified with these active peptides can induce tissue formation conforming to the cell type seeded on the material. Therefore extensive research over

Table 1 Chemical bond forces, energies and corresponding adhesion forces

Bond forces	Dipole (Keesom force)	Induction (Debye force)	Dispersion (London force)	Hydrogen	Covalent	Ionic	Metallic
Energies [kJ/mol]	<30	<10	<10	<50	60–800	600–1,000	100–800
Adhesion [N/mm^2]	$\approx 10^2$	$\approx 10^2$	$\approx 10^2$	$\approx 10^2$–10^3	$\approx 10^4$–10^5	$\approx 10^4$–10^5	$\approx 10^4$–10^5

the last decade has been performed on the incorporation of adhesion-promoting oligopeptides into biomaterial surfaces [106, 127–132]. In this way, engineered microenvironments have been designed for controlled SC differentiation including biomimetics and controlled release [120].

Recent studies on biomimetic materials include the following: bioceramic implants with drug and protein controlled delivery capability [133]; surface modification with fibrin/hyaluronic acid hydrogel on solid freeform-based scaffolds followed by bone morphogenic protein-2 (BMP-2) loading to enhance bone regeneration [134]; synergistic effects of the dual release of stromal cell-derived factor-1 and BMP-2 from hydrogels on bone regeneration [135]; spatial control of gene expression within a scaffold by localized inducer release [136]; bio-activation via glycosaminoglycans as regulators for SC differentiation [137]; and biomimetic properties of an injectable chitosan/nano-HA/collagen composite [138]. In vivo studies demonstrated that chondrogenic differentiation of MSCs is induced by collagen-based hydrogels [139].

So it is possible to individually control the release of several agents by biomaterial drug release systems. Nevertheless, the appropriate combination of bioactive factors needed at different time points during tissue regeneration has still to be studied in more detail. Furthermore, the therapeutic application of growth factors can be accompanied by undesirable side effects due to the difficulty in controlling the release in an appropriate dose-dependent manner. Bioactive factors have been extensively investigated for their effects on angiogenesis, cell growth or SC differentiation [140]. Both native and artificial receptor ligands, i.e., extracellular nucleotides, are known to induce SC differentiation or growth, e.g., BMPs and transforming growth factor-β used in bone and cartilage regeneration. BMPs have been shown to recruit MSCs from bone marrow and periosteum to the site of repair and to support proliferation and differentiation of these cells. Furthermore, they have been shown to induce vascularization, bone formation, remodeling, and marrow differentiation. In vitro studies have shown that in articular cartilage, transforming growth factor- β (TGF-β) induces MSC differentiation to chondrocytes and promotes cell proliferation [72].

Wei and Ma studied poly(lactides)/apatite composite scaffolds prepared by a biomimetic approach. The poly(lactide) scaffolds were prepared by conventional phase separation in dioxane. Nanofibrous scaffolds were prepared by sugar template leaching and phase separation in tetrahydrofuran. Nanosphere drug release systems are immobilized on nanofiber composites; invitro release kinetics of rhBMP-7 from nanosphere-immobilized nanofibrous scaffolds were studied. Three distinct release profiles were achieved from three different co-polymers (different co-monomer ratios, or different molecular weights). New bone formation in rhBMP-7 incorporated PLA nanofibrous scaffolds retrieved six weeks after subcutaneous implantation in rats [117].

So-called "integrated biomimetic systems" combine conventional materials and fabrication methods, respectively, with nanomaterials and nanotechnologies: e.g., chitosan/hyaluronic acid composites [138], PLA/hyaluronic acid [102], and calcium phosphate/polymer composites (Wagoner et al. 2009).

Biomimetic composite coating is performed to improve the functional performance of rapid prototyped scaffolds for bone tissue engineering. Thus, rapid prototyped poly(ε-caprolactone)/tri-calcium phosphate (PCL/TCP) scaffolds were fabricated using the screw extrusion system (SES). The fabricated PCL/TCP scaffolds were coated with a carbonated hydroxyapatite–gelatin composite via biomimetic co-precipitation. The cell–scaffold interaction was studied by culturing porcine BMSCs on the scaffolds and assessing the proliferation and bone-related gene and protein expression capabilities of the cells. Confocal laser microscopy and scanning electron microscopy (SEM) images of the cell–scaffold constructs showed a uniformly distributed cell sheet and accumulation of ECM in the interior of carbonated hydroxyapatite–gelatin composite-coated PCL/TCP scaffolds. The proliferation rate of BMSCs on gelatin composite–coated PCL/TCP scaffolds was about 2.3 and 1.7 times higher than that on PCL/TCP scaffolds and gelatin-coated PCL/TCP scaffolds, respectively, by day ten. Furthermore, reverse transcription polymerase chain reaction and Western blot analysis revealed that gelatin composite-coated PCL/TCP scaffolds stimulated osteogenic differentiation of BMSCs [141].

The effect of surface-modified collagen on the adhesion, biocompatibility, and differentiation of BMSCs has been studied in poly(lactides–co-glycolide)/chitosan scaffolds. The scaffold containing type I collagen (640 µg/ml) had about 1.2 times the cell adhesion efficiency of the corresponding unmodified scaffold. In addition, the modification of type I collagen increased the cell viability about 1.3-fold and the biodegradation 1.2-fold. The differentiation of BMSCs in PLGA/chitosan scaffolds produced osteoblasts with mineral deposition on the substrate. Moreover, the surface collagen promoted the formation of mineralized tissue and reduced the amount of phenotypic BMSCs in the constructs [142].

Nanoscaled drug release systems incorporated into nanostructured biomaterials represent a novel and promising strategy for tissue regeneration [104]. Biomaterials used as matrices for controlled drug release include hyaluronic acid, acrylic acid, dextran methacrylic acid, polyethylene glycol acrylate/methacrylate, and polyethylene glycol diacrylate/dimethacrylate [143]. BMP-2 loading to enhance bone regeneration was studied using solid-freeform-based polymer scaffolds with fibrin hyaluronic acid hydrogel-modified surfaces [134]. Novel dendron-like nanoparticles have been investigated by Oliveira and colleagues, including in vivo studies of SC differentiation into osteoblasts. Biodegradable dexamethasone-loaded dendron-like nanoparticles of carboxymethylchitosan/poly(amido amine) dendrimer have been synthesized and used as intracellular drug delivery systems. Results proved a noncytotoxic in vitro behavior, supporting cell attachment and incorporation. In vivo experiments using rats could demonstrate a good performance of dendron-like nanoparticles for intracellular delivery of dexamethasone [114].

In addition, conventional surface coating can be performed via chemical and physical vapour deposition (CVD, PVD) methods. CVD and PVD are well-known technologies mainly used in the microelectronics industry to modify surfaces. Werner and co-workers used ammonia plasma treatment and studied maleinic acid-based co-polymer surfaces immobilized with amines. Plasma-immobilized hydrogels of poly(N-alkylacrylamide)-g-poly(ethylene glycol) were prepared using ammonia plasma treatment of poly(3-hydroxybutyrate). A platform of thin polymer coatings was introduced for the functional modulation of immobilized bioactive molecules at solid–liquid interfaces. The approach is based on covalently attached alternating maleic acid anhydride copolymers with a variety of co-monomers and extended through conversion of the anhydride moieties by hydrolysis, reaction with functional amines, and other conversions of the anhydride moieties. We demonstrated that these options permit control of the physicochemical constraints for bioactive molecules immobilized at interfaces to influence important performance characteristics of biofunctionalized materials for medical devices and molecular diagnostics. Examples concern the impact of the substrate-anchorage of fibronectin on the formation of cell–matrix adhesions, the orientation of endothelial cells according to lateral anti-adhesive micropatterns using grafted PEO, and the spacer-dependent activity of immobilized synthetic thrombin inhibitors [144].

4.1.2 Modification of Surface Topography

Cellular behavior can be influenced and even dictated in a controlled manner by topographically patterned surfaces [145]. Surface roughness at the micro- and even nanoscale is known to influence biocompatibility of synthetic materials used for tissue engineering applications. Furthermore, adhesion and alignment strongly depends on micro- and nanotopographical features. Symmetry and regularity of surface patterns (isotropic versus anisotropic grinding) causes differences in cell responses. Conventional surface modification strategies can be divided into two groups. The first one covers methods which changing surface chemistry and topography: e.g., chemical adsorption, plasma treatment methods, and chemical etching. The second group alters surface topography: mechanical roughening, the so-called substrate templating methods (e.g., lithography), electro and vapour deposition methods, and novel moulding processes [146]. In the last decade, material surfaces used for tissue engineering applications have been micro- and nanostructured during scaffold fabrication via solid freeform techniques, e.g., 3D printing, 3D plotting [117, 147]. Rapid prototyping processes do show differences in resolution and all are characterized by advantages and certain limitations.

The highest resolutions can be realized via SLA (70–250 μm). However, SLA requires appropriate liquid photopolymers that are still limited in availability, whereas 3D printing is a rather fast process but characterized by weak bonding between powder particles [122].

Surface patterning of novel polyesters to be used in bone tissue engineering is realized via the 3D printing process [123]. Novel biocompatible polyacrylate-based photopolymers have been synthesized for scaffold fabrication via stereolithography [148]. Whitesides and co-workers first described the patterning of proteins and cells using so-called soft lithography [149, 150]. Patterning via electrospinning was used to study the role of nanostructured mesoporous silicon in discriminating invitro calcification for electrospun scaffold composites [118]. Electrospun scaffold composites consisting of PCL/gelatine nanofiber have been produced and micropatterned by femtosecond laser ablation for tissue engineering applications [116].

The geometry and size of ECM structures do have significant effects on various cell properties including attachment/adhesion, migration, and proliferation. Differences in the height of nanotopographic features influence cell behavior through secondary effects, such as alterations in the effective substrate stiffness [151].

Gerecht and co-workers could demonstrate that nanotopographic-structured ECM alters the morphology and proliferation of human ESCs through cytoskeletal-mediated mechanisms. Poly(dimethylsiloxane) gratings with 600-nm features and spacing have been designed that are able to induce ESC alignment and elongation [152]. In addition, they could also show that nanotopographic features altered the organization of various cytoskeletal components such as F-actin, vimentin, γ-tubulin, and α-tubulin. Changes in proliferation and morphology were abolished by the effect of actin-disrupting agents. Furthermore, the influence of nanotopographic features may be mediated through secondary effects such as alterations in the effective stiffness perceived by the cell or differences in protein adsorption caused by ECM nanotopographics.

Surface patterning via *self-assembled composites*. In addition to functionalization via chemical reactions, solid substrates (scaffolds) can be covered by ultrathin films, single monolayers or multilayers using different methods: (a) self-assembling methods, (b) the co-called layer-by-layer (LbL) method, or (c) the Langmuir–Blodgett (LB) technique. Both surface chemistry and topography can be varied in a very controlled manner via film-coating [153, 154]. Whitesides and colleagues have reviewed the application of so-called soft lithography in biology and biochemistry using self-assembling processes [150].

Chen and co-workers studied a hybrid system consisting of a self-assembled composite matrix in a hierarchical 3D scaffold to be used for bone tissue engineering. The effects of the PCL-based hybrid scaffold on hMSC seeding efficiency, proliferation, distribution, and differentiation were investigated. Porous PCL meshes prepared by FDM were embedded in a matrix of hyaluronic acid, methylated collagen, and terpolymer via polyelectrolyte complex coacervation. Studies showed clearly that embedded scaffolds provided a higher cell seeding efficiency, a more homogeneous cell distribution, and more osteogenically differentiated cells, verified by a more pronounced gene expression of the bone markers alkaline phosphatase, osteocalcin, bone sialoprotein I, and bone sialoprotein II. In addition, dynamic culture resulted in higher amounts of deoxyribonucleic acid (DNA) and calcium. Embedding synergistically enhanced the calcium deposition of hMSC [155]. Further studies recently reported include the design of novel 2D and 3D biointerfaces using self-organization to control cell behavior [156], concave pit-containing scaffold surfaces that improve SC-derived osteoblast performance and lead to significant bone tissue formation [157], polymer thin films for biomedical applications designed via nanotechnologies [158], and direct patterning of protein- and cell-resistant polymeric monolayers and microstructures [159].

LbL technique using polyelectrolytes. Picart and co-workers studied new methods for positioning and anchoring of biomolecules onto scaffold surfaces using multiple functionalities of polyelectrolyte multilayer films [153]. The LbL technique using polyelectrolytes was first described by Decher [160]. A variety of depositing methods have since been developed for LbL formation including dip coating, spin coating, and spraying. Entcheva and colleagues developed a new dewetting method, which appears to be efficient, economical, and fast and could be used to create unique adsorption topographies, including fractal networks and aligned fibers [161]. For future use and industrial applications of LbL films, the total time required for film preparation and the anchorage of the layer to the underlying substrate are probably important constraints. Rapid methods such as spraying are being further developed. In addition, anchorage to the underlying substrate was improved, in particular for hydrophobic surfaces like poly(tetrafluoroethylene) and poly(ethylene), which often require priming methods. Assembling polyelectrolyte multilayers and their effects on self-assembly of particles in a so-called bottom-up approach is reported for polymers, particles, nanoparticles, and carbon nanotubes [162]. In another approach, polyelectrolyte multilayer films have been designed for vascular tissue engineering applications. Human mesenchymal SC differentiation into endothelial-like cells could be observed on surfaces coated with polyelectrolyte multilayer films [163].

LB technique. Beside the LbL method, mainly limited to polyelectrolytes, the LB technique can be used to design mono- and multilayers as coating materials, as illustrated in Fig. 4.

LB monolayers can be modified in a very controlled manner to obtain tailor-made surfaces. The surface roughness is limited to a nanometer scale, depending on the chemical structure of the monomers and polymers used for film formation. Appropriate polymers for mono- and multilayer formation via LB thin film technology are rigid rod-like polymers. Rigidity can be caused by different structural reasons: rigid monomer units (e.g., aromatic rings in poly-p-phenylenes), supramolecular structures (e.g., helical structures of polypeptides, DNA, polyglutamates, cellulose derivatives) or specific packing resulting in rod-like systems, e.g., phthalocyaninato poly(siloxanes). In Fig. 4b the regeneration of cellulose ethers is illustrated. This reaction can be performed after film transformation directly on the scaffold (c). Part D in Fig. 4 illustrates the so-called reaction zones for surface modification reactions. Thus, cell-attracting functionalities can

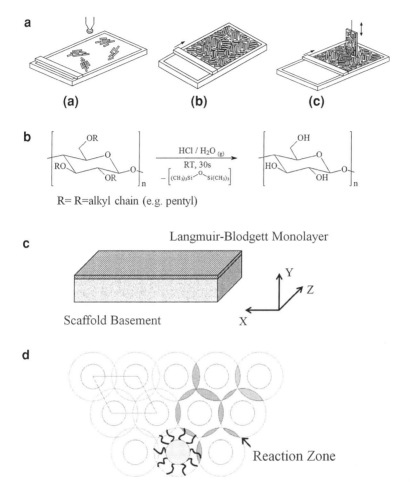

Fig. 4 Surface feature of scaffold. **a** Monolayer formation via LB technique including three steps: (**a**) spreading of polymer solution; (**b**) compression to single monolayer; (**c**) film transformation onto substrate. **b** Regeneration of cellulose LB-monolayer; **c** scaffold coated with a LB-monolayer; **d** reaction zones of functionalized polymers within a multilayer system

be introduced as well as crosslinkable groups to introduce surface patterns via lithographic processes [164]. Furthermore, bioactive factors and the corresponding ligands, such as puriregic (P) ligands could be connected via non-covalent or covalent bonds. LB films based on cellulose, PLA, and polypeptides have been studied as well as films prepared from polyelectrolytes [165]. L-Carnitine is a natural betaine with vitamin properties playing an essential role in fatty acid metabolism. Polymer synthesis routes are designed to maintain the primary structure and specific properties of carnitine, such as hydrophilicity and stiffening. Poly(carnitine) co-polymers such as poly(croton betain) and poly(carnitine ally-lester) have been synthesized that can be used for microemulsion and microcapsule preparation [166]. Poly(carnitines) are not yet used in thin film technology; however, these polycationic structures could be combined with polyanions resulting in polyelectrolyte systems analogous to LbL films [163].

LB multilayers have been designed mainly to study surface processes such as adhesion for sensor and membrane applications and integrated optics [167]. Lenhert and colleagues used LB lithography to pattern polystyrene surfaces and investigate osteoblast alignment, elongation, and migration. This so-called *nanoimprinting* enables the fabrication of nanostructured surface areas on a wide spectrum of different biomaterials. For many biomaterial applications, relatively large surface areas are required—they are beyond the limits of traditional lithography. LB lithography, a recently developed method, was used to fabricate regularly spaced grooves of different depths (50 and 150 nm) with a periodicity of 500 nm over several square centimeters on silicon surfaces. These topographies were transferred onto polystyrene surfaces by means of nanoimprinting. Primary osteoblasts were cultured on the patterned polymer surfaces, and were observed to align, elongate, and migrate parallel to the grooves. Osteoblasts show a significant anisotropic behavior on these surfaces, which can enhance cell settlement on the surface or be used to direct tissue generation on the biomaterial interface [168].

4.2 MSCs on Artificial Surfaces

Artificial surfaces should be constructed to provide an environment mimicking the ECM and the correct microenvironment for the SCs, similar to the postulated SC niche [169], to construct an environment favorable for SC maintenance or differentiation. To do so the first step is to understand this microenvironment, which is not only very complex, but new data suggest that it seems to be different for different kinds of SCs.

SCs have by definition the ability for self-renewal and differentiation towards specific lineages. However the mechanism by which they regulate these two characteristics is poorly understood. Two major hypotheses have been suggested. In the asymmetrical system, the SC will divide and one daughter cell will remain a SC, whereas the second cell will start to differentiate and leave the SC niche. The question is, does this SC leave the niche and by doing so, due to the changed microenvironment, start differentiating? Or does the SC start differentiation due to the asymmetrical cleavage and then leave the niche? The second hypothesis is that after an external trigger, presumably from within the SC niche, the SC will start proliferating and thus lose the SC characteristics. Both daughter cells will leave the SC niche which ultimately leads to a depletion of the SC pool. Which of these hypotheses is true cannot yet be decided. There is data supporting both of them and it is not unlikely that both systems exist, depending on the SC type. What is already clear is that the microenvironment plays a key role in this event.

In the following, the major components characterizing the SC microenvironment are discussed to provide an overview of the subject, focusing on the SC niche of adult SCs within the human body.

4.2.1 The SC Microenvironment or the SC Niche

Mechanical and Physicochemical Factors of the SC Microenvironment

The SC niche is the microenvironment which physically surrounds the SCs within a given tissue and actively influences SC fate either to promote self-renewal and the maintenance of the SC character or to promote the differentiation needed for tissue repair or regeneration. Mechanical and physicochemical factors positively or negatively regulate SC responses within the niche.

Mechanical triggers are derived from the ECM components, which form the local structural geometry and topography. The elasticity of matrices constructed to mimic the softness or rigidity of different tissues has been shown to directly influence the differentiation of SCs into specific lineages in vitro [170–172]. The mechanotransduction can also influence the SC shape via a physical control. This effect seems to be activated by the adhesive interactions between the cell and its substrate [151]. Cell shape, cytoskeletal tension, and the small GTPase RhoA regulate SC lineage commitment [173]. Effects of the material on the cell response are also dependent on the stage of cell commitment. It has been shown that more differentiated cells are also more responsive to a model cell adhesion ligand [174].

Physicochemical factors such as pH, oxygen tension, and ionic strength also regulate the fate of adult SCs. (see Fig. 5 for an overview). The pH is known to be a modulator of cell proliferation. A reduced pH (7.1 instead of 7.6) increased the cloning efficiency of progenitor cells and a pH between 7.2 and 7.4 was optimal for their differentiation, at least in vitro [175]. Low oxygen concentration maintained the cells in their undifferentiated and multipotent state [176]. The Ca^{2+} content of the niche seems to influence the favored localization of adult mammalian hematopoiesis in bone [177].

As well as the mechanical and physicochemical nature of the microenvironment, cell–cell interactions between SCs, exchanges between SCs and neighboring differentiated cells, and interactions between SCs and adhesion molecules characterize the SC niche.

Cell–Cell and Cell–ECM Interactions

The interaction of SCs with their particular microenvironment is thought to be responsible for SCs' fate, maintaining their potential and quiescent state and the regulation of their specific differentiation properties. This is achieved next to the solid-state signals (see Sect. 4.2.1) of cell–ECM interactions through a complex

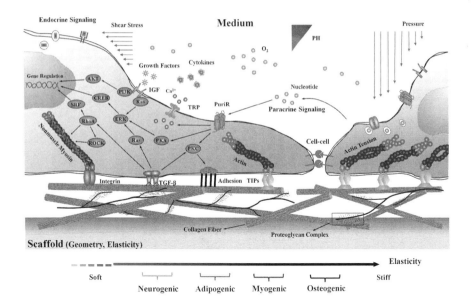

Fig. 5 SCs and their natural microenvironment. SCs are affected by their microenvironment which is defined by ECM properties such as elasticity and geometry, molecules which connect to the ECM, for instance TGF-β, TIPs, integrins, and transient receptor potential (TRP) which can regulate cytoskeleton tension, successively followed by gene expression and focal adhesion though the activation of a series of mechanical transduction events. Various soluble factors such as extracellular nucleotide, growth factors, and cytokines also influence SC fate. Mechanical forces such as shear stress and blood pressure influence SC proliferation and differentiation from the media side of the niche as well as chemical and physical factors like pH or oxygen

system of paracrine signals via growth factors and cytokines as well as through autocrine neighboring cell–SC and SC–SC interactions.

A typical example of cell–ECM interactions is the action via the transmembrane adhesion receptor family of the integrins bound to insoluble ECM proteins. However, the ligand-activated induction of these transmembrane receptors is not the only effect of the integrins. There is also a structural component via tractional forces, which leads to a rearrangement of the actin and thus a cell shape change, which both together influence proliferation [178]. Other pathways with a similar effect involve TGF-β, tension-induced proteins (TIPs), and non-muscle myosin, all ultimately influencing SC shape and fate (see Fig. 5 for an overview) [179, 180].

The soluble factors can be classified into those which act on a more general basis and those which have a more specific effect, but both will ultimately determine SC behavior. Several chemicals seem to be more universal differentiation-inducing factors. A typical example is retinoic acid (RA), which can change the differentiation status of the epidermis [181] and also induce differentiation in neuronal and hematopoietic cells [182, 183]. Another molecule group very broadly affecting SCs is the purine derivatives. ATP itself is not only an energy source, it is also a substrate in signal transduction pathways, and a major factor in the SC niche for self-renewal [184] and survival under ischemic stress by maintaining the multipotency of MSCs [185]. Cyclic adenosine monophosphate (cAMP) can induce differentiation in epidermal and neuronal cells similarly to RA, and combined with cytokines, cAMP seem to improve viability and function of hematopoietic cells [186]. ATP and UTP, as well as their breakdown products, signal through a family of so-called P receptors, which is composed of four P1 and 15 P2 receptors. Recently it has been shown that the P2 receptors in particular play a major role in MSC proliferation and differentiation into the adipogenic and osteogenic lineages [187]. Exploring this receptor family can also be expected to shed more light on the known effect of calcium ions within the SC microenvironment, since the subfamily of P2X receptors are ligand-gated ion channels as well [188]. To round off the story, it should be mentioned that they also are a source of cytokine secretion (see Fig. 5).

Cell–cell interactions seem to be fundamental for differentiation control. An example is Notch signaling, which has been described to influence asymmetric cell division [189]. The interactions between SCs and neighboring cells are best understood for the bone marrow-located niche of the HSCs. An overview of the cells contributing to the maintenance of a functional HSC niche, which are composed of sinusoidal endothelial cells, macrophages, perivascular MSCs, sympathetic nerve fibers, and of course the cells of the osteoblastic lineage regulating the behavioral control of HSCs, is given in the review of Ehninger and Trumpp [190]. The network of cytokine interactions and enzymes yielding the particular response of these SCs is summarized in the review of Visigalli and Biffi [191]. Another fairly well characterized niche is the hair follicle. Wnt/β-catenin is required for follicle SC maintenance and β-catenin activation is crucial for inducing quiescent follicle SCs to proliferate and terminally differentiate along the hair cell lineage [192].

For other SC niches, various components have been described [179], but the overall picture is as yet unclear. It is clear that several cytokines such as interleukin 11 (IL-11) in the HSCs' fate are another group of key players in SCs' fate. Another example is IL-6, which triggers the maintenance of human limbal epithelial cells in a progenitor-like state [193]. However, knowing the components is not enough. The local concentration might cause a threshold-based reaction and the soluble factors interact with the solid-state signals. Much work must be devoted in the future to elucidating in more detail the concerted action of the multiple temporal and spatial factors, which often cause the SCs to react in a synergistic way with a specific cell fate response that ultimately leads SCs to organize into tissues which organize into organs. Microscale engineering strategies for systematically examining and reconstructing individual niche components might be a way to shed more light on the complex composition of the SC microenvironment [194]. There is no doubt that synthetic SC niche engineering may form a new foundation for regenerative therapies.

4.2.2 SCs on Artificial Surfaces Mimicking the SC Microenvironment

Cells, including SCs, seem to have a robust intrinsically determined set of cell fates. They can be quiescent or proliferative, they can differentiate or die, but the choices are very selective. Waddington's epigenetic landscape has been an accepted model for this since 1956 [195]. It describes a landscape of valleys and hills and a marble (the cell) which ultimately rolls downhill and the only choice is through which valley, meaning which lineage-specific differentiation pathway or other fate the cell can have. This model was greatly challenged when Yamanaka introduced his iPS cells [27]. This would show cells rolling *up-hill* in Waddington's landscape, a reaction which was thought it be impossible. The question now arises whether the model is wrong or whether it can be altered. An revised model has been suggested by Huang and Ingber [196]. It reminds the authors of Einstein's space-time bending where gravity effects are often pictured as dents in a layer and the deepness of the dent is dependent on the size of the star or planet. This model seems to be a good new representation of (stem) cell fate with the dents being proliferation, differentiation or apoptosis. The lack of a down-hill feature as in the Waddington model also allows the explanation of reprogramming processes as in iPS, but the limited number of possible fates (number of dents) is preserved by this picture, too. It also allows an explanation of the hotly discussed topic of whether a SC can differentiate invitro into lineages which in nature would not be an expected fate of this specific SC, a feature which is called transdifferentiation. This means the differentiation of a SC which should be lineage-committed to the cells of a specific germinal layer (ectoderm, endoderm or mesoderm), but can be differentiated into one of the other two layers.

SCs' fate can be affected by ECM properties. Therefore an artificial surface cannot only be used to systematically examine and reconstruct individual niche components for a better understanding, as already discussed in the paragraph above [194], it could also be altered to trigger the SCs into a fate which was not planned for this SC in nature. There is increased scientific effort to use artificial surfaces to influence the biological system [189, 197]. One focus is the use of bioactive native or artificial ligands which influence the fate of SCs bound or loosely integrated into the matrix, because these factors are known to be key players in the SC niche (see Sect. 4.2.1). Another focus of research to influence SCs' fate on artificial scaffolds is the nanostructure of the surface [198]. It has been shown that the surface structure directly influences the lineage-specific differentiation of SCs, but the thinness of the matrix can also trigger the differentiation process. The tension caused by size variations and physical deformation within the matrix influencing cell shape and cell distortion seems to be a dominant control element to guide cell fate. Naive MSCs have been shown to commit to phenotypes with high sensitivity to tissue-level elasticity. Soft matrices mimicking brain are neurogenic, stiffer matrices mimicking muscle are myogenic, and rather rigid matrices mimicking collagenous bone have been proven to trigger the osteogenic lineage commitment (see Fig. 5) [170].

In general, an artificial substrate should provide an environment similar to the ECM and the SC niche [94, 199, 200]. Three-dimensional culture systems, which generally induce a more rounded, spheroidal cell morphology in comparison to standard 2D culture systems, are by themselves already influencing SC fate [151], and it is the beauty of the system that even our very limited understanding of the SC microenvironment already leads to reasonable results due to the intrinsic set of possible SC fates. Nevertheless that should not obviate further efforts towards a better understanding of the SC and its fate within its natural or artificial microenvironment, for a future use of the system in regenerative medicine.

5 Scaffold–Cell Interaction in Hard Tissue Engineering

5.1 MSCs on Scaffolds for Bone Defects

Bone defects must be divided into small bone defects and critical-size defects (CSD) of the bone that will not heal during the lifetime of an animal, when used in the preclinical field of orthopedic and trauma surgery. CSD need cells and scaffold material to bridge the gap. These bone substitutes must be evaluated for their biocompatibility and preferably they should also allow osteoinductivity and osteoconductivity. To achieve this, the cell type (SCs or differentiated cells) and the physicochemical and mechanical factors of the SC environment must be considered.

An overview of the SCs used to generate the major cells of a bone tissue, namely osteoblasts, osteoclasts, and chondrocytes, can be found in zur Nieden's publication [201]. But the cell type alone does not give the full story. Osteoinductivity and osteoconductivity must match the state of the respective cell with which the scaffold is loaded, because a SC or a differentiating cell will react differently to an inducing signal. Hsiong and colleagues showed that SCs were less sensitive in their uncommitted state to a model cell adhesion ligand (arginine-glycine-aspartic acid [RGD]-containing peptide) presented from hydrogels of varying stiffness than cells differentiated into the osteoblast lineage [174].

Several mechanical factors and chemicals such as oxygen or calcium, with or without additional enzymes, are known to be key players in influencing SC fate towards osteogenesis. The regulation of osteogenesis and chondrogenesis during skeletogenesis in regulating limb development and regenerative events such as fracture repair are dependent on mechanical signals influencing MSCs [202]. The effect of matrix stiffness on osteo-differentiation already mentioned (see Sect. 4.2.1) often shows an additive or even synergistic effect with soluble factors. The effect of matrix stiffness on the differentiation of MSCs in response to TGF-β can promote MSC differentiation into either smooth muscle cells (SMCs) or chondrogenic cells [203]. The small GTPase RhoA and its effector protein ROCKII regulate fluid-flow-induced osteogenic differentiation via isometric tension within the actin cytoskeleton, simultaneously being negative regulators of both adipogenic and chondrogenic differentiation [204]. Low oxygen throughout the cultivation time of mesenchymal progenitor cells elevates culture markers for osteogenesis, including alkaline phosphatase activity, calcium content, and von Kossa staining [205]. Local Ca^{2+} concentration influences cell morphology through the cell–cell or cell–matrix interactions of osteoblasts, thus affecting osteogenic differentiation while not influencing proliferation [206]. Calcium and purine derivates can also directly influence the lineage commitment of MSCs to the adipo- or osteogenic lineages via P receptors. Several specific P2 receptors are involved in this process. P2Y5 and P2Y14 influence proliferation of MSCs, and P2X6 up- or down-regulation is plays a key role in the lineage commitment of SCs between the adipogenic or osteogenic lineages, respectively [187].

Another unsolved problem of large bone defects is the necessity of angiogenesis. MSCs have been shown to differentiate in vitro not only into the osteogenic lineages and others, but also into endothelial cells, which are the fundamental cells for angiogenesis. Although the sprouting of endothelial cells to form new vessels has already been demonstrated [207], and although the size of the interconnecting pores needed within the scaffold for the vessel is known as well as major inducing factors such as vascular endothelial growth factor (VEGF), this final step towards an artificial scaffold for large bone defects needs more basic research efforts before a breakthrough achievement permits clinical trials with such an approach. Until now the sprouting new vessels are too small and the inducing factors have severe side effects such as an increase in carcinogenesis.

5.2 MSCs for Dental Tissues

Several different problems must be faced in dentistry and efforts are being made to create new solutions for old problems on the basis of SCs used for regenerative approaches. One major problem is the tooth decay formerly known as dental caries. Here replacement of the enamel with or without dentine would be beneficial. Even the re-growth of full teeth is already being considered and researched. The second problem is periodontitis, which represents a major health problem worldwide. Periodontal defects primarily evolve as a consequence of inflammation and do not heal spontaneously. Root cementum is the major hard tissue which must be considered if this problem is to be addressed. Last but not least, tooth loss due to tooth decay or periodontitis creates a need for bone reconstruction and implant stabilization. Taken together, dental materials contain all calcified tissues of the body: enamel and dentine within the teeth and root cementum and alveolar bone in the periodontium. For an overview of tooth organogenesis and regeneration see Thesleff and Tummers [208].

Considerable efforts have been made to identify SCs in dental tissues (e.g., SCAPs, see "MSCs and Ectomesenchymal SCs" and [82]). However, the availability of dental SCs is limited as it is restricted to specific time points. SCs derived from the tooth buds of wisdom teeth are such a source. These cells are a specific type of MSCs, the ectomesenchymal SCs. They are already more committed towards hard tissues since they develop from oral ectoderm and neural crest-derived mesenchyme and form the periodontium during tooth development. SCs from dental pulp, the dental papilla, and apical papilla progenitor cells are another widely used source of SCs with a fairly good differentiation potential (see [209] for a review). Thus, the tooth seems to have several different SCs with various lineage potentials, but the limited availability nevertheless makes MSCs a promising source [210], because they are easily accessible at nearly all times and capable of differentiating into calcified tissue-forming cells.

Adipose tissue-derived SCs have been used in vitro to differentiate into a 3D dental bud structure. The cells were positive for ameloblastic and odontoblastic markers after four weeks [211]. Thetooth-specific hard tissues, enamel and dentin, are secreted by ameloblasts and odontoblasts, respectively. However, the bud is only the first step in tooth development, followed by the cap and bell stages, and after the crown is complete, root formation must be initiated. Thus, there is a long way to go before a new tooth can be grown and transplanted. Light might be shed on the basics for the development by considering the epithelial SC niche existing for continuously growing teeth such as in rodents, where epithelial SCs are maintained in a cervical loop.

For defects caused by peritonitis, other difficulties must be faced. The periodontium is a tissue surrounding and supporting the tooth. It is composed of alveolar bone, which is bordered on the upper part by the connective tissue of the gums (the gingiva). In the lower part of the tooth, the connection of the tooth root to the alveolar bone is via periodontal ligament and cementum. Defects in the periodontium are therefore easily overgrown by fibroblast from the gingiva. Collagen-based tissues can be attached to the tooth to prevent this ingrowth and give the SCs time to fill in the gap by differentiating into the lost tissue type. It would be beneficial to load the scaffold with bioactive molecules such as ligand for specific P receptors to accelerate the differentiation of the ectomesenchymal SCs [187]. Furthermore, materials for a site-specific differentiation of MSCs within the same defect in two different lineages, such as the osteogenic and cementogenic lineages, to form the periodontium would be favorable.

After tooth loss, the alveolar bone is quickly reduced in material due to the lower mechanical forces on the bone. If an implant is not set in time, replacement material for the bone is needed to stabilize the artificial tooth. For this, bone material can be taken from the hip, but severe side effects such as infection risk, additional pain, and reduced motility have to be considered. Artificial material such as HA and others are an alternative, as are MSCs differentiated into bone (see Sect. 5.1). If the bone itself is strong enough, there is still the question of stabilizing the implant. For this, bone chips, a waste product derived from drilling the hole into the bone for the implant, can be used. This material is discussed widely among dentists, because it is in contact with the bacteria from the oral cavity during the drilling process and it was unclear until recently whether this material contains living cells. New data shows that this is indeed the case: the bone chips contain not only living cells, but bone precursor cells and SCs depending from where in the alveolar bone the material is taken (Tobiasch, unpublished). In addition, pre-treatment of the patient with antibiotics, as is common anyway, reduces the contamination of the material to levels found in other tissues as well. Thus, this material is a good and autologous source for stem and precursor cells to stabilize implants without additional surgical interventions.

Even modern approaches such as guided tissue regeneration, bioactive proteins alone or loaded on scaffolds, and replacement materials with or without cells, are only moderately successful. Depending on which of the questions given above should be addressed, specific lineage-committed SCs and innovative scaffolds must be developed, in addition considering clinical handling and surgical procedures for an implementation in clinical periodontology and dental surgery in the future.

6 Future Developments

Great hopes have arisen in the last couple of years for SC-derived new therapies for various diseases of genetic or degenerative or even infectious origin. The question is whether this hype has an underlying hope and what can be expected in the future.

The authors are reminded of a comparable hype 20 years ago: gene therapy. There, a similar excitement suddenly vanished when the death of a patient, Jesse Gelsinger, in a clinical trial caused the cessation of clinical trials worldwide, followed by a huge general decline in interest and therefore also in money for scientific research in this field. What was the cause? Big promises had been made that this technique could, for the first time, lead not only to the treatment of genetic diseases but also their cure. This statement, although true in general, was followed by enormous excitement and hope in the population, which led to more money for this research, but also to more pressure for fast results. But results cannot be forced in science and the delay in positive outcomes of trials ultimately led to a disappointment. A similar story can be seen at present in SC research. A lot of effort and money is spent and the population expects visible effects soon. More than 3,000 clinical trials, mainly in phase I or II, are being performed at the moment for the use of SCs to treat various diseases. Most of them are for cancer ($>2,500$) but all kinds of illnesses such as cardiovascular diseases, diabetes, stroke, multiple sclerosis, amyotrophic lateral sclerosis (ALS), and others are addressed as well. In general, they mostly show a clinical improvement, often with an early onset of positive progress, but more often than not these positive results are only short-term and even when they lead to a prolongation of life they do not show the long-term effects which scientists have been hoping for. A closer look at the effects reveals that the SCs mainly have a diffuse stimulatory effect on the neighboring cells, whose underlying mechanism(s) cannot be pinpointed exactly at the moment.

Other effects are due to the use of SCs as a vehicle for gene transfer and other outcomes seem to be due to immunomodulatory mechanisms.

Thus, surprisingly few data support the theory that SCs back up to repair the damaged tissue in this approaches, although this is the typical characteristic and theory of the mechanism of action of adult SCs in vivo. Obviously, there is still a lack of knowledge of how to treat and stimulate differentiation of SCs in vitro to achieve the desired effect in tissue or organ regeneration. Nevertheless, good results can already be achieved in some approaches and these are, as well as the hematopoietic SC transplantation mentioned above, the use of SCs in less complex tissues such as hard tissues for bone or dental replacements. Small bone defects such as the osseointegration of implants can already be treated with reasonably good results and the next successful step can be expected to be the treatment of large bone defects, once the regulation of angiogenesis has been solved by supportive scaffolds allowing the in-growth of vessels and the regulated replacement of the artificial material or the long-term integration of the tissue replacement into the natural bone.

It could be the missing or incorrect microenvironment, mimicking the SC niche in vitro, which is the cause of the shortfall in SC-derived therapies for more complex tissues or even organs.The use of scaffolds might therefore be the solution for this problem in the future. However, fundamental questions have first to be answered: how does the structure of hMSCs control the interfacial cell–scaffold architecture and how might this be influenced by the scaffold manufacturing process? For this, model interfaces on the nanometer scale have to be created to understand the processes at the cell–scaffold interface *in detail* and finally develop approaches for quantitative control.

Acknowledgments We would like to acknowledge Yu Zhang for his great help in drawing the beautiful pictures for this work. Without his highly appreciated input, this chapter would have been less vivid. The results summarized in this work were supported by BMBF-AIF, AdiPaD; FKZ: 1720X06.

References

1. Bartold PM (2000) Periodontology 40:164–172
2. Aejaz HM, Aleem AK, Parveen N et al (2007) Stem cell therapy-present status. Transplant Proc 39:694–699
3. McKay R (2000) Stem cells—hype and hope. Nature 406:361–364
4. Chung Y, Klimanskaya I, Becker S et al (2005) Embryonic and extraembryonic stem cell lines derived from single mouse blastomeres. Nature 439: 216–219
5. Bladé J, Samson D, Reece D et al (1998) Criteria for evaluating disease response and progression in patients with multiple myeloma treated by high-dose therapy and haemopoietic stem cell transplantation. Br J Haematol 102:1115–1123
6. Pavletic S, Khouri I, Haagenson M et al (2005) Unrelated donor marrow transplantation for B-cell chronic lymphocytic leukemia after using myeloablative conditioning: results from

the Center for international blood and marrow transplant research. J Clin Oncol 23: 5788–5794
7. Crisostomoto PR, Wang Y, Markel TA et al (2008) Human mesenchymal stem cells stimulated by TNF-alpha, LPS, or hypoxia produce growth factors by an NF kappa B— but not JNK-dependent mechanism. Am J Physiol Cell Physiol 294:675–682
8. Slayton WB, Spangrude GJ (2004) Adult stem cell plasticity. In: Turksen K (ed) Adult stem cells. Humana Press, New Jersey, pp 1–3
9. Forte E, Chimenti I, Barile L et al (2011) Cardiac cell therapy: the next (re)generation. Stem Cell Rev Rep. doi:10.1007/s12015-011-9252-8 [Epub ahead of print]
10. Mozid AM, Arnous S, Sammut EC et al (2011) Stem cell therapy for heart diseases. Br Med Bull 98:143–159
11. Vanikar AV, Dave SD, Thakkar UG et al (2010) Cotransplantation of adipose tissue-derived insulin-secreting mesenchymal stem cells and hematopoietic stem cells: a novel therapy for insulin-dependent diabetes mellitus. Stem Cells Int 2010:582382. 20 Dec 2010
12. Meng J, Muntoni F, Morgan JE (2011) Stem cells to treat muscular dystrophies—where are we? Neuromuscul Disord 1:4–12
13. Illes J, Reimer JC, Kwon BK (2011) Stem cell clinical trials for spinal cord injury: readiness, reluctance, redefinition. Stem Cell Rev Rep. doi:10.1007/s12015-011-9259-1 [Epub ahead of print]
14. Joe AW, Gregory-Evans K (2010) Mesenchymal stem cells and potential applications in treating ocular disease. Curr Eye Res 35:941–52
15. Barranco C (2011) Stem cells: mesenchymal stem cells from adipose tissue could be used to deliver gene therapy to the liver. Nat Rev Gastroenterol Hepatol 8:64
16. Liu T, Wang Y, Wen C, Zhang S et al (2011) Stem cells or macrophages, which contribute to bone marrow cell therapy for liver cirrhosis? Hepatology. doi:10.1002/hep.24431
17. Rhee YH, Ko JY, Chang MY et al (2011) Protein-based human iPS cells efficiently generate functional dopamine neurons and can treat a rat model of Parkinson disease. J Clin Invest. doi:10.1172/JCI45794
18. Lindvall O, Kokaia Z (2010) Stem cells in human neurodegenerative disorders—time for clinical translation. J Clin Invest 120:29–40
19. Mucke L (2009) Neuroscience: Alzheimer's disease. Nature 461:895–897
20. Ringe J, Kaps C, Burmester G-R et al (2002) Stem cells for regenerative medicine: advances in the engineering of tissues and organs. Naturwissenschaften 89:338–351
21. Schaefer D, Klemt C, Zhang X et al (2000) Tissue engineering with mesenchymal stem cells for cartilage and bone regeneration. Chirurg 71:1001–1008
22. Breitbach M, Bostani T, Roell W et al (2007) Potential risks of bone marrow cell transplantation into infarcted hearts. Blood 110:1362–1369
23. Thomson JA, Itskovitz-Eldor J, Shapiro SS et al (1998) Embryonic stem cell lines derived from human blastocysts. Science 282:1145–1147
24. Whittaker PA (2005) Therapeutic cloning: the ethical limits. Toxicol Appl Pharmacol 207:689–691
25. Peljto M, Wichterle H (2011) Programming embryonic stem cells to neuronal subtypes. Curr Opin Neurobiol 21:43–51 Feb 2011 Epub 20 Oct 2010
26. Canaari J, Kollet O, Lapidot T et al (2011) Neural regulation of bone, marrow, and the microenvironment. Front Biosci (Schol Ed) 3:1021–1031, June 1
27. Takahashi K, Yamanaka S (2006) Induction of pluripotent stem cells from mouse embryonic and adult fibroblast cultures by defined factors. Cell 126:663–676
28. Takahashi K, Okita K, Nakagawa M et al (2007) Induction of pluripotent stem cells from fibroblast cultures. Nat Protoc 2:3081–3089
29. Wernig A, Schäfer R, Knauf U et al (2005) On the regenerative capacity of human skeletal muscle. Artif Organs 29:192–198
30. Yu J, Vodyanik M, Smuga-Otto K et al (2007) Induced pluripotent stem cell lines derived from human somatic cells. Science 318:1917–1920

31. Eminli S, Foudi A, Stadtfeld M et al (2009) Differentiation stage determines potential of hematopoietic cells for reprogramming into induced pluripotent stem cells. Nat Genet 41:968–976
32. Kim J, Greber B, Araúzo-Bravo M et al (2009a) Direct reprogramming of human neural stem cells by OCT4. Nature 461:643–649
33. Kim J, Sebastiano V, Wu G et al (2009b) Oct4-induced pluripotency in adult neural stem cells. Cell 136:411–419
34. Kaji K, Norrby K, Paca A et al (2009) Virus-free induction of pluripotency and subsequent excision of reprogramming factors. Nature 458:771–775
35. Woltjen K, Michael I, Mohseni P et al (2009) piggyBac transposition reprograms fibroblasts to induced pluripotent stem cells. Nature 458:766–770
36. Hentze H, Soong PL, Wang ST et al (2009) Teratoma formation by human embryonic stem cells: evaluation of essential parameters for future safety studies. Stem Cell Res 2:198–210
37. Locke M, Ussher JE, Mistry R et al (2011) Transduction of human adipose-derived mesenchymal stem cells by recombinant adeno-associated virus vectors. Tissue Eng Part C Methods 17:949–959
38. Ohi Y, Qin H, Hong C et al (2011) Imcomplete DNA methylation underlines a transcriptional memory of somatic cells in human iPS cells. Nat Cell Biol 5:541–549
39. Zhao T, Zhang ZN, Rong Z et al (2011) Immunogenicity of induced pluripotent stem cells. Nature 474:212–215
40. Bilousova G, Hyun JD, King KB et al (2011) Osteoblasts derived from induced pluripotent stem cells from calcified structures in scaffolds both in vitro and in vivo. Stem Cells 29:206–216
41. Ye JH, Xu YJ, Gao J et al (2011) Critical size calvarial bone defects healing in a mouse model with silk scaffolds and SATB2-modified iPSCs. Biomaterials 32:5065–5076
42. Goodman J, Hodgson G (1962) Evidence for stem cells in the peripheral blood of mice. Blood 19:702–714
43. Blanpain C (2010) Stem cells. Skin regeneration and repair. Nature 464:686–687
44. Toma J, Akhavan M, Fernandes K et al (2001) Isolation of multipotent adult stem cells from the dermis of mammalian skin. Nat Cell Biol 3:778–784
45. Marshman E, Booth C, Potten CS (2002) The intestinal epithelial stem cell. Bioessays 24:91–98, Review
46. Becker AJ, McCulloch EA et al (1963) Cytological demonstration of the clonal nature of spleen colonies derived from transplanted mouse marrow cells. Nature 2:452–454
47. Bianco P, Riminucci M, Gronthos S et al (2001) Bone marrow stromal stem cells: nature, biology, and potential applications. Stem Cells 19:180–192 (Review)
48. Cudkowicz G, Upton A, Smith L et al (1964) An approach to the characterization of stem cells in mouse bone marrow. Ann NY Acad Sci 31:571–585
49. Michalopoulos GK, DeFrances MC (1997) Liver regeneration. Science 276:60–66
50. Xiao JC, Jin XL, Ruck P et al (2004) Hepatic progenitor cells in human liver cirrhosis: immunohistochemical, electron microscopic and immunofluorencence confocal microscopic findings. World J Gastroenterol 10:1208–1211
51. Miura M, Gronthos S, Zhao M et al (2003) SHED: stem cells from human exfoliated deciduous teeth. Proc Natl Acad Sci USA 100:5807–5812
52. Morsczeck C, Götz W, Schierholz J et al (2005) Isolation of precursor cells (PCs) from human dental follicle of wisdom teeth. Matrix Biol 24:155–165
53. Murphy M, Reid K, Dutton R et al (1997) Neural stem cells. J Investig Dermatol Symp Proc 2:8–13 (Review)
54. Schultz SS, Lucas PA (2006) Human stem cells isolated from adult skeletal muscle differentiate into neural phenotypes. J Neurosci Methods 152:144–155
55. Tedesco FS, Dellavalle A, Diaz-Manera J et al (2010) Repairing skeletal muscle regenerative potential of skeletal muscle stem cells. J Clin Invest 120:11–9
56. Rodriguez A-M, Elabd C, Amri E-Z et al (2005) The human adipose tissue is a source of multipotent stem cells. Biochimie 87:125–128

57. Zuk P, Zhu M, Mizuno H et al (2001) Multilineage cells from human adipose tissue: implications for cell-based therapies. Tissue Eng 7:211–228
58. Leri A, Hosoda T, Kajstura J et al (2011) Identification of a coronary stem cell in the human heart. J Mol Med 89:947–959
59. Kajstura J, Rota M, Hall SR et al (2011) Evidence for human lung stem cells. N Engl J Med 364:1795–806
60. McQualter J, Yuen K, Williams B et al (2010) Evidence of an epithelial stem/progenitor cell hierarchy in the adult mouse lung. Proc Natl Acad Sci USA 4:1414–1419
61. Dominici M, Le Blanc K, Mueller I et al (2006) Minimal criteria for defining multipotent mesenchymal stromal cells. The international society for cellular therapy position statement. Cytotherapy 8:315–317
62. De Ugarte DA, Morizono K, Elbarbary A et al (2003) Comparison of multi-lineage cells from human adipose tissue and bone marrow. Cells Tissues Organs 174:101–109
63. Izadpanah R, Trygg C, Patel B et al (2006) Biologic properties of mesenchymal stem cells derived from bone marrow and adipose tissue. J Cell Biochem 99:1285–1297
64. Kern S, Eichler H, Stoeve J et al (2006) Comparative analysis of mesenchymal stem cells from bone marrow, umbilical cord blood, or adipose tissue. Stem Cells 24:1294–1301
65. Minguell J, Erices A, Conget P (2001) Mesenchymal stem cells. Exp Biol Med 226: 507–520 (Review)
66. Psaltis PJ, Zannettino ACW, Worthley SG et al (2008) Concise review: mesenchymal stromal cells: potential for cardiovascular repair. Stem Cells 26:2201–2210
67. Zuk P, Zhu M, Ashjian P et al (2002) Human adipose tissue is a source of multipotent stem cells.Mol Biol Cell 13:4279–4295
68. Shay JW, Wright WE (2010) Telomeres and telomerase in normal and cancer stem cells. FEBS Lett 584(17):3819–3825
69. Zuk PA (2010) The adipose-derived stem cell: looking back and looking ahead. Mol Biol Cell 21:1783–1787
70. Caplan AI (1991) Mesenchymal stem cells. J Orthop Res 9:641–650
71. Dicker A, Le Blanc K, Aström G et al. (2005) Functional studies of mesenchymal stem cells derived from adult human adipose tissue. Exp Cell Res 308:283–290
72. Lee J, Kim Y, Kim S et al (2004) Chondrogenic differentiation of mesenchymal stem cells and its clinical applications. Yonsei Med J 30:41–47
73. Pansky A, Roitzheim B, Tobiasch E (2007) Differentiation potential of adult human mesenchymal stem cells. Clin Lab 53:81–84
74. Pittenger MF, Mackay AM, Beck SC et al (1999) Multilineage potential of adult human mesenchymal stem cells. Science 284:143–147
75. Pacary E, Legros H, Valable S et al. (2006) Synergistic effects of CoCl2 and ROCK inhibition on mesenchymal stem cell differentiation into neuron-like cells. J Cell Sci 119:2667–2678
76. Liu M, Han Z (2008) Mesenchymal stem cells: biology and clinical potential in type 1 diabetes therapy. J Cell Mol Med 12:1155–1168
77. Xie QP, Huang H, Xu B et al (2009) Human bone marrow mesenchymal stem cells differentiate into insulin-producing cells upon microenvironmental manipulation in vitro. Differentiation 77:483–491
78. Saulnier N, Lattanzi W, Puglisi MA et al (2009) Mesenchymal stromal cells multipotency and plasticity: induction toward the hepatic lineage. Eur Rev Med Pharmacol Sci 13:71–78
79. Planat-Benard V, Silvestre J, Cousin B et al (2004) Plasticity of human adipose lineage cells toward endothelial cells: physiological and therapeutic perspectives. Circulation 109:656–663
80. De Francesco F, Tirino V, Desiderio V et al (2009) Human CD34+/CD90+ASCs Are Capable of Growing as Sphere Clusters, producing high levels of VEGF and forming capillaries. PLoS One 4:6537
81. Porada C, Zanjani E, Almeida-Porad G (2006) Adult mesenchymal stem cells: a pluripotent population with multiple applications. Curr Stem Cell Res Ther 1:365–369

82. Morsczeck C, Reichert TE, Vollner F et al (2007) The state of the art in human dental stem cell research. Mund Kiefer Gesichtschir 11:259–266
83. Williams DF (1999) Williams dictionary of biomaterials. Liverpool University Press, Liverpool
84. Griffith L, Naughton G (2002) Tissue engineering—current challenges and expanding opportunities. Science 295:1009–1014
85. Khademhosseini A, Vacanti J, Langer R (2009) Progress in tissue engineering. Sci Am 300:64–71
86. Langer R, Vacanti J (1993) Tissue engineering. Science 260:920–926
87. Park J, Lakes R (2007) Biomaterials: an introduction. Springer Science and Business Media, New York
88. Hutmacher D, Schantz J, Lam C et al (2007) State of the art and future directions of scaffold-based bone engineering from a biomaterials perspective. J Tissue Eng Reg Med 1:245–260
89. Ogushi H, Caplan A (1999) Stem cell technology and bioceramics: from cell to gene engineering. J Biomed Mater Res 48:913–927
90. Kon E, Muraglia A, Corsi A et al (2000) Autologous bone marrow stromal cells loaded onto porous hydroxyapatite ceramic accelerate bone repair in critical-size defects of sheep long bones. J Biomed Mater Res 49:328–337
91. Hoppe A, Guldal NS, Boccaccini AR (2011) A review of the biological response to ionic dissolution products from bioactive glasses and glass–ceramics. Biomaterials 32:2757–2774
92. Hofmann I, Haas D, Eckert A et al (2008) Mechanical properties of cellulose-apatite composite fibers for biomedical applications. Adv Appl Ceramics 107:293–297
93. Dvir T, Tsur-Gang O, Cohen S (2005) Designer-scaffolds for tissue engineering and regeneration. Israel J Chem 45:487–494
94. Kim I, Seo S, Moon H et al (2008) Chitosan and its derivatives for tissue engineering applications. Biotech Adv 26:1–21
95. Segura T, Anderson B, Chung P et al (2005) Crosslinked hyaluronic acid hydrogels: a strategy to functionalize and pattern. Biomaterials 26:359–371
96. Sachlos E, Czernuszka J (2003) Making tissue engineering scaffolds work. Eur Cell Mater 30:29–39
97. Ragetly G, Griffon DJ, Chung YS (2010) The effect of type II collagen coating of chitosan fibrous scaffolds on mesenchymal stem cell adhesion and chondrogenesis. Acta Biomater 6:3988–3997
98. Mauney J, Jaquiéry C, Volloch V et al (2005) In vitro and in vivo evaluation of differentially demineralized cancellous bone scaffolds combined with human bone marrow stromal cells for tissue engineering. Biomaterials 26:3173–3185
99. Ragetly G, Ratanavaraporn J, Damrongsakkul S et al (2011) Osteogenic differentiation of bone-marrow-derived stem cells cultured with mixed gelatine and chitosanoligosaccharide scaffolds. J Biomater Sci Polym Ed 22:1083–1098
100. Zippel N, Schulze M, Tobiasch E (2010) Biomaterials and mesenchymal stem cells for regenerative medicine. Recent Pat Biotech 4:1–22
101. Moroni L, Wijn J, Van Blitterswijck A (2008) Integrating novel technologies to fabricate smart scaffolds. J Biomater Sci Polymer Ed 19:543–572
102. Kim SH, Oh SA; Lee WK et al (2011a) Poly(lactic acid) porous scaffold with calcium phosphate mineralized surface and bone marrow mesenchymal stem cell growth and differentiation. Mater Sci Eng C-Mater Biol Appl 31:612–619
103. Li Y, Danmark S, Edlund U et al (2011) Resveratrol-conjugated poly-epsilon-caprolactone facilitates in vitro mineralization and in vivo bone regeneration. Acta Biomater 7:751–758
104. Wan ACA, Ying JY (2010) Nanomaterials for in situ cell delivery and tissue regeneration. Adv Drug Deliv Rev 62:731–740
105. Goldstein A, Zhu G, Morris G et al (1999) Effect of osteoblastic culture conditions on the structure of poly(D,L-lactic-co-glycolic acid) foam scaffolds. Tissue Eng 5:421–434
106. Ma PX (2008) Biomimetic materials for tissue engineering. Adv Drug Deliv Rev 60:184–98

107. Nasibulin AG, Anisimov AS, Pikhitsa PV et al (2007) Investigations of NanoBud formation. Chem Phys Letters 446:109–114
108. Wang J, Yu X (2010) Preparation, characterization and in vitro analysis of novel structured nanofibrous scaffolds for bone tissue engineering. Acta Biomater 6:3004–3012
109. Kim K, Dean D, Lu AQ et al (2011b) Early osteogenic signal expression of rat bone marrow stromal cells is influenced by both hydroxyapatite nanoparticles content and initial cell seeding density in biodegradable nanocomposites scaffolds. Acta Biomater 7:1249–1264
110. Zhang L, Webster T (2009) Nanotechnology and nanomaterials: promises for improved tissue regeneration. Nano Today 4:66–80
111. Bauer S, Park J, von der Mark K et al (2008) Improved attachment of mesenchymal stem cells on super hydrophobic TiO_2 nanotubes. Acta Biomater 4:1576–1582
112. Park J, Bauer S, Schmuki P et al (2009) Narrow window in nanoscale dependent activation of endothelial cell growth and differentiation on TiO_2 nanotube surfaces. Nano Lett 9:3157–3164
113. Sahithi K, Swetha M, Ramasamy K et al (2010) Polymeric composites containing carbon nanotubes for bone tissue engineering. Int J Biol Macromol 46:281–283
114. Oliveira JM, Sousa RA, Malafaya PB et al (2011) In vivo study of dendron-like nanoparticles for stem cells "tune-up": from nano to tissues. Nanomedicine. doi:10.1016/j.nano.2011.03.002 [Epub ahead of print]
115. Sitharaman B, Avti PK, Schaefer K et al (2011) A novel nanoparticle-enhanced photoacoustic stimulus for bone tissue engineering. Tissue Eng A 17:1851–1858
116. Lim YC, Johnson J, Fei ZZ et al (2011) Micropatterning and characterization of electrospun poly(epsilon-caprolactone)/gelatin nanofiber tissue scaffolds by femtosecond laser ablation for tissue engineering applications. Biotechnol Bioeng 108:116–126
117. Wei G, Ma PX (2008) Nanostructured biomaterials for regeneration, nano-scaled drug release systems incorporated into nanostructured biomaterials represents a novel and promising strategy to tissue regeneration. Adv Funct Mater 18:3568–3582
118. Fan DM, Akkaraju GR, Couch EF et al (2011) The role of nanostructured mesoporous silicon in discriminating in vitro calcification for electrospun composite tissue engineering scaffolds. Nanoscale 3:354–361
119. Poursamar SA, Azami M, Mozafari M (2011) Controllable synthesis and characterization of porous polyvinyl alcohol/hydroxyapatite nanocomposite scaffolds via an in situ colloidal technique. Colloids Surf B Biointerfaces 84:310–316
120. Burdick JA, Vunjak-Novakovic G (2009) Engineered microenvironments for controlled stem cell differentiation. Tissue Eng A15:205–219
121. Kretlow J, Mikos A (2008) From material to tissue: biomaterial development. Scaffold fabrication, and tissue engineering. AIChE J 54:3048–3067
122. Peltola S, Sanna M, Melchels F et al (2008) A review of rapid prototyping techniques for tissue engineering purposes. Ann Med 40:268–280
123. Seyednejad H, Gawlitta D, Dhert WJ et al (2011) Preparation and characterization of a three-dimensional printed scaffold based on a functionalized polyester for bone tissue engineering applications. Acta Biomater 7:1999–2006
124. De Gennes PG, Brochard-Wyart F, Quéré D (2002) Capillary and wetting phenomena—drops, bubbles, pearls, waves. Springer, New York
125. He L, Dexter AF, Middelberg APJ (2006) Biomolecular engineering at interfaces. Chem Eng Sci 61:989–1003
126. Chew SY, Low WC (2011) Scaffold-based approach to direct stem cell neural and cardiovascular differentiation: an analysis of physical and biochemical effects. J Biomed Mater Res A 29:355–374
127. Biondi M, Ungaro F, Quaglia F et al (2008) Controlled drug delivery in tissue engineering. Adv Drug Deliv Rev 60:229–242
128. Cartmell S (2009) Controlled release scaffolds for bone tissue engineering. J Pharm Sci 98:430–441

129. Quaglia F (2008) Bioinspired tissue engineering: the great promise of protein delivery technologies. Int J Pharm 364:281–297
130. Richardson TP, Peters MC, Ennett AB et al (2001) Polymeric system for dual growth factor delivery. Nature Biotech 19:1029–1034
131. Sokolsky-Papkov M, Agashi K, Olaye A et al (2007) Polymeric carriers for drug delivery in tissue engineering. Adv Drug Deliv Rev 59:187–206
132. Ubersax L, Merkle H, Meinel L (2009) Biopolymer based growth factor delivery for tissue repair: from natural concepts to engineered systems. Tissue Eng Part B Rev 15:263–289
133. Vallet-Regi M, Balas F, Colilla M et al (2008) Bone-regenerative bioceramic implants with drug and protein controlled delivery capability. Prog Solid State Chem 36:163–91
134. Kong SW, Kim JS, Park KS et al (2011) Surface modification with fibrin hyaluronic acid hydrogel on solid-free form-based scaffolds followed by BMP-2 loading to enhance bone regeneration. Bone 48:298–306
135. Ratanavaraporn J, Furuya H, Kohara H et al (2011) Synergistic effects of the dual release of stromal cell-derived factor-1 and bone morphogenic protein-2 from hydrogels on bone regeneration. Biomaterials 32:2797–2811
136. Baraniak PR, Nelson DM, Leeson CE et al (2011) Spatial control of gene expression within a scaffold by localized inducer release. Biomaterials 32:3062–3071
137. Smith RA, Meade K, Pickford CE et al (2011) Glycosaminoglycans as regulators of stem cell differentiation. Biochem Soc Trans 39:383–387
138. Huang Z, Feng QL, Yu B et al (2011) Biomimetic properties of an injectable chitosan/nano-hydroxyapatite/collagen composite. Mater Sci Eng C-Mater Biolog Appl 31:683–687
139. Zheng L, Fan H, Sun J et al (2010) Chondrogenic differentiation of mesenchymal stem cells induced by collagen-based hydrogel: an in vivo study. J Biomed Mater Res A 93:783–792
140. Fischbach C, Mooney DJ (2006) Polymeric systems for bioinspired delivery of angiogenetic molecules. Polym Reg Med Adv Polym Sci 203:191–221
141. Arafat MT, Lam CXF, Ekaputra AK et al (2011) Biomimetic composite coating on rapid prototyped scaffolds for bone tissue engineering. Acta Biomater 7:809–820
142. Kuo YC, Yeh CF (2011) Effect of surface-modified collagen on the adhesion, biocompatibility and differentiation of bone marrow stromal cells in poly(lactides-co-glycolide)/chitosan scaffolds. Colloids Surf B Biointerfaces 82:624–631
143. Lin CC, Metters AT (2006) Hydrogels in controlled release formulations: network design and mathematical modelling. Adv Drug Deliv Rev 58:1379–1408
144. König U, Nitschke M, Menning A et al (2002) Durable surface modification of poly(tetrafluoroethylene) by low pressure H_2O plasma treatment followed by acrylic acid graft polymerization. Colloids Surf B Biointerfaces 24:63–71
145. Curtis A, Wilkinson C (1997) Topographical control of cells. Biomaterials 18:1573–1583
146. Schaefer D, Klemt C, Zhang X et al (2000) Tissue engineering with mesenchymal stem cells for cartilage and bone regeneration. Chirurg 71:1001–1008
147. Faid K, Voicu R, Bani-Yaghoub M et al (2005) Rapid fabrication and chemical patterning of polymer microstructures and their applications as a platform for cell cultures. Biomed Microdevices 7:179–184
148. Bens A, Bermes G, Emons M et al (2007) Non-toxic flexible photopolymers for medical stereolithography technology. Rapid Prot J 13:38–47
149. Kane RS, Takayama S, Ostuni E et al (1999) Patterning proteins and cells using soft lithography. Biomaterials 20:2363–2376
150. Whitesides GM, Ostuni E, Takayama S et al (2001) Soft lithography in biology and biochemistry. Ann Rev Biomed Eng 3:335–373
151. Guilak F, Cohen DM, Estes BT et al (2009) Control of stem cell fate by physical interactions with the extracellular matrix. Cell Stem Cell 5:17–26
152. Gerecht S, Vunjak-Novakovic G, Langer R (2007) Engineering biomaterials for vascular differentiation and regeneration. Circulation 116:235
153. Boudou T, Crouzier T, Ren K et al (2010) Multiple functionalities of polyelectrolyte multilayer films: new biomedical applications. Adv Mater 22:441–467

154. Whitesides GM, Grzybowski B (2002) Self-assembly at all scales. Science 295:2418–2421
155. Chen M, Le DQ, Baatrup A et al (2011) Self-assembled composite matrix in a hierarchical 3-D scaffold for bone tissue engineering. Acta Biomater 7:2244–2255
156. Tanaka M (2011) Design of novel 2D and 3D biointerfaces using self-organization to control cell behaviour. Biochim Biophys Acata-Gen Subjects 1810:251–258
157. Graziano A, d'Aquino R, Cusella-De Angelis MG et al (2007) Concave pit-containing scaffold surfaces improve stem cell-derived osteoblast performance and lead to significant bone tissue formation. PLoS One 6:e496
158. Vendra VK, Wu L, Krishnan S (2007) Polymer thin films for biomedical applications in nanotechnologies for the life sciences. Wiley–VCH, New York
159. Khademhosseini A, Jon S, Suh KY et al (2003) Direct patterning of protein- and cell-resistant polymeric monolayers and microstructures. Adv Mater 15:1995–2000
160. Decher G (1997) Fuzzy nanoassemblies: toward layered polymeric multicmposites. Science 277:1232–1237
161. Entcheva E, Bien H, Yin L et al (2004) Functional cardiac cell constructs on cellulose-based scaffolding. Biomaterials 25:5753–5762
162. Marchenko I, Yashchenok A, German S et al (2010) Polyelectrolytes: influence on evaporative self-assembly of particles and assembly of multilayers. Polymers 2:690–708
163. Moby V, Labrude P, Kadi A et al (2011) Polyelctrolyte multilayer film and human mesenchymal stem cells: An attractive alternative in vascular engineering approaches. J Biomed Mater Res 96A:313–319
164. Schulze M (1997) Supramolecular architectures from cellulose materials. Macromol Chem Macromol Symp 120:237–242
165. Rulkens R, Wegner G, Enkelmann V et al (1996) Synthesis and properties of rigid polyelectrolytes based on sulfonated poly-p-phenylenes. Ber Bunsenges Phys Chem 100:707–715
166. Kamm B, Kamm M, Kiener A et al (2005) Polycarnitine—a new biomaterial. Appl Microbiol Biotechnol 67:1–7
167. Gupta VK, Kornfield JA, Ferencz A et al. (1994) Controlling molecular order in "hairy-rod" Langmuir–Blodgett films: a polarization-modulation microscopy study. Science 265:940–942
168. Lenhert S, Meier MB, Meyer U et al (2005) Osteoblast alignment, elongation and migration on grooved polystyrene surface patterned by Langmuir–Blodgett lithography. Biomaterials 26:563–570
169. Schreiber TD, Steinl C, Essl M et al (2009) The integrin {alpha}9{beta}1 onhaematopoietic stem and progenitor cells: involvement in cell adhesion, proliferation anddifferentiation. Haematologica 94:1493–1501
170. Engler AJ, Sen S, Sweeney HL et al (2006) Matrix elasticity directs stem cell lineage specificationn. Cell 126:677–689
171. Terraciano V, Hwang N, Moroni L et al (2007) Differential response of adult and embryonic mesenchymal progenitor cells to mechanical compression in hydrogels. Stem Cells 25:2730–2738
172. Yim EKF, Pang SW, Leong KW (2007) Synthetic nanostructures inducing differentiation of human mesenchymal stem cells into neuronal lineage. Exp Cell Res 313:1820–1829
173. McBeath R, Pirone DM, Nelson CM et al (2004) Cell shape, cytoskeletal tension, and RhoA regulate stem cell lineage commitment. Dev Cell 6:483–495
174. Hsiong SX, Carampin P, Kong HJ et al (2008) Differentiation stage alters matrix control of stem cells. J Biomed Mater Res A 85:145–156
175. McAdams TA, Miller WM, Papoutsakis ET (1997) Variations in culture pH affect the cloning efficiency and differentiation of progenitor cells in ex vivo haemopoiesis. Br J Haematol 97:889–95
176. Basciano L, Nemos C, Foliguet B et al (2011) Long term culture of mesenchymal stem cells in hypoxia promotes a genetic program maintaining their undifferentiated and multipotent status. BMC Cell Biol 12:12

177. Adams GB, Chabner KT, Alley IR et al (2006) Stem cell engraftment at the endosteal niche is specified by the calcium-sensing receptor. Nature 439:599–603
178. Tanentzapf G, Devenport D, Godt D et al (2007) Integrin-dependent anchoring of a stem-cell niche. Nat Cell Biol 9(12):1413–1418
179. Dennis E, Discher D, Mooney DJ et al (2009) Growth factors, matrices, and forces combine and control stem cells. Science 324:1673–1677
180. Vicente-Manzanares M, Ma X, Adelstein RS et al (2009) Non-muscle myosin II takes centre stage in cell adhesion and migration. Nat Rev Mol Cell Biol 10:778–790
181. Tobiasch E, Winter H, Schweizer J (1992) Structural features and sites of expression of a new murine 65 kD and 48 kD hair-related keratin pair, associated with a spezial type of parakeratotic epithelial differentiation. Differentiation 50:163–178
182. Gurumurthy S, Xie SZ, Alagesan B (2010) The Lkb1 metabolic sensor maintains haematopoietic stem cell survival. Nature 468:659–663
183. Jia C, Doherty JP, Crudgington S et al (2009) Activation of purinergic receptors induces proliferation and neuronal differentiation in Swiss Webster mouse olfactory epithelium. Neuroscience 163:120–128
184. Xi R, Xie T (2005) Stem cell-renewal controlled by chromatin remodelling factors. Science 310:1487–1489
185. Mylotte LA, Duffy AM, Murphy M et al (2008) Metabolic flexibility permits mesenchymal stem cell survival in an ischemic environment. Stem Cells 5:1325–1336
186. D'Atri LP, Etulain J, Romaniuk MA et al (2011) The low viability of human CD34+ cells under acidic conditions is improved by exposure to thrombopoietin, stem cell factor, interleukin-3, or increased cyclic adenosine monophosphate levels. Transfusion. doi:10.1111/j.1537-2995.2010.03051
187. Zippel N, Scholze NJ, Müller CA et al (2011) Purinergic receptors influence the differentiation of human mesenchymal stem cells. Stem Cells Develop. doi:10.1089/scd.2010.0576 [Epub ahead of print]
188. De Proost I, Pintelon I, Wilkinson WJ et al (2009) Purinergic signalling in the pulmonary neuroepitelial body microenvironment unravelled by live cell imaging. FASEB J 4:1153–1160
189. Williams SE, Beronja S, Pasolli HA et al (2011) Asymmetric cell divisions promote Notch-dependent epidermal differentiation. Nature 470:353–358
190. Ehninger A, Trumpp A (2011) The bone marrow stem cell niche grows up: mesenchymal stem cells and macrophages move. J Exp Med 208:421–428
191. Visigalli I, Biffi A (2011) Maintenance of a functional hematopoietic stem cell niche through galactocerebrosidase and other enzymes. Curr Opin Hematol 18:214–219
192. Yang L, Peng R (2010) Unweiling hair follicle stem cells. Stem Cell Rev 4:658–664
193. Notara M, Shortt AJ, Galatowicz G et al (2010) IL6 and the human limbal stem cell niche: a mediator of epithelial–stromal interaction. Stem Cell Res 3:188–200
194. Peerani R, Zandstra PW (2010) Enabling stem cell therapies through synthetic stem cell-niche engineering. J Clin Invest 120:60–70
195. Waddington CH (1956) Principles of embryology. Allen & Unwin, London
196. Huang S, Ingber DE (2004) From stem cells to functional tissue architecture. In: Sell S (ed) Stem cells handbook. Humana press, New Jersey
197. Curran JM, Chen R, Hunt JA (2006) The guidance of human mesenchymal stem cell differentiation in vitro by controlled modifications to the cell substrate. Biomaterials 27:4783–4793
198. Cavalcanti-Adam EA, Aydin D, Hirschfeld-Warneken VC et al (2008) Cell adhesion and response to synthetic nanopatterned environments by steering receptor clustering and spatial location. HFSP J 2:276–285
199. Catledge SA, Vohra YK, Bellis SL et al (2004) Mesenchymal stem cell adhesion and spreading on nanostructured biomaterials. J Nanosci Nanotech 4:986–989
200. Pennisi CP, Sevencu C, Dolatshahi-Pirouz A et al (2009) Responses of fibroblasts and glial cells to nanostructured platinum surfaces. Nanotechnology 20:1–9

201. zur Nieden NI (2011) Embryonic stem cells for osteo-degenerative diseases. Methods Mol Biol 690:1–30
202. Kelly DJ, Jacobs CR (2010) The role of mechanical signals in regulating chondrogenesis and osteogenesis of mesenchymal stem cells. Birth Defects Res C Embryo Today A 90: 75–85
203. Park JS, Chu JS, Tsou AD et al (2011) The effect of matrix stiffness on the differentiation of mesenchymal stem cells in response to TGF-β. Biomaterials 32:3921–3930
204. Arnsdorf EJ, Tummala P, Kwon RY et al (2009) Mechanically induced osteogenic differentiation—the role of RhoA, ROCKII and cytoskeletal dynamics. J Cell Sci 122: 546–553
205. Lennon DP, Edmison JM, Caplan AI (2001) Cultivation of rat marrow-derived mesenchymal stem cells in reduced oxygen tension: effects on in vitro and in vivo osteochondrogenesis. J Cell Physiol 187:345–55
206. Nakamura S, Matsumoto T, Sasaki J et al (2010) Effect of calcium ion concentrations on osteogenic differentiation and hematopoietic stem cell niche-related protein expression in osteoblasts. Tissue Eng Part A 16:2467–2473
207. Gong Z, Niklason LE (2008) Small-diameter human vessel wall engineered from bone marrow-derived mesenchymal stem cells (hMSCs). FASEB J 22:1635–1648
208. Thesleff I, Tummers M (2008) Tooth organogenesis and regeneration. StemBook [Internet]. Harvard Stem Cell Institute, Cambridge, 31 Jan 2008–2009
209. Tziafas D, Kodonas K (2010) Differentiation potential of dental papilla, dental pulp, and apical papilla progenitor cells. J Endod 36:781–789
210. Estrela C, Alencar AH, Kitten GT et al (2011) Mesenchymal stem cells in the dental tissues: perspectives for tissue regeneration. Braz Dent J 22:91–8
211. Ferro F, Spelat R, Falini G et al (2011) Adipose tissue-derived stem cell in vitro differentiation in a three-dimensional dental bud structure. Am J Pathol 178:2299–2310

Adv Biochem Engin/Biotechnol (2012) 126: 195–226
DOI: 10.1007/10_2011_106
© Springer-Verlag Berlin Heidelberg 2011
Published Online: 16 November 2011

Bioactive Glass-Based Scaffolds for Bone Tissue Engineering

Julia Will, Lutz-Christian Gerhardt and Aldo R. Boccaccini

Abstract Originally developed to fill and restore bone defects, bioactive glasses are currently also being intensively investigated for bone tissue engineering applications. In this chapter, we review and discuss current knowledge on porous bone tissue engineering scaffolds made from bioactive silicate glasses. A brief historical review and the fundamental requirements in the field of bone tissue engineering scaffolds will be presented, followed by a detailed overview of recent developments in bioactive glass-based scaffolds. In addition, the effects of ionic dissolution products of bioactive glasses on osteogenesis and angiogenic properties of scaffolds are briefly addressed. Finally, promising areas of future research and requirements for the advancement of the field are highlighted and discussed.

Keywords Bioactive glasses · Scaffolds · Bone tissue engineering · Angiogenesis

Contents

1 Introduction	196
2 Scaffold Requirements	199
3 Bioactive Glass Processing	200
4 Bioactive Glass–Ceramic Scaffolds	202
4.1 Fabrication and Microstructures	202
4.2 Mechanical Properties	204
4.3 In-Vitro and In-Vivo Studies	208
5 Bioactive Glass Containing Composite Scaffolds	208
6 Effect of Bioactive Glass on Angiogenesis	213
7 Conclusions and Future Work	215
References	217

J. Will · A. R. Boccaccini (✉)
Department of Materials Science and Engineering,
Institute of Biomaterials, University of Erlangen-Nuremberg,
Cauerstr. 6, 91058 Erlangen, Germany
e-mail: aldo.boccaccini@ww.uni-erlangen.de

L.-C. Gerhardt
Biomedical Engineering, Soft Tissue Biomechanics and Engineering,
Technische Universiteit Eindhoven, PO Box 513, 5600 MB,
Eindhoven, The Netherlands

1 Introduction

Tissue engineering (TE) and regenerative medicine aim to restore diseased or damaged tissue using combinations of functional cells, bioactive molecules and biodegradable scaffolds made from engineered biomaterials [1, 2]. Some of the most promising biomaterials for application in bone TE are bioceramics such as hydroxyapatite (HA), calcium phosphates, bioactive silicate glasses and related composite materials combining bioactive inorganic materials with biodegradable polymers [3, 4]. Bioactive inorganic materials are capable of reacting with physiological fluids to form strong bonds to bone through the formation of bone-like hydroxyapatite layers, leading to effective biological interaction and fixation of bone tissue with the implanted material surface [5, 6]. Moreover, in the case of silicate bioactive glasses, such as 45S5 Bioglass® [5], reactions on the material surface induce the release and exchange of critical concentrations of soluble Si, Ca, P and Na ions, which can lead to favorable intracellular and extracellular responses promoting rapid bone formation [7–11].

In 1971, Hench and colleagues discovered that rat bone can bond chemically to certain silicate-based glass compositions [12]. This group of glasses was later termed "bioactive", meaning "a material that elicits a specific biological response at the material surface which results in the formation of a bond between the tissues and the materials" [5, 13]. Hench [13] has published the history of the development of bioactive glass (BG), focusing on the breakthrough discovery of the classical 45S5 Bioglass® composition. This oldest BG composition consists of a silicate network (45 wt% SiO_2) incorporating 24.5 wt% Na_2O, 24.5 wt% CaO and 6 wt% P_2O_5. The high amounts of Na_2O and CaO, as well as the relatively high CaO/P_2O_5 ratio, make the glass surface highly reactive in physiological environments [5]. A schematic diagram showing the series of events that occur on the surface of BG in contact with a biological environment, as proposed in the literature [5], is presented in Fig. 1.

Other bioactive glass compositions developed over the years have additional elements incorporated in the silicate network, such as fluorine [14], magnesium [15, 16], strontium [17–19], iron [20], silver [21–24], boron [25–28], potassium [29], or zinc [30, 31]. The biological response to the different ion dissolution products released from BG has recently been reviewed by Hoppe et al. [9].

The typical characteristic of all bioactive glasses, which are usually fabricated by melting or sol–gel methods (see Sect. 3), is the ability to form a strong bond to bone and in some cases soft tissues [32, 33]. It is now widely accepted that for establishing a bond with bone, a biologically active apatite surface layer must form at the material/bone interface [1, 5, 12, 34–36]. Early clinical applications of bioactive glasses were in the form of solid pieces for small bone replacement, e.g. in middle ear surgery [1, 5, 13]. Later, other clinical applications of bioactive glasses were proposed, for example as coatings on metallic orthopedic implants or in periodontology [5, 13, 32].

Bioactive Glass-Based Scaffolds for Bone TE

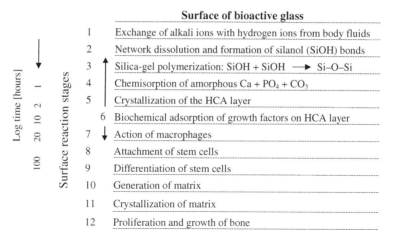

Fig. 1 Sequence of interfacial reactions involved in forming a bond between bone and a bioactive glass (modified from Ref. [5])

Since the late 1990s and the beginning of the new millennium, great potential has been attributed to the application of bioactive glasses in TE and regenerative medicine [1, 7, 9, 13, 34–39]. The application involves both micron-sized and nanoscale bioactive glass particles of different compositions [24, 40, 41] as well as the fabrication of composite materials which are developed by combining biodegradable polymers and bioactive glass particles or fibres [34, 42–47].

Based on the attractive osteogenic and angiogenic properties of bioactive glasses, bone TE is one of the most exciting future clinical applications of these materials. Both micron-sized and nanoscale particles [40, 43] are considered in this application field. Bioactive silicate glasses exhibit three major advantages for bone TE applications over other conventional non-degradable (insoluble) bioceramics such as TiO_2, Al_2O_3, ZrO_2, or sintered hydroxyapatite (Fig. 2). Firstly, chemical reactions on the material surface lead to a strong bond to bone by means of a hydroxyl carbonate apatite (HCA) layer [5]. Secondly, ion release and dissolution products from the bioactive glass activate and up-regulate gene expression in osteoprogenitor cells that give rise to rapid bone regeneration, which explains the higher rate of bone formation in comparison to other inorganic ceramics such as hydroxyapatite [7, 9–11, 13, 48]. Thirdly, recent studies (reviewed in Ref. [49]) have demonstrated angiogenic effects of 45S5 Bioglass®, i.e., increased secretion of vascular endothelial growth factor (VEGF) and VEGF gene expression in fibroblasts, the proliferation of endothelial cells and formation of endothelial tubules in vitro, as well as enhancement of vascularization in vivo [49–53]. Figure 2 summarizes schematically these three effects of bioactive glasses in the context of tissue engineering.

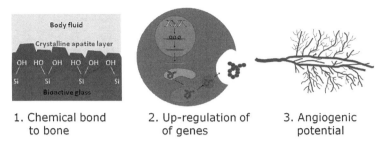

Fig. 2 Schematic representation of the main effects of bioactive glasses in the context of bone tissue engineering

In addition, the incorporation of particular ions into the silicate network, such as silver [21–23] and boron [27, 28], has been investigated in order to develop antibacterial and antimicrobial materials. Bioactive glasses can also serve as carriers for the local delivery of selected ions and drugs to control specific cell functions [9, 31, 54–60]. For example, mesoporous BG microspheres have demonstrated enhanced haemostatic activity, as well as reduced clot detection times and increased coagulation rates compared to nonporous microspheres [61].

Bioactive glasses belong to the group of Class A bioactive materials which are characterized by both osteoconduction (i.e., growth of bone at the implant surface) and osteoinduction (i.e., activation and recruitment of osteoprogenitor cells by the material itself stimulating bone growth on the surface of the material) [5, 60, 62]. Differences between Class A and B bioactive materials are discussed elsewhere [5, 13, 62]. As indicated above, the range of bioactive glasses exhibiting these attractive properties has been extended over the years, in terms of both chemical composition and morphology, as new preparation methods have become available. A recent review summarizes these latest developments [60]. At this point, for completeness, it has to be mentioned that an early significant modification of bioactive silicate glasses was the development of apatite/wollastonite (A/W) bioactive glass–ceramics [63, 64]. A recent review summarizing research on Ca–Si-based ceramics is available [65].

The present chapter covers specifically the field of bioactive glass-derived scaffolds for bone TE. In Sect. 2, the essential requirements for bone TE scaffolds are highlighted. Section 3 covers fabrication technologies of bioactive glasses. Sections 4 and 5 summarize the latest developments of bioactive glass–ceramic and BG-containing composite scaffolds, respectively, including an overview of both materials science aspects and in vitro/in vivo studies. Section 6 discusses the angiogenic properties of BG. Finally, in Sect. 7, remaining challenges in the field are discussed, and areas where further research is needed are identified.

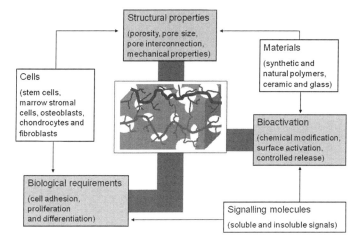

Fig. 3 Schematic diagram of key factors involved in the design of optimal scaffolds for bone tissue engineering (modified from Ref. [3])

2 Scaffold Requirements

The success of a bone tissue scaffold is determined by its ability to stimulate and aid in both the onset and completion of bone regeneration. Thus, the most important function of a bone TE scaffold is its role as a template that allows cells to attach, proliferate, differentiate and organize into normal, healthy bone as the scaffold degrades. Figure 3 illustrates the most important factors involved in the design of TE scaffolds and their interdependencies, according to Guarino et al. [3]. Depending on the final application, scaffold requirements include matching the structural and mechanical properties with those of the recipient tissue and optimization of the microenvironment to support cell integration, adhesion and growth, issues that have become known as structural and surface compatibility of biomaterials [66].

Considering the complexity of the TE task, scaffolds are subjected to many interrelated biological and structural requirements which must be taken into consideration when selecting a suitable biomaterial, fabrication procedure and final scaffold structure and surface condition. Firstly, scaffolds need to promote cell attachment, differentiation and proliferation, which are cell functions highly dependent on substrate material properties. For bone engineering, scaffolds should be osteoconductive, which is important not only to avoid the formation of encapsulating tissue but also to induce a strong bond between the scaffold and host bone [3, 4, 34]. The rate of biodegradation in vivo is another criterion for selection of biomaterials for fabricating scaffolds. The scaffold degradation rate must be tailored to match the rate of regeneration of new tissue. Further requirements are related to the scaffold architecture. An ideal bone tissue scaffold should possess interconnected porosity, i.e., it should be highly permeable with porosity and pore diameters in a wide range (e.g., 10–500 µm) for cell seeding, tissue ingrowth and vascularization, as well as for nutrient delivery and cellular waste removal [3, 4, 34, 67, 68]. A particular design criterion is the mimicry and implementation of the hierarchical porosity of cancellous bone tissue, which is an important factor for effective scaffold vascularization and for bone ingrowth [69]. Microporosity (≈ 2–10, <50 µm) is essential for immediate protein and cell adhesion, cell migration and osteointegration [15, 67, 69, 70]. Higher pore sizes (>300 µm) are required for enhanced new bone formation, larger bone ingrowth and the formation of vascular capillaries. Because of vascularization, pore size has been shown to affect the progression of osteogenesis. Small pores favor hypoxic conditions and induce osteochondral formation before osteogenesis, while large pores that are well-vascularized lead to direct osteogenesis (without preceding cartilage formation) [67]. However, higher scaffold porosity results in reduced mechanical properties. Therefore, the design of the scaffold pore structure must consider the need for an optimal porosity enabling sufficiently high permeability for waste removal and nutrient supply and the required stiffness and strength to sustain the loads transferred to the scaffold from the surrounding tissue [71]. Finally, it should be possible to fabricate scaffolds in complex or irregular shapes in order to match specific defect morphologies in bone. In addition, the material of the scaffold should be suitable for sterilization by clinically approved methods and the scaffold technology should be advantageous for commercialization, i.e., the scaffold production must be scalable and cost-effective.

3 Bioactive Glass Processing

Bioactive glasses can be fabricated using two different methods: the traditional melt-derived approach and the sol–gel process. Each technique yields different structures and properties.

In the melting process conventional glass technology is used [5]. The glass components in the form of grains of oxides or carbonates are mixed and then melted and homogenized at high temperatures, i.e., 1250–1400 °C. The molten glass is then cast into steel or graphite molds to make bulk implants by subsequent grinding or polishing. If bioactive glass powder is required (e.g., for treatment of periodontal lesion or for fabricating scaffolds by sintering), it is made by pouring the molten glass into a liquid medium, such as water, thus fracturing the frozen glass into small fragments (quenching). Subsequent grinding and size separation steps are necessary to achieve powders with specific size ranges. However, there may be some disadvantages of these conventional glass-derived methods for bioactive glasses.

For example, it might be difficult to maintain the very high purity required for optimal bioactivity due to the high melting temperatures and the process steps of grinding and polishing. The melting method is limited by the evaporation of the volatile component P_2O_5 during high-temperature processing. This is due to the extremely high equilibrium liquidus temperature of SiO_2, (1713 °C), and the extremely high viscosity of silicate melts with high SiO_2 content. The processing costs are considerable, due to the energy costs and use of platinum crucibles.

Melt-derived bioactive glasses have been used successfully as bone-filling materials in orthopedic and dental surgery but their poor mechanical strength and low toughness limit their application in load-bearing positions. However, one method suggested to improve the mechanical strength of these bioactive glasses is their transformation into glass–ceramics [72]. Glass–ceramics are partially crystallized glasses produced by heating the parent bioactive melt-derived glass powder above its crystallization temperature, usually at about 610–630 °C [73, 74]. Sintering of glass powder is one way to fabricate glass–ceramic scaffolds. During the occurrence of crystallization and densification, the microstructure of the parent glass shrinks, porosity is reduced and the solid structure gains mechanical strength with increasing crystallization [62].

Low-temperature sol–gel processing offers an alternative to the conventional glass and melting process [75]. This process involves the synthesis of an inorganic network by mixing the metal alkoxides in solution, followed by hydrolysis, gelation, and low-temperature firing to produce a glass. The sol–gel processing of a silicate glass involves hydrolysis of alkoxide precursors, such as tetraethylorthosilicate (TEOS), to form a colloidal solution (sol). Polycondensation of silanol (Si–OH) groups continues after hydrolysis is complete, beginning the formation of the silicate (–Si–O–Si–) network [76]. As the network connectivity increases, viscosity increases and a gel is formed. The gel is then subjected to controlled thermal processes of aging to strengthen the gel, drying to remove the liquid byproduct of the polycondensation reaction and thermal stabilization/sintering to remove organic species from the surface of the material [77]. After these thermal treatments, the sol–gel powder is derived via crushing and milling. Inherent in this process is the ability to modify the network structure through controlled hydrolysis and polycondensation reactions. Structural variation can thus be obtained without compositional changes. Because the glasses can be prepared from gels by heat treatment at relatively low temperatures (600–700 °C), most of the disadvantages of high-temperature processing can be eliminated with much higher control over purity. Also, sol–gel processing offers the potential advantages of ease of powder production, a broader range of bioactivity, and a better control of bioactivity by changing microstructure through processing parameters. Sol–gel-derived bioactive glasses provide excellent matrices for entrapping a variety of organic and inorganic compounds and biologically important molecules [78].

In addition, sol–gel bioactive glasses with compositions varying over a wide range have demonstrated bioactivity in vitro and in vivo, because they usually have high specific surface area and Si–OH groups which could accelerate the surface crystallization of HCA. Moreover, bioactive glasses prepared via sol–gel always have an interconnected mesoporous structure, with pores of about 5–10 nm in diameter. A macroporous sol–gel bioactive glass with two simultaneous pore classes: i.e., larger than 100 μm and about 5–10 nm, has been proposed as a suitable bone scaffold material [33]. However, it is very difficult to produce macroporous sol–gel glasses with a pore size larger than 100 μm because of the large shrinkage during sol–gel processing. Jie et al. [79] reported a foaming method to prepare macroporous sol–gel bioactive glasses with pores larger than 100 μm. In the last few years, the successful application of high relative humidity during gel drying has made it possible to fabricate macroporous structures using a pore former such as polyvinyl alcohol (PVA) [80]. Jones [81] has carried out extensive work developing sol–gel derived bioactive scaffolds exhibiting nano-structured topography.

Recent key papers [82–88] and an informative review [78] highlighting the potential of the sol–gel technology in the field of bone-tissue scaffold development can be consulted for completeness.

4 Bioactive Glass–Ceramic Scaffolds

4.1 Fabrication and Microstructures

The limited strength, brittleness and low fracture toughness (i.e., ability to resist fracture when a crack is present) of bioactive glasses obtained either via the melting route or sol–gel processes have so far prevented their use for load-bearing implants [13, 62, 73, 109], and thus the repair and regeneration of large bone defects in load-bearing anatomical sites (e.g., limbs) remain a clinical/orthopedic challenge [110]. Recent developments related to bone TE try to overcome this problem by fabricating architectures and components carefully designed on different length scales, i.e., from the macroscale, mesoscale, and microscale down to the nanometer scale [60, 111], including both multifunctional bioactive glass composite structures and advanced bioactive glass–ceramic scaffolds exhibiting oriented microstructures, controlled porosity and directional mechanical properties [60, 91, 93, 94, 98, 102], as discussed in the following paragraphs. Most studies (summarized in Table 1) have mainly investigated the mechanical properties, in-vitro and cell biological behavior of glass–ceramic scaffolds. Scaffolds exhibiting compressive strength [91, 94] and elastic modulus values [93, 94] above those of cancellous bone and close to the lower limit of cortical bone have been developed.

The "replication" or "polymer-sponge" fabrication process is one of the successful methods introduced for producing bioactive glass–ceramic scaffolds [72]. The foams are manufactured by coating a polyurethane or polyester foam with a glass particle slurry. The polymer foam determines the final scaffold macrostructure, and thus serves as a sacrificial substrate for the glass coating. The slurry infiltrates the polymer structure and adheres to the surface of the polymer. Excess slurry is squeezed out leaving a glass coating on the struts of the foam. After drying, the polymer is burned out and the glass is sintered to the desired density. The process replicates the macrostructure of the sacrificial polymer, and results in a distinctive pore microstructure within the macrostructure (Fig. 4). In addition to

Table 1 Overview of recent studies performed on silicate bioactive glass–ceramic scaffolds

Glass composition/system	Particle size of starting glass powder	Fabrication technique/process	References
45S5	<5 μm	Polymer foam replication	[72]
SiO_2–CaO–CaF_2–Na_2O–K_2O–P_2O_5–MgO	<32 μm	Polymer foam replication	[14]
SiO_2–P_2O_5–CaO–MgO–Na_2O–K_2O	<30 μm	Polymer foam replication	[16]
SiO_2–P_2O_5–CaO–MgO–Na_2O–K_2O	<30 μm	Polymer foam replication	[89]
45S5	10–20 μm	Polymer foam replication	[38]
SiO_2–Na_2O–CaO–MgO	<100 μm	Starch consolidation	[15]
SiO_2–P_2O_5–B_2O_3–CaO–MgO–K_2O–Na_2O	75 μm[a]	Compaction and sintering of melt-spun fibers	[90]
SiO_2–CaO–Na_2O–K_2O–P_2O_5–MgO–CaF_2	<106 μm	Polymer porogen bake-out	[91]
45S5	20–50 μm	Polymer foam replication	[74]
SiO_2–Na_2O–K_2O–MgO–CaO–P_2O_5	255–325 μm	Slip casting	[92]
SiO_2–Na_2O–K_2O–MgO–CaO–P_2O_5	<5–10 μm	Polymer foam replication	[93]
SiO_2–Na_2O–K_2O–MgO–CaO–P_2O_5	<5 μm	Freeze casting	[94, 95]
SiO_2–CaO–K_2O	<106 μm	Polymer porogen burn-off	[96]
SiO_2–TiO_2–B_2O_3–P_2O_5–CaO–MgO–K_2O–Na_2O	75 μm[a]	Compaction and sintering of melt-spun fibers	[31]
45S5	45–90 μm	Polymer porogen bake-out	[97]
45S5	<5 μm	Polymer foam replication	[98]
SiO_2–Na_2O–K_2O–MgO–CaO–P_2O_5; 45S5	25–40 μm[a]	Densification and sintering of melt-spun fibers	[99]
45S5	≈5 μm	Polymer foam replication	[72]
45S5	5–10 μm	Polymer foam replication	[100]
45S5	≈10 μm	Polymer foam replication	[101]
SiO_2–P_2O_5–CaO–MgO–Na_2O–K_2O	n.a.	Polymer burn-off, foam replication	[102]
45S5	<5 μm	Polymer foam replication	[103]
SiO_2–Na_2O–CaO–P_2O_5–B_2O_3–TiO_2	n.a.	Solution combustion	[48]
SiO_2–Na_2O–CaO–P_2O_5–B_2O_3–TiO_2	n.a.	Solution combustion	[104]
SiO_2–CaO–P_2O_5–Al_2O_3	8–30 μm[a]	Manual free-forming of melt-spun fibers	[105]
SiO_2–CaO–Na_2O–P_2O_5–K_2O–MgO–B_2O_3	n.a.	Polymer foam replication	[106]
SiO_2–CaO–Na_2O–K_2O–MgO–P_2O_5–B_2O_3	75 μm[a]	Densification and sintering of melt-spun fibers	[107]
SiO_2–Na_2O–K_2O–MgO–CaO–P_2O_5	0.5–4 μm	Ink-write assembling	[108]

[a] fiber diameter
n.a. not available

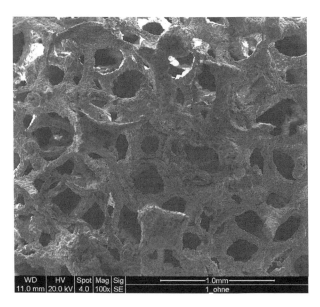

Fig. 4 Scanning electron microscopy (SEM) image of the surface of a 45S5 Bioglass®-derived scaffold fabricated by the foam replication method similar to that reported in Ref. [72]

the foam replica method, other techniques have been considered for fabricating porous glass–ceramic scaffolds. For example, organic particles such as starch, rice, potato, or corn grains [15] swell in water and leave a porous and highly interconnected structure following burn-out from the glass slurry. Porosity can also be introduced by addition of thermally removable phases such as polyethylene particles [102]. Sugar or salt leaching [29, 44] is another common method of producing porous scaffolds. Particles are incorporated into the slurry and leached out upon sintering, leaving an interconnected pore network. The compaction and sintering of melt-spun fibers from bioactive glass is another method of producing scaffolds [31, 90, 99]. After glass production, fibers can be manufactured by melt spinning and packed in a ceramic mould and sintered. It is also possible to manually form melt-spun fibers [105]. Freeze casting techniques uses camphene, ice or water and glycerol as freeze vehicles [112]. After mixing the glass powder with the relevant vehicles, the slurries are cast and frozen at temperatures between −20 and −70 °C, followed by a sintering process.

4.2 Mechanical Properties

In a recent study, Fu et al. [94] fabricated bioactive glass (13–93) scaffolds with oriented (i.e., columnar and lamellar) microstructures and found that at an equivalent porosity of 55–60%, the columnar scaffolds had a compressive strength

Table 2 Mechanical properties of human cancellous and cortical bone in comparison to dense bioactive glass (45S5 Bioglass®)

Material property	Trabecular bone	Cortical bone	Bioglass®45S5
Compressive strength (MPa)	0.1–16 [114, 115]	130–200 [34, 114]	500 [34]
Tensile strength (MPa)	n.a.	50–151 [34]	42 [62]
Compressive modulus (GPa)	0.12–1.1 [116, 117]	11.5–17 [67]	n.a.
Young's modulus (GPa)	0.05–0.5 [34, 118]	7–30 [6, 34, 118]	35 [62]
Fracture toughness (MPa m$^{1/2}$)	n.a.	2–12 [34, 62]	0.9 [119, 120]

n.a. not available

of 25 ± 3 MPa, compressive modulus of 1.2 GPa, and pore width of 90–110 µm, compared to values of 10 ± 2 MPa, 0.4 GPa, and 20–30 µm, respectively, for the lamellar scaffolds. The compressive strength of these columnar bioactive glass scaffolds is >1.5 times higher than the highest strength reported for trabecular bone (0.1–16 MPa, see Table 2). In addition, the cellular response of murine post-osteoblasts/pre-osteocytes to columnar scaffolds indicated that these structures were the most favorable for cell proliferation, migration and mineralization (e.g., bone nodule formation, alkaline phosphatase activity). From the results of their study [94], the authors claimed that 13–93 bioactive glass scaffolds with columnar microstructure are promising candidate materials for the repair and regeneration of load-bearing bones in vivo. It is interesting to note in this regard that highly porous lamellar HA scaffolds (porosity ≈ 50–70%) fabricated by freeze casting exhibited 2.5–4 times higher compressive strength (≈20–140 MPa) than conventional porous HA [111].

Multi-directional, anisotropic mechanical properties of scaffolds have been also reported by Baino et al. [91]. They prepared glass–ceramic scaffolds containing fluoroapatite and investigated their mechanical, structural and bioactive properties upon soaking in simulated body fluid (SBF). The scaffolds had interconnected macropores (porosity = 23.5–50%) and orthotropic mechanical properties, with compressive strength values in the range 20–150 MPa. Thick hydroxyapatite layers were formed on the surface of the scaffolds after 7 days of immersion in SBF, demonstrating the scaffold's excellent bioactivity. Compressive strength values reported in Ref. [91] are considerably higher than those found for bioactive glass–ceramic scaffolds with similar porosities (porosity = 54–73%) prepared by the foam replication technique [113]. The latter scaffolds formed from SiO_2–P_2O_5–CaO–MgO–Na_2O–K_2O bioactive glass had a compressive strength of 1.3–5.4 MPa [113].

Ideally, the elastic modulus of the scaffold should be comparable to that of the tissue to be replaced in order to promote load transfer and minimize stress shielding, reducing the problems of bone resorption [121]. Stress shielding describes the mismatch in elastic moduli between biomaterial and the adjacent/surrounding bone. In cases of large elastic mismatch, bone becomes "stress shielded", which is undesirable since living bone must be under some stress to

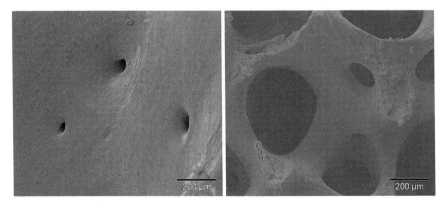

Fig. 5 SEM micrographs showing the structure of bone. Specimens are taken from human mandible showing high-density structure (*left*), and a low-density, sponge-like structure (*right*). (Figures courtesy of R. Detsch, University of Erlangen-Nuremberg, Germany)

remain healthy. In the literature, depending on the measurement technique and parameters used, the source of bone and the structural variation in bone from a given source, a wide range of values has been reported for the compressive modulus of trabecular (0.12–1.1 GPa) and cortical bone (11.5–17 GPa) (Table 2).

Fu et al. [93] reported a compressive strength of 11 ± 1 MPa and compressive modulus of 3.0 ± 0.5 GPa for magnesium- and potassium-substituted bioactive glass–ceramic scaffolds (porosity = 85 ± 2%, pore size = 100–500 µm), which match the highest values reported for human trabecular bone (Table 2). Interestingly, these values are more than ten times higher than compressive strengths reported for 45S5 Bioglass®-based scaffolds [72] of similar porosity and prepared by the same foam replication method. This finding confirms that glass composition and sintering parameters also affect the mechanical properties of glass–ceramic scaffolds. Upon immersion in SBF, Fu et al. [93] observed a nanostructured HA layer formed on the surface of the porous scaffolds within 7 days, indicating the in-vitro bioactivity of the scaffolds. Such HA nanocrystals are found in human bone and are believed to be beneficial for increased cell adhesion, proliferation and greater tissue growth into the scaffold [122–124]. Cell culture results and scanning electron microscopy (SEM) observations presented in Ref. [93] confirmed an excellent attachment and subsequent proliferation of osteoblastic cells.

Engineering constructs with graded porosity represent an interesting approach to the development of bone TE scaffolds. Vitale-Brovarone et al. [102] and Bretcanu et al. [98] manufactured highly porous bioactive glass–ceramic scaffolds with tailored porosity gradients in order to mimic the morphology and lightweight structure of human bone, formed by cortical (compact bone with dense structure) and cancellous bone (trabecular bone with highly porous structure) (Fig. 5). Trabecular bone represents only about 20 wt% of the skeletal mass, but has a nearly ten times greater surface-to-volume ratio (100–300 cm^2/cm^3) than compact

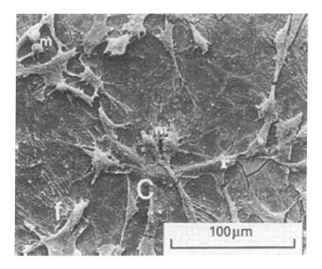

Fig. 6 SEM image of osteoblasts cultured for 2 days on bioactive glass stored under dry conditions [136]. The osteoblasts mainly exhibit a "stand-off" morphology (C) with many dorsal ruffles (r) and filapodia (f). Cell divisions (m) are also relatively often seen. Cracks on the glass surface are due to dehydration of the top layer during critical point drying. (Figure reprinted with permission of Springer)

bone [125, 126]. Therefore, trabecular bone is more important in phosphate and calcium homeostasis than compact bone. The unique hierarchical structure of bone enables its self-repairing properties; bone can alter its geometry and material properties in response to changing external stimuli (e.g., mechanical stresses), and it undergoes a continuous remodeling process [121, 127]. Bone grows in response to load, so that the density of trabecular bone depends on the magnitude of the loads and the orientation of the trabeculae depends on the loading direction.

Analyzing experimental results from the literature [72, 91, 94, 113], a linear relationship between scaffold porosity and compressive strength is found, with coefficients of determination R^2 between 0.80 and 0.99, as reported elsewhere [128].

For human bone, different functional relationships between bone volume fraction (i.e., porosity) and mechanical properties have been observed. On the basis of image-guided failure assessment (IGFA), Nazarian et al. [117] found highly positive linear correlations (R^2= 0.8–0.9) between bone volume fraction and compressive yield strength as well as between bone volume fraction and elastic modulus. Other authors reported quadratic [129] or power-law relationships [114] between bone volume fraction (relative density) and compressive strength, as well as between bone volume fraction and Young's modulus of human bone [129, 130]. Moreover, a second-order polynomial relationship between porosity and Young's modulus has been found in the modeling of the mechanical properties of a face-centered cubic scaffold microstructure [131, 132].

For human bone, the microstructure–property relationship and the relative importance of bone mineral density and bone architecture on fracture behavior need further investigation [117, 133–135].

4.3 In-Vitro and In-Vivo Studies

In an early study by Vrouwenvelder et al. [136] osteoblasts were seeded on polished bioactive glass (45S5) slides and cultured for several days. Figure 6 shows the morphology of osteoblasts after 2 days' cultivation. The cells in the center show a well-developed morphology; they are well spread and tend to form a monolayer.

In addition to providing excellent in vitro bioactivity, suitable cellular behavior and favorable mechanical properties, bioactive glass–ceramic scaffolds have shown also superior in vivo behavior (e.g., bone formation, mineralization, higher interfacial strength between implant and bone) compared to the glass in particulate form [90] or compared to other bioactive materials (e.g., HA, tricalcium phosphate) [48].

For example, Wang et al. [137] implanted sol–gel-derived porous bioactive glass discs of 1 cm diameter in New Zealand rabbits. After 5 weeks, histological results showed newly formed tissue in the form of widely distributed collagen fiber strands (Fig. 7). San Miguel et al. [107] reported superior osteoconductive behavior using a rabbit calvarial bone model (i.e., significantly higher bone formation, bone deposition) of SBF-pretreated scaffolds (BG fiber constructs) compared with non-treated porous BG scaffolds, bioactive glass granules (PerioGlas®) and empty bone defects.

Fu et al. [138] reported on the implantation of a 13–93 bioactive glass scaffold into subcutaneous pockets in the dorsum of rats. Scaffolds with both a "trabecular" microstructure (processed via the polymer sponge method, porosity 65%, pore size between 100–500 µm), and with a columnar microstructure (processed via unidirectional freezing of the suspension, porosity 65%, pore size between 90 and 110 µm) were investigated. After 4 weeks the columnar scaffolds showed abundant tissue ingrowth whereas the trabecular scaffolds showed only limited tissue infiltration.

5 Bioactive Glass Containing Composite Scaffolds

Most glass–ceramic scaffolds analyzed in the literature show a suitable interconnected macroporous network and compressive strengths >2 MPa, which is in the range of the compressive strength of cancellous bone (Table 2). The scaffolds can therefore fulfill the requirement in terms of compressive strength. However, load-bearing bone defect sites are usually under cyclic loading and, as the scaffolds are

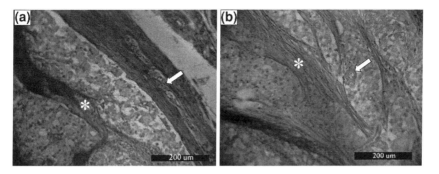

Fig. 7 Histology of porous BG scaffolds after 5 weeks of implantation in New Zealand rabbits [137]. **a** New vasularization formed with the fibrous tissue band of the capsule (*arrow*). **b** Collagen fiber strands (*asterisk*) are distributed throughout the scaffold material showing new blood vessel formation (*arrow*). (Figures reprinted with permission of Springer)

made from porous glass, they are normally inherently brittle and have poor fracture toughness (Table 2).

Fracture toughness values in the range reported for cortical bone (2–12 MPa m$^{1/2}$) are required for load-bearing applications and therefore toughening effects must be introduced into this type of scaffold, which can be achieved by producing composites [34]. Polymer/bioceramic composite scaffolds thus represent a convenient alternative due to the possibility of tailoring their various properties (e.g., mechanical and structural behavior, degradation kinetics and bioactivity) [34, 139]. Composites made of polymers and bioceramics combine the advantages of their individual components [3, 34, 109]. Polymers exhibit generally high ductility, flexibility and favorable formability as well as processibility and plasticity. The glass or glass–ceramic phase adds stiffness and adequate mechanical strength to the composite. In particular, composites based on biodegradable polymers and bioactive glasses are being increasingly studied as bone TE materials because this particular combination does not require a revision surgery for their removal, since newly formed bone gradually substitutes the implanted scaffold during degradation [34, 37]. Much current research is therefore focused on the fabrication of bioactive composite materials with bioactive glass incorporated either as filler or coating (or both) into the bioresorbable polymer matrix [34]. Effort is devoted in particular to the development of porous, high-strength composite structures for the regeneration of human bone at load-bearing sites. A comprehensive general review on bone TE scaffolds based on composites with inorganic bioactive fillers has been published by Rezwan et al. [34]. The state of knowledge on polymer–bioceramic composites with focus on polymer coatings and interpenetrating polymer–bioceramic structures for bone TE has been summarized by Yunos et al. [140]. Polymer/bioactive glass nanocomposites, based on bioactive glass nanoparticles and nanofibres, have been reviewed by Boccaccini et al. [42].

Many studies have been carried out in the last 10 years to optimize and investigate bone TE composite scaffolds concerning material combinations,

Table 3 Overview of studies performed on BG containing composite scaffolds for bone TE

Bioactive glass	wt%	Matrix	Fabrication technique/process	Study
45S5 m-BG	5, 29, 40	PDLLA	Co-extrusion+compaction; TIPS	[143]
45S5 m-BG	4.8, 28.6	PDLLA	TIPS	[144]
45S5 m-BG	10	P(3HB)	ST/PL	[44]
45S5 n-BG	10	P(3HB)	ST/PL	[44]
S53P4 m-BG	20, 50	P(CL/DLLA)	ST/PL	[145]
S53P4 m-BG	30	P(CL/DLLA)	ST/PL	[146]
45S5 m-BG	10, 30	PLGA	Microsphere emulsification	[147]
45S5 m-BG	10	PDLG	TIPS	[148]
45S5 m-BG	25, 50	PLA	Freeze extraction technique	[149]
45S5 m-BG	5, 40	PDLLA	Solvent casting	[150]
45S5 m-BG	10, 25, 50	PDLLA	TIPS	[151]
45S5 m-BG	10, 25, 50	PLGA	TIPS	[151]
45S5 m-BG	5, 10, 40	PDLLA	TIPS	[152]
45S5 m-BG	5, 40	PDLLA	TIPS	[153]
45S5 m-BG	10, 25, 50	PLGA	TIPS	[37]
45S5 m-BG	25	PLGA	Solvent casting	[47]
45S5 m-BG	20	P(3HB)	Solvent casting	[154]
45S5 m-BG	20	P(3HB)	Solvent casting	[155]
45S5 n-BG	10, 20	P(3HB)	Solvent casting	[43]
45S5 m-BG	10, 20, 30	P(3HB)	Solvent casting	[45]
45S5 n-BG	10, 20, 30	P(3HB)	Solvent casting	[45]
45S5 m-BG	5, 30	PDLLA	TIPS	[156]
45S5 m-BG	5, 30	PDLLA	TIPS	[157]
45S5 m-BG	5, 40	PDLLA	TIPS	[158]
SiO$_2$–3CaO–P$_2$O$_5$–MgO	10, 30, 50	PLA	TIPS	[159]

PDLLA poly(d,l-lactide), *P(3HB)* poly(3-hydroxybutyrate), *P (CL/DLLA)* poly(ε-caprolactone/d,l-lactide), *PLGA* poly(lactic-co-glycolic acid), *PDLG* poly(d,l-lactide-co-glycolide), *PLA* poly(l-lactide), *S53P4* 53 wt% SiO$_2$, 23 wt% Na$_2$O, 20 wt% CaO, 4 wt% P$_2$O$_5$, *m-BG* micron-sized bioactive glass, *n-BG* nano-sized bioactive glass, *ST/PL* sugar template/particulate leaching, *TIPS* thermally induced phase separation bioactive properties, degradation characteristics in vitro and in vivo behavior, as well as mechanical properties (Table 3).

The mechanical strength of most of today's available porous polymer/BG composite scaffolds is inadequate for bone substitution because they are at least one order of magnitude weaker than natural cancellous bone and orders of magnitude weaker than cortical bone. Moreover there is still limited understanding of how microstructure features (e.g., geometry of struts, pore size distribution, pore orientation, interconnectivity, morphology and distribution of the BG filler) affect the scaffold's mechanical response and its functional performance [141]. In addition, insufficient particle–matrix bonding is considered a possible reason for the low mechanical properties of these composites. With regard to the latter, two key issues have to be solved to effectively improve the material properties of scaffolds by adding bioactive glass particles as filler: (1) interfacial bonding and (2) the proper, homogeneous dispersion of the individual particles in the matrix (e.g., by particle surface functionalization). According to the concepts of the composites theory [142], load transfer at the filler/matrix interface is key to achieving strengthening and stiffening, which depends on the quality of interfacial bonding between the two phases (filler and matrix). Strong interfacial bonding is therefore a significant requirement for improving the mechanical properties of biodegradable polymer composite scaffolds.

Blaker et al. [144] have developed highly porous (porosity $\approx$ 94%) poly (d,l-lactide) (PDLLA)/Bioglass® foams using thermally induced phase separation (TIPS). The scaffolds exhibited a bimodal and anisotropic pore structure, with tubular micropores of $\approx$ 100 µm in diameter, and with interconnected micro-pores of $\approx$ 50–10 µm, along with anisotropic mechanical properties. With respect to the direction of the tubular pores, similar axial yield strengths of about 0.08 MPa were found for all composites (0, 4.8, 28.6 wt% Bioglass®), whereas a higher axial compressive modulus (1.2 MPa) was obtained for 28.6 wt% Bioglass® containing scaffolds compared to the pure PDLLA constructs (0.89 MPa). The yield strength values reported in Ref. [144] are considerably lower than those for cancellous bone [117], so a further improvement is necessary to increase the mechanical performance up to the levels required for bone TE applications. The compressive moduli are in the range of those determined for trabecular bone, but lower than those for cortical bone (see Table 2).

Other authors have found, however, considerably higher mechanical strength for their composite scaffolds [47, 151]. Maquet et al. [151], for example, have reported highly porous (porosity > 90%) PDLLA and PLGA scaffolds containing 50 wt% Bioglass®, exhibiting compressive moduli of about 21 and 26 MPa, respectively: a factor 1.5–2.5 higher than the values of the pure polymer scaffolds. Lu et al. [47] determined for PLGA scaffolds incorporated with 25 wt% Bioglass® (porosity = 43%, pore diameter = 89 µm) a compressive modulus of about 51 MPa, and compressive strength of about 0.42 MPa, which are in the range of values reported for trabecular bone (Table 2), but at the cost of porosity (43%).

Interestingly, numerical analyses presented in Ref. [144] showed that the compressive modulus of the composite foams can be well predicted by micromechanic theories based on the combination of the Ishai–Cohen [160] and Gibson–Ashby models [161]. The modulus–density (pore volume fraction) relationship was characterized by a power-law function with exponents between 2 and 3. This is close to the exponents found for similar relationships valid for human bone (2–3.2) [129, 130].

Extensive work has been also carried out to investigate the cellular response to polymer/bioactive glass composites concerning composition, particle concentration and particle size effect in vitro and in vivo, as discussed in Ref. [128]. Some key findings are included here. For example, Lu et al. [162] showed that for PLGA/bioactive glass films (0, 10, 25, 50 wt%), the growth, mineralization and differentiation of human osteoblast-like SaOS-2 cells as well as the kinetics of Ca–P layer formation and the resulting Ca–P chemistry were dependent on BG content. The 10 and 25 wt% BG composite supported greater osteoblast growth and differentiation than the 50 wt% BG group. Such bioactive glass dose-dependent cell proliferation and alkaline phosphatase (ALP) synthesis were also reported by Yang et al. [158], Verrier et al. [153] and Tsigkou et al. [150]. Tsigkou et al. [150] observed, for example, that human fetal osteoblasts were less spread and elongated on PDLLA and PDLLA/5 wt% BG, whereas cells on PDLLA/40 wt% BG were elongated but with multiple protrusions spreading over the BG particles. However, when differentiation and maturation of fetal osteoblasts were examined, incorporation of 45S5 Bioglass® particles within the PDLLA matrix was found to significantly enhance ALP and osteocalcin protein synthesis compared to PDLLA alone. Alizarin red staining indicated extracellular matrix mineralization on 5 and 40 wt% BG-containing films, with significantly more bone nodules formed than on neat PDLLA films. Yang et al. [158] pre-treated 45S5 BG containing (0, 5, 40 wt%) PDLLA scaffolds with serum and found a significant increase in ALP activity in human bone marrow mesenchymal stem cells in 5 wt% Bioglass® composites relative to the 0 and 40 wt% Bioglass® groups, whereas in vivo studies indicated significant new bone formation throughout all the scaffolds. The results of numerous studies [150, 153, 158, 162] have confirmed the osteogenic potential of BG-containing scaffolds and suggest that there is a critical threshold range of BG content (5–40 wt%) which is optimal for osteoblast growth and Ca–P formation. This finding is also relevant for the vascularization and angiogenic properties of composite scaffolds, as discussed in Sect. 6.

Several cell culture studies (see Ref. [49, 128]) have demonstrated the pro-angiogenic potential of bioactive glass over a limited range of concentrations, implying that dose-dependent effects are also involved in angiogenesis similar to those shown for osteogenic differentiation (e.g., ALP synthesis) and cellular behavior (adhesion, proliferation, mineralization). BG has been reported to have pro-angiogenic potential over a limited range of lower concentrations and greater osteogenic potential at higher concentrations [50, 163]. To our knowledge, Misra et al. [45] were the first to incorporate bioactive glass nanoparticles (30–50 nm) of composition matching the 45S5 BG composition into degradable matrices (in their case P(3HB) was used) and compared their thermal, mechanical, microstructural, bioactive and cell biological properties with those of conventional, micron-sized

(5 μm) BG-containing composites. The addition of bioactive glass nanoparticles (n-BG) enhanced the Young's modulus by 50–100% to values of 1.2 and 1.6 GPa, compared to both pure polymer films and the corresponding micro-sized BG (m-BG)-containing films (10, 20, 30 wt%). The nanostructured surface topography induced by n-BG considerably improved protein adsorption on the n-BG composites compared to the unfilled polymer and the m-BG composites, whereas no substantial differences in the proliferation of MG-63 osteoblasts were observed between the different surfaces. It was thus confirmed that the addition of nanosized bioactive glass particles had a more significant effect on the mechanical and structural properties of a composite system in comparison with microparticles. The addition of nanoparticles also enhanced protein adsorption, a desirable effect for the application of composites in bone TE.

It has been also reported that tailoring porosity (e.g., nano- or mesoporosity) and surface topography (e.g., by the incorporation of nanophase bioactive glass particles into degradable polymer matrices) can favor protein adsorption and cellular interactions [45, 155], as well as improve the bioactive behavior [43], antimicrobial/antibacterial effect [164–166] and mechanical properties [45] of bioactive glass and related (composite) scaffolds.

6 Effect of Bioactive Glass on Angiogenesis

The ability of BG to stimulate the release of pro-angiogenic factors (e.g., VEGF) from transplanted and/or host cells that have migrated into the scaffold might be extremely beneficial in inducing neo-vascularization and rapid vascular in-growth sufficient to meet the metabolic requirements of new bone. Enhancement of the angiogenic potential of implantable biomaterials and scaffolds is crucial for the success of tissue engineering approaches.

Despite increased evidence relating bioactive glass to angiogenic effects both in vitro and in vivo [49–53], there has been limited research to date on understanding the specific role of bioactive glass in vascularization of bone scaffolds. There is for example limited quantitative data regarding how the shape, size and concentration of BG particles (e.g., as inclusion in polymer matrices) affect angiogenesis. In addition, the influence of specific ions (as dissolution products of bioactive glasses) on dissolution products of bioactive glasses on angiogenesis has not been widely investigated. Moreover, the design criteria and the specific requirements of the geometry and morphology of the scaffold (pore size distribution, shape, interconnectivity) to achieve tailored scaffolds with angiogenic properties are unknown [128]. A detailed overview of the available literature investigating bioactive glasses with respect to angiogenesis has been published by Gorustovich et al. [49], and the effect of ion release products concerning angiogenic response has been reviewed by Hoppe et al. [9]. An important conclusion of the literature analyzed is that cell culture studies demonstrated the pro-angiogenic potential of BG over a limited range of BG concentrations, implying that dose-dependent effects

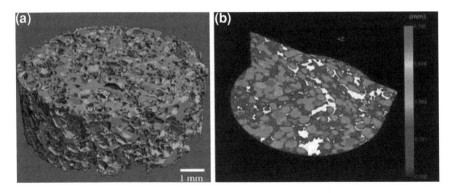

Fig. 8 a Micro-computed tomography image of a PDLLA scaffold fabricated by a sugar template particulate leaching technique [50]. (Figure courtesy of R. Stämpfli, Empa St Gallen, Switzerland). **b** Color-coded scaffold micro-architecture. The intermeshing, overlapping color circles, i.e., color-coded pores (200–750 μm, 90th percentile: 600 μm), indicate high pore interconnectivity and high porosity, which are both essential for cell penetration, proper vascularization, bone tissue in-growth, waste removal, oxygen and nutrient delivery. The white regions around the pores are PDLLA scaffold matrix areas. (Figure kindly provided by J.A. Sanz-Herrera and I. Ochoa, University of Zaragoza, Spain)

are involved in angiogenesis. It has been shown, for example, that bioactive glass stimulates the secretion of angiogenic growth factors in fibroblasts [52, 128, 148, 167, 168], the proliferation of endothelial cells [50, 53], and the formation of endothelial tubules [53].

In addition, in vivo results confirmed that BG is able to stimulate and promote neo-vascularization [49, 103–105, 169]. For example, Leu et al. [53, 170] filled calvarial defects in Sprague–Dawley rats with 45S5 Bioglass®-impregnated (1.2 mg) collagen sponges (volume = 0.05 cm^3), using unloaded, empty sponges as control. After 2 weeks of implantation, histological analyses of calvaria demonstrated significantly greater neo-vascularization and vascular density within defects treated with 45S5 BG (35 ± 16 vessels/mm^2) than with collagen controls alone (12 ± 2 vessels/mm^2).

The angiogenic effect of bioactive glass has been shown to be much more pronounced in bioactive glass-based scaffolds (i.e., loaded sponges [170], discs [171], meshes [169], tubes [172], and porous glass–ceramic scaffolds [49, 104, 105] than in composite structures incorporating and fully embedding bioactive glass particles in polymer matrices (e.g., microsphere composites [148] or foams [52, 173]).

Day et al. [52], for example, found favorable angiogenic properties (i.e., greater tissue infiltration and higher blood vessel formation) for compression-molded BG composites compared to the corresponding unfilled polymer scaffolds. Interestingly, the same authors found no difference in the number of blood vessels formed in scaffolds prepared by thermally-induced phase separation technology. These findings indicate that the geometry and morphology of the scaffold (pore orientation, pore size, interconnectivity, strut thickness) can be used to control the

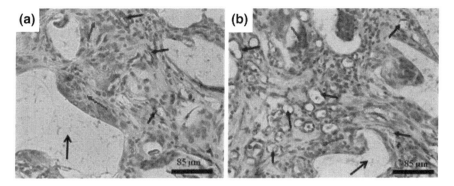

Fig. 9 Histology of PDLLA/BG scaffolds after 8 weeks of implantation in Sprague–Dawley rats [50]. **a** PDLLA and **b** scaffold containing 20 wt% micron-sized BG stained with hematoxylin and Factor VIII (*brown rings*). Scaffolds were well interspersed with newly formed tissue and blood vessels. Key: *black arrow*: scaffold (opaque material), *red arrow*: blood vessel immunolocalized for Factor VIII, *yellow arrow*: cellular infiltrate. (Micrographs taken with assistance of Dr T. Ansari's research group, Northwick Park Institute for Medical Research, UK)

complex mechanisms of angiogenesis. The scaffold microstructure also affects the angiogenic response of the construct in vivo. Consequently, a "dimension response element" has been suggested to be involved in the regulation of osteogenic and angiogenic gene expression [3].

Recently, PDLLA/bioactive glass containing composite scaffolds (Fig. 8) have been fabricated, which exhibit high porosities high porosities (81–93%), suitable permeability ($k = 5.4$–8.6×10^{-9} m^2) and compressive strength values (0.4–1.6 MPa) in the lower range of trabecular bone [50]. After 8 weeks of implantation, scaffolds containing m-sized BG and nano-sized BG were infiltrated with newly formed tissue (Fig. 9) and demonstrated higher vascularization and percentage of blood vessel formation (11.6–15.1%) than neat PDLLA scaffolds (8.5%). This work thus showed the potential for the regeneration of hard-soft tissue defects and increased bone formation arising from enhanced vascularization of PDLLA/bioactive glass constructs. Advanced functional bioactive glass scaffolds showing osteogenic and angiogenic properties, i.e., the ability to become tightly attached to the host tissue, mineralized and vascularized, represent an attractive solution for the regeneration of complex tissue structure defects, for example at soft–hard tissue interfaces (e.g., the scaffold could be used as a "plug" at the cartilage–bone interface).

7 Conclusions and Future Work

One of the most significant challenges in bone tissue engineering remains the fabrication of scaffolds exhibiting mechanical, structural, surface-chemical, topographical and biological properties suitable for regenerating large (critical size) cortical bone defects and capable of functioning under relevant loads. Although a number of bioactive glass and glass–ceramic scaffolds with favorable properties are available, as comprehensively discussed in this chapter, several issues need to be addressed prior to clinical application, such as mechanical reliability of scaffolds, tailored degradability, and induction of vascularization. A major challenge remains the proper cellularization and controlled vascularization of 3D scaffolds. For successful bone regeneration, there is a need for functional, mature vessels promoting functionality to the intrinsically "inactive" man-made TE constructs. Angiogenesis requires that capillaries develop and stabilize before differentiating into arterioles and venules by the appearance of circumferentially located smooth muscle cells, and may ultimately mature into arteries and veins [174]. A stable vasculature is important for the long-term success of TE constructs and bone regeneration [175].

One alternative to accelerate osteogenesis and angiogenesis is the incorporation of active biomolecules such as growth factors into the scaffold structure [176–181]. However, the short half-life and uncontrolled release of growth factors from scaffolds associated with possible toxicity effects may be a problem or limitation of current drug delivery scaffolds. The use of bioactive glass as a filler in degradable matrices might offer a promising strategy for the regulated in situ secretion/expression of angiogenic growth factors (e.g., VEGF) and osteogenic markers (e.g., ALP) at therapeutic levels, leading to successful vascularization and bone formation (mineralization) of TE scaffolds.

Further improvement in scaffold function is related to surface modification, e.g., through the control of specific/non-specific protein adsorption [182], plasma treatment [183, 184] or enzyme grafting [185], to provide biofunctional groups for cell attachment and response, thus making the scaffold more surface-compatible. There is still limited understanding of the long-term in-vivo behavior of bioactive glass-based scaffolds and polymer/BG composite scaffolds, particularly regarding their degradation rate, ion release kinetics, variation in mechanical properties and angiogenic effect. In this context, it has to be pointed out that the influence of sterilization on the cytotoxic, mechanical (e.g., compressive strength, fracture toughness) and physical properties (glass transition temperature, crystallinity) of biodegradable composites has often been overlooked in the past. This is particularly important for scaffolds incorporating a polymeric phase. Sterilization issues have to be considered and monitored in parallel to the design and development stages of scaffolds because standard medical product sterilization techniques (gamma irradiation, ethylene oxide gas exposure) have shown to reduce the molecular weight of resorbable polymers by a factor of 2–3 [186–188].

Moreover, more focus on in-vivo studies is required and there is need for further research on the evaluation of scaffolds in realistic biological systems. Engineered scaffolds from silicate amorphous or partially crystallized glasses, combined with biodegradable polymers, will continue to be improved and optimized. These scaffolds constitute attractive alternative approaches in future developments and their combination with stem cells is of great interest [57, 189–191]. The use of bioactive glass and glass–ceramic nanoparticles and carbon nanotubes (CNTs)

[192, 193] as well as their combination with bioresorbable polymers may also improve the local environment in order to enhance cell attachment, proliferation and angiogenic and osteogenic properties as well as adding extra functionalities to the base scaffold. However, possible toxicity issues associated with nanoparticles and CNTs remain to be properly investigated [194].

References

1. Hench LL, Polak JM (2002) Third-generation biomedical materials. Science 295: 1014–1017
2. Williams D (2004) Benefit and risk in tissue engineering. Mater Today 7:24–29
3. Guarino V, Causa F, Ambrosio L (2007) Bioactive scaffolds for bone and ligament tissue. Expert Rev Med Dev 4:405–418
4. Hutmacher DW, Schantz JT, Lam CXF, Tan KC, Lim TC (2007) State of the art and future directions of scaffold-based bone engineering from a biomaterials perspective. J Tissue Eng Regen Med 1:245–260
5. Hench LL (1998) Bioceramics. J Am Ceram Soc 81:1705–1728
6. Kokubo T, Kim HM, Kawashita M (2003) Novel bioactive materials with different mechanical properties. Biomaterials 24:2161–2175
7. Xynos ID, Edgar AJ, Buttery LDK, Hench LL, Polak JM (2001) Gene-expression profiling of human osteoblasts following treatment with the ionic products of Bioglass®45S5 dissolution. J Biomed Mater Res 55:151–157
8. Hench LL (2009) Genetic design of bioactive glass. J Eur Ceram Soc 29:1257–1265
9. Hoppe A, Güldal NS, Boccaccini AR (2011) A review of the biological response to ionic dissolution products from bioactive glasses and glass–ceramics. Biomaterials 32(11): 2757–2774
10. Jell G, Stevens MM (2006) Gene activation by bioactive glasses. J Mater Sci Mater Med 17:997–1002
11. Tsigkou O, Jones JR, Polak JM, Stevens MM (2009) Differentiation of fetal osteoblasts and formation of mineralized bone nodules by 45S5 Bioglass (R) conditioned medium in the absence of osteogenic supplements. Biomaterials 30:3542–3550
12. Hench LL, Splinter RJ, Allen WC, Greenlee TK (1971) Bonding mechanisms at the interface of ceramic prosthetic materials. J Biomed Mater Res 5:117–141
13. Hench LL (2006) The story of Bioglass®. J Mater Sci Mater Med 17:967–978
14. Vitale-Brovarone C, Miola M, Balagna C, Verné E (2008) 3D-glass–ceramic scaffolds with antibacterial properties for bone grafting. Chem Eng J 137:129–136
15. Vitale-Brovarone C, Verne E, Bosetti M, Appendino P, Cannas M (2005) Microstructural and in vitro characterization of SiO_2-Na_2O-CaO-MgO glass–ceramic bioactive scaffolds for bone substitutes. J Mater Sci Mater Med 16:909–917
16. Vitale-Brovarone C, Verne E, Robiglio L, Appendino P, Bassi F et al (2007) Development of glass–ceramic scaffolds for bone tissue engineering: characterisation, proliferation of human osteoblasts and nodule formation. Acta Biomater 3:199–208
17. Gentleman E, Fredholm YC, Jell G, Lotfibakhshaiesh N, O'Donnell MD et al (2010) The effects of strontium-substituted bioactive glasses on osteoblasts and osteoclasts in vitro. Biomaterials 31:3949–3956
18. Pan HB, Zhao XL, Zhang X, Zhang KB, Li LC et al (2010) Strontium borate glass: potential biomaterial for bone regeneration. J R Soc Interface 7:1025–1031
19. O'Donnell MD, Hill RG (2010) Influence of strontium and the importance of glass chemistry and structure when designing bioactive glasses for bone regeneration. Acta Biomater 6:2382–2385

20. Hsi C-S, Cheng H-Z, Hsu H-J, Chen Y-S, Wang M-C (2007) Crystallization kinetics and magnetic properties of iron oxide contained $25Li_2O-8MnO_2-20CaO-2P_2O_5-45SiO_2$ glasses. J Eur Ceram Soc 27:3171–3176
21. Balamurugan A, Balossier G, Laurent-Maquin D, Pina S, Rebelo AHS et al (2008) An in vitro biological and anti-bacterial study on a sol–gel derived silver-incorporated bioglass system. Dent Mater 24:1343–1351
22. Bellantone M, Williams HD, Hench LL (2002) Broad-spectrum bactericidal activity of Ag_2O-doped bioactive glass. Antimicrob Agents Chemother 46:1940–1945
23. Blaker JJ, Nazhat SN, Boccaccini AR (2004) Development and characterisation of silver-doped bioactive glass-coated sutures for tissue engineering and wound healing applications. Biomaterials 25:1319–1329
24. Delben JRJ, Pimentel OM, Coelho MB, Candelorio PD, Furini LN et al (2009) Synthesis and thermal properties of nanoparticles of bioactive glasses containing silver. J Therm Anal Calor 97:433–436
25. Liu X, Huang W, Fu H, Yao A, Wang D et al (2009) Bioactive borosilicate glass scaffolds: improvement on the strength of glass-based scaffolds for tissue engineering. J Mater Sci Mater Med 20:365–372
26. Liu X, Huang W, Fu H, Yao A, Wang D et al (2009) Bioactive borosilicate glass scaffolds: in vitro degradation and bioactivity behaviors. J Mater Sci Mater Med 20:1237–1243
27. Munukka E, Lepparanta O, Korkeamaki M, Vaahtio M, Peltola T et al (2008) Bactericidal effects of bioactive glasses on clinically important aerobic bacteria. J Mater Sci Mater Med 19:27–32
28. Gorriti MF, Porto López JM, Boccaccini AR, Audisio C, Gorustovich AA (2009) In vitro study of the antibacterial activity of bioactive glass–ceramic scaffolds. Adv Engin Mater 11:B67–B70
29. Cannillo V, Sola A (2009) Potassium-based composition for a bioactive glass. Ceram Int 35:3389–3393
30. Aina V, Malavasi G, Fiorio Pla A, Munaron L, Morterra C (2009) Zinc-containing bioactive glasses: surface reactivity and behaviour towards endothelial cells. Acta Biomater 5:1211–1222
31. Haimi S, Gorianc G, Moimas L, Lindroos B, Huhtala H et al (2009) Characterization of zinc-releasing three-dimensional bioactive glass scaffolds and their effect on human adipose stem cell proliferation and osteogenic differentiation. Acta Biomater 5:3122–3131
32. Wilson J, Pigott GH, Schoen FJ, Hench LL (1981) Toxicology and biocompatibility of bioglasses. J Biomed Mater Res 15:805–817
33. Hench LL, Xynos ID, Polak JM (2004) Bioactive glasses for in situ tissue regeneration. J Biomat Sci Polym Ed 15:543–562
34. Rezwan K, Chen QZ, Blaker JJ, Boccaccini AR (2006) Biodegradable and bioactive porous polymer/inorganic composite scaffolds for bone tissue engineering. Biomaterials 27:3413–3431
35. Ylanen HO, Helminen T, Helminen A, Rantakokko J, Karlsson KH et al (1999) Porous bioactive glass matrix in reconstruction of articular osteochondral defects. Ann Chir Gynaecol 88:237–245
36. Garcia AJ, Ducheyne P, Boettiger D (1998) Effect of surface reaction stage on fibronectin-mediated adhesion of osteoblast-like cells to bioactive glass. J Biomed Mater Res 40:48–56
37. Boccaccini AR, Maquet V (2003) Bioresorbable and bioactive polymer/Bioglass®composites with tailored pore structure for tissue engineering applications. Compos Sci Technol 63:2417–2429
38. Chen QZ, Efthymiou A, Salih V, Boccaccini AR (2008) Bioglass-derived glass–ceramic scaffolds: study of cell proliferation and scaffold degradation in vitro. J Biomed Mater Res Part A 84:1049–1060
39. Xynos ID, Hukkanen MVJ, Batten JJ, Buttery LD, Hench LL et al (2000) Bioglass®45S5 stimulates osteoblast turnover and enhances bone formation in vitro: Implications and applications for bone tissue engineering. Calcif Tissue Int 67:321–329

40. Brunner TJ, Grass RN, Stark WJ (2006) Glass and bioglass nanopowders by flame synthesis. Chem Commun 1384–1386
41. Vollenweider M, Brunner TJ, Knecht S, Grass RN, Zehnder M et al (2007) Remineralization of human dentin using ultrafine bioactive glass particles. Acta Biomater 3:936–943
42. Boccaccini AR, Erol M, Stark WJ, Mohn D, Hong Z et al (2010) Polymer/bioactive glass nanocomposites for biomedical applications: a review. Compos Sci Technol 70:1764–1776
43. Misra SK, Ansari T, Mohn D, Valappil SP, Brunner TJ et al (2010) Effect of nanoparticulate bioactive glass particles on bioactivity and cytocompatibility of poly(3-hydroxybutyrate) composites. J R Soc Interface 7:453–465
44. Misra SK, Ansari TI, Valappil SP, Mohn D, Philip SE et al (2010) Poly(3-hydroxybutyrate) multifunctional composite scaffolds for tissue engineering applications. Biomaterials 31:2806–2815
45. Misra SK, Mohn D, Brunner TJ, Stark WJ, Philip SE et al (2008) Comparison of nanoscale and microscale bioactive glass on the properties of P(3HB)/Bioglass composites. Biomaterials 29:1750–1761
46. Liu A, Hong Z, Zhuang X, Chen X, Cui Y et al (2008) Surface modification of bioactive glass nanoparticles and the mechanical and biological properties of poly(l-lactide) composites. Acta Biomater 4:1005–1015
47. Lu HH, El-Amin SF, Scott KD, Laurencin CT (2003) Three-dimensional, bioactive, biodegradable, polymer-bioactive glass composite scaffolds with improved mechanical properties support collagen synthesis and mineralization of human osteoblast-like cells in vitro. J Biomed Mater Res Part A 64:465–474
48. Ghosh SK, Nandi SK, Kundu B, Datta S, De DK et al (2008) In vivo response of porous hydroxyapatite and beta-tricalcium phosphate prepared by aqueous solution combustion method and comparison with bioglass scaffolds. J Biomed Mater Res B Appl Biomater 86:217–227
49. Gorustovich A, Roether J, Boccaccini AR (2010) Effect of bioactive glasses on angiogenesis: in vitro and in vivo evidence. A review. Tissue Eng Part B Rev 16:199–207
50. Gerhardt LC, Widdows KL, Erol MM, Burch CW, Sanz JA et al (2011) The pro-angiogenic properties of multifunctional bioactive glass composite scaffolds. Biomaterials 32:4096–4108
51. Day RM (2005) Bioactive glass stimulates the secretion of angiogenic growth factors and angiogenesis in vitro. Tissue Eng 11:768–777
52. Day RM, Maquet V, Boccaccini AR, Jerome R, Forbes A (2005) In vitro and in vivo analysis of macroporous biodegradable poly(D, L-lactide-co-glycolide) scaffolds containing bioactive glass. J Biomed Mater Res Part A 75:778–787
53. Leu A, Leach JK (2008) Proangiogenic potential of a collagen/bioactive glass substrate. Pharm Res 25:1222–1229
54. Cauda V, Fiorilli S, Onida B, Vernè E, Vitale Brovarone C et al (2008) SBA-15 ordered mesoporous silica inside a bioactive glass–ceramic scaffold for local drug delivery. J Mater Sci Mater Med 19:3303–3310
55. El-Ghannam AR (2004) Advanced bioceramic composite for bone tissue engineering: design principles and structure-bioactivity relationship. J Biomed Mater Res Part A 69:490–501
56. Habraken WJ, Wolke JG, Jansen JA (2007) Ceramic composites as matrices and scaffolds for drug delivery in tissue engineering. Adv Drug Deliv Rev 59:234–248
57. Juan Z, Min W, Jae Min C, Athanasios M (2009) The incorporation of 70s bioactive glass to the osteogenic differentiation of murine embryonic stem cells in 3D bioreactors. J Tissue Eng Regen Med 3:63–71
58. Mortera R, Onida B, Fiorilli S, Cauda V, Brovarone CV et al (2008) Synthesis and characterization of MCM-41 spheres inside bioactive glass–ceramic scaffold. J Chem Eng 137:54–61

59. Zhu Y, Kaskel S (2009) Comparison of the in vitro bioactivity and drug release property of mesoporous bioactive glasses (MBGs) and bioactive glasses (BGs) scaffolds. Microporous Mesoporous Mater 118:176–182
60. Rahaman MN, Day DE, Bal BS, Fu Q, Jung SB, et al (2011) Bioactive glass in tissue engineering. Acta Biomater 7:2355–2373
61. Ostomel TA, Shi QH, Tsung CK, Liang HJ, Stucky GD (2006) Spherical bioactive glass with enhanced rates of hydroxyapatite deposition and hemostatic activity. Small 2:1261–1265
62. Thompson ID, Hench LL (1998) Mechanical properties of bioactive glasses, glass–ceramics and composites. Proc Inst Mech Eng Part H-J Eng Med 212:127–136
63. Kokubo T, Ito S, Shigematsu M, Sakka S, Yamamuro T (1985) Mechanical-properties of a new type of apatite-containing glass–ceramic for prosthetic application. J Mater Sci 20:2001–2004
64. Nakamura T, Yamamuro T, Higashi S, Kokubo T, Itoo S (1985) A new glass–ceramic for bone replacement-evaluation of its bonding to bone. J Biomed Mater Res 19:685–698
65. Wu C (2009) Methods of improving mechanical and biomedical properties of Ca-Si-based ceramics and scaffolds. Expert Rev Med Devices 6:237–241
66. Ramakrishna S, Mayer J, Wintermantel E, Leong KW (2001) Biomedical applications of polymer-composite materials: a review. Compos Sci Technol 61:1189–1224
67. Karageorgiou V, Kaplan D (2005) Porosity of 3D biomaterial scaffolds and osteogenesis. Biomaterials 26:5474–5491
68. Tabata Y (2009) Biomaterial technology for tissue engineering applications. J R Soc Interface 6:S311–S324
69. Smith IO, Ren F, Baumann MJ, Case ED (2006) Confocal laser scanning microscopy as a tool for imaging cancellous bone. J Biomed Mater Res B Appl Biomater 79:185–192
70. Woodard JR, Hilldore AJ, Lan SK, Park CJ, Morgan AW et al (2007) The mechanical properties and osteoconductivity of hydroxyapatite bone scaffolds with multi-scale porosity. Biomaterials 28:45–54
71. Ochoa I, Sanz-Herrera JA, García-Aznar JM, Doblaré M, Yunos DM et al (2009) Permeability evaluation of 45S5 Bioglass®-based scaffolds for bone tissue engineering. J Biomech 42:257–260
72. Chen QZ, Thompson ID, Boccaccini AR (2006) 45S5 Bioglass®-derived glass–ceramic scaffolds for bone tissue engineering. Biomaterials 27:2414–2425
73. Boccaccini AR (2005) Ceramics. In: Hench LL, Jones JR (eds) Biomaterials, artificial organs and tissue engineering, 1st edn. Woodhead Publishing Limited CRC Press, Cambridge, pp 26–36
74. Boccaccini AR, Chen Q, Lefebvre L, Gremillard L, Chevalier J (2007) Sintering, crystallisation and biodegradation behaviour of Bioglass-derived glass–ceramics. Faraday Discuss 136:27–44
75. Li R, Clark AE, Hench LL (1991) An investigation of bioactive glass powders by sol–gel processing. J Appl Biomater 2:231–239
76. Jones JR (2007) Bioactive ceramics and glasses. In: Boccaccini AR, Gough JE (eds) Tissue engineering using ceramics and polymers, 1st edn. Woodhead Publishing Limited CRC Press, Cambridge, pp 52–71
77. Pereira MM, Jones JR, Orefice RL, Hench LL (2005) Preparation of bioactive glass-polyvinyl alcohol hybrid foams by the sol–gel method. J Mater Sci Mater Med 16:1045–1050
78. Gupta R, Kumar A (2008) Bioactive materials for biomedical applications using sol–gel technology. Biomed Mater 3:034005
79. Jie Q, Lin K, Zhong J, Shi Y, Li Q et al (2004) Preparation of macroporous sol–gel bioglass using PVA particles as pore former. Sci Technol 30:49–61
80. Costa HS, Mansur AAP, Barbosa-Stancioli EF, Pereira MM, Mansur HS (2008) Hybrid bioactive glass-polyvinyl alcohol prepared by sol–gel. Mater Sci Forum 587–588:62–64

81. Jones JR (2009) New trends in bioactive scaffolds: the importance of nanostructure. J Eur Ceram Soc 29:1275–1281
82. Ravarian R, Moztarzadeh F, Hashjin MS, Rabiee SM, Khoshakhlagh P et al (2010) Synthesis, characterization and bioactivity investigation of bioglass/hydroxyapatite composite. Ceram Int 36:291–297
83. Mansur HS, Costa HS (2008) Nanostructured poly(vinyl alcohol)/bioactive glass and poly(vinyl alcohol)/chitosan/bioactive glass hybrid scaffolds for biomedical applications. J Chem Eng 137:72–83
84. Mishra R, Basu B, Kumar A (2009) Physical and cytocompatibility properties of bioactive glass-polyvinyl alcohol-sodium alginate biocomposite foams prepared via sol–gel processing for trabecular bone regeneration. J Mater Sci Mater Med 20:2493–2500
85. Hong Z, Reis RL, Mano JF (2008) Preparation and in vitro characterization of scaffolds of poly(l-lactic acid) containing bioactive glass ceramic nanoparticles. Acta Biomater 4:1297–1306
86. Peter M, Sudheesh Kumar PT, Binulal NS, Nair SV, Tamura H et al (2009) Development of novel α-chitin/nanobioactive glass ceramic composite scaffolds for tissue engineering applications. Carbohydr Polym 78:926–931
87. El-Kady AM, Ali AF, Farag MM (2010) Development, characterization, and in vitro bioactivity studies of sol–gel bioactive glass/poly(l-lactide) nanocomposite scaffolds. Mater Sci Eng C 30:120–131
88. Xie E, Hu Y, Chen X, Bai X, Li D et al (2008) In vivo bone regeneration using a novel porous bioactive composite. Appl Surf Sci 255:545–547
89. Renghini C, Komlev V, Fiori F, Verne E, Baino F et al (2009) Micro-CT studies on 3-D bioactive glass–ceramic scaffolds for bone regeneration. Acta Biomater 5:1328–1337
90. Moimas L, Biasotto M, Di Lenarda R, Olivo A, Schmid C (2006) Rabbit pilot study on the resorbability of three-dimensional bioactive glass fibre scaffolds. Acta Biomater 2:191–199
91. Baino F, Verné E, Vitale-Brovarone C (2009) 3-D high-strength glass–ceramic scaffolds containing fluoroapatite for load-bearing bone portions replacement. Math Sci Eng C 29:2055–2062
92. Fu Q, Rahaman MN, Bal BS, Huang W, Day DE (2007) Preparation and bioactive characteristics of a porous 13–93 glass, and fabrication into the articulating surface of a proximal tibia. J Biomed Mater Res Part A 82:222–229
93. Fu Q, Rahaman MN, Bal BS, Brown RF, Day DE (2008) Mechanical and in vitro performance of 13–93 bioactive glass scaffolds prepared by a polymer foam replication technique. Acta Biomater 4:1854–1864
94. Fu Q, Rahaman MN, Bal BS, Brown RF (2010) Preparation and in vitro evaluation of bioactive glass (13–93) scaffolds with oriented microstructures for repair and regeneration of load-bearing bones. J Biomed Mater Res A 93A:1380–1390
95. Liu X, Ramahan MN, Fu Q (2011) Oriented bioactive glass (13–93) scaffolds with controllable pore size by unidirectional freezing of camphene-based suspensions: microstructure and mechanical response. Acta Biomater 7:406–416
96. Brovarone CV, Verne E, Appendino P (2006) Macroporous bioactive glass–ceramic scaffolds for tissue engineering. J Mater Sci Mater Med 17:1069–1078
97. Deb S, Mandegaran R, Di Silvio L (2010) A porous scaffold for bone tissue engineering/ 45S5 Bioglass®derived porous scaffolds for co-culturing osteoblasts and endothelial cells. J Mater Sci Mater Med 21:893–905
98. Bretcanu O, Samaille C, Boccaccini AR (2008) Simple methods to fabricate Bioglass®-derived glass–ceramic scaffolds exhibiting porosity gradient. J Mater Sci 43:4127–4134
99. Brown RF, Day DE, Day TE, Jung S, Rahaman MN et al (2008) Growth and differentiation of osteoblastic cells on 13–93 bioactive glass fibers and scaffolds. Acta Biomater 4:387–396
100. Chen QZ, Rezwan K, Armitage D, Nazhat SN, Boccaccini AR (2006) The surface functionalization of 45S5 Bioglass®-based glass–ceramic scaffolds and its impact on bioactivity. J. Mater Sci Mater Med 17:979–987

101. Chen QZ, Rezwan K, Francon V, Armitage D, Nazhat SN et al (2007) Surface functionalization of Bioglass®-derived porous scaffolds. Acta Biomater 3:551–562
102. Vitale-Brovarone C, Baino F, Verne E (2010) Feasibility and tailoring of bioactive glass-ceramic scaffolds with gradient of porosity for bone grafting. J Biomater Appl 24:693–712
103. Vargas GE, Mesones RV, Bretcanu O, López JMP, Boccaccini AR et al (2009) Biocompatibility and bone mineralization potential of 45S5 Bioglass®-derived glass–ceramic scaffolds in chick embryos. Acta Biomater 5:374–380
104. Nandi SK, Kundu B, Datta S, De DK, Basu D (2009) The repair of segmental bone defects with porous bioglass: an experimental study in goat. Res Vet Sci 86:162–173
105. Mahmood J, Takita H, Ojima Y, Kobayashi M, Kohgo T et al (2001) Geometric effect of matrix upon cell differentiation: BMP-induced osteogenesis using a new bioglass with a feasible structure. J Biochem 129:163–171
106. Mantsos T, Chatzistavrou X, Roether JA, Hupa L, Arstila H et al (2009) Non-crystalline composite tissue engineering scaffolds using boron-containing bioactive glass and poly(D,L-lactic acid) coatings. Biomed Mater 4:55002
107. San Miguel B, Kriauciunas R, Tosatti S, Ehrbar M, Ghayor C et al (2010) Enhanced osteoblastic activity and bone regeneration using surface-modified porous bioactive glass scaffolds. J Biomed Mater Res A 94A:1023–1033
108. Fu Q, Saiz E, Tomsio AP (2011) Bioinspired strong and highly porous glass scaffolds. Adv Funct Mater 21
109. Thompson ID (2005) Biocomposites. In: Hench LL, Jones JR (eds) Biomaterials, artificial organs and tissue engineering, 1st edn. Woodhead Publishing Limited CRC Press, Cambridge, pp 48–58
110. Kanczler JM, Oreffo RO (2008) Osteogenesis and angiogenesis: the potential for engineering bone. Eur Cells Mater 15:100–114
111. Deville S, Saiz E, Nalla RK, Tomsia AP (2006) Freezing as a path to build complex composites. Science 311:515–518
112. Mallik KK (2008) Freeze Casting of Porous Bioactive Glass and Bioceramics. J Am Ceram Soc 92:S85–S94
113. Vitale-Brovarone C, Baino F, Verné E (2009) High strength bioactive glass–ceramic scaffolds for bone regeneration. J Mater Sci Mater Med 20:643–653
114. Hernandez CJ, Beaupre GS, Keller TS, Carter DR (2001) The influence of bone volume fraction and ash fraction on bone strength and modulus. Bone 29:74–78
115. Keaveny TM, Morgan EF, Niebur GL, Yeh OC (2001) Biomechanics of trabecular bone. Annu Rev Biomed Eng 3:307–333
116. Sun SS, Ma HL, Liu CL, Huang CH, Cheng CK et al (2008) Difference in femoral head and neck material properties between osteoarthritis and osteoporosis. Clin Biomech 1(Suppl 23):S39–S47
117. Nazarian A, von Stechow D, Zurakowski D, Muller R, Snyder BD (2008) Bone volume fraction explains the variation in strength and stiffness of cancellous bone affected by metastatic cancer and osteoporosis. Calcif Tissue Int 83:368–379
118. Hench LL, Wilson J (1993) Introduction. In: Hench LL, Wilson J (eds) An introduction to bioceramics, 1st edn. World Scientific Publishing, Singapore, pp 1–24
119. Clupper DC, Gough JE, Embanga PM, Notingher I, Hench LL et al (2004) Bioactive evaluation of 45S5 bioactive glass fibres and preliminary study of human osteoblast attachment. J Mater Sci Mater Med 15:803–838
120. Clupper DC, Hench LL, Mecholsky JJ (2004) Strength and toughness of tape cast bioactive glass 45S5 following heat treatment. J Eur Ceram Soc J 24:2929–2934
121. Frost HM (1998) Could some biomechanical effects of growth hormone help to explain its effects on bone formation and resorption? Bone 23:395–398
122. Meyers MA, Chen P-Y, Lin AY-M, Seki Y (2008) Biological materials: Structure and mechanical properties. Prog Mater Sci 53:1–206
123. Webster TJ, Ahn ES (2007) Nanostructured biomaterials for tissue engineering bone tissue engineering II. Springer, Berlin, pp 275–308

124. Zhang X, Fu H, Liu X, Yao A, Wang D et al (2009) In vitro bioactivity and cytocompatibility of porous scaffolds of bioactive borosilicate glasses. Chin Sci Bull 54:3181–3186
125. O'Flaherty EJ (1991) Physiologically based models for bone-seeking elements : I. Rat skeletal and bone growth. Toxicol Appl Pharmacol 111:299–312
126. O'Flaherty EJ (1991) Physiologically based models for bone-seeking elements : III. Human skeletal and bone growth. Toxicol Appl Pharmacol 111:332–341
127. Woźniak P, El Haj AJ (2007) Bone regeneration and repair using tissue engineering. In: Boccaccini AR, Gough JE (eds) Tissue engineering using ceramics and polymers, 1st edn. Woodhead Publishing Limited CRC Press, Cambridge, pp 294–318
128. Gerhardt LC, Boccaccini AR (2010) Bioactive glass and glass–ceramic scaffolds for bone tissue engineering. Materials 3:3867–3910
129. Gibson LJ (2005) Biomechanics of cellular solids. J Biomech 38:377–399
130. Jacobs CR (1994) Numerical simulation of bone adaptation to mechanical loading, Ph.D. Dissertation, Stanford University, California
131. Diego RB, Estelles JM, Sanz JA, Garcia-Aznar JM, Sanchez MS (2007) Polymer scaffolds with interconnected spherical pores and controlled architecture for tissue engineering: fabrication, mechanical properties, and finite element modeling. J Mater Res Part B Appl Biomater 81:448–455
132. Sanz-Herrera JA, Garcia-Aznar JM, Doblare M (2008) A mathematical model for bone tissue regeneration inside a specific type of scaffold. Biomech Model Mechanobiol 7:355–366
133. Müller R (2003) Bone microarchitecture assessment: current and future trends. Osteoporosis Int 5(Suppl 14):S89–S95
134. Müller R, Bosch T, Jarak D, Stauber M, Nazarian A, et al (2002) Micro-mechanical evaluation of bone microstructures under load. In: Bonse U (ed) Developments in X-Ray tomography III, pp 189–200
135. Beaupied H, Lespessailles E, Benhamou CL (2007) Evaluation of macrostructural bone biomechanics. Joint Bone Spine 74:233–239
136. Vrouwenvelder WC, Groot CG, Groot K (1995) Preliminary ageing study of bioactive glass in a cell culture model. J Mater Sci Mater Med 6:144–149
137. Wang S, Falk MM, Rashad A, Saad MM, Marques AC, et al (2011) Evaluation of 3D nano-macro porous bioactive glass scaffold for hard tissue engineering. J Mater Sci Mater Med 22:1195–1203
138. Fu Q, Ramahan MN, Bal BS, Kuroki K, Brown RF (2010) In vivo evaluation of 13–93 bioactive glass scaffolds with trabecular and oriented microstructures in a subcutaneous rat implantation model. Biomed Mater Res A 95A:235–244
139. Navarro M, Aparicio C, Charles-Harris M, Ginebra MP, Engel E et al (2006) Development of a biodegradable composite scaffold for bone tissue engineering: physicochemical topographical, mechanical, degradation, and biological properties. Ordered polymeric nanostructures at surfaces. Springer, Berlin, pp 209–231
140. Yunos DM, Bretcanu O, Boccaccini AR (2008) Polymer-bioceramic composites for tissue engineering scaffolds. J Mater Sci 43:4433–4442
141. Sanz-Herrera JA, Doblaré M, García-Aznar JM (2010) Scaffold microarchitecture determines internal bone directional growth structure: a numerical study. J Biomech 43:2480–2486
142. Matthews FL, Rawlings RD (eds) (1994) Composite materials: engineering and science, Woodhead Publishing Limited, CRC Press, Cambridge
143. Blaker JJ, Bismarck A, Boccaccini AR, Young AM, Nazhat SN (2010) Premature degradation of poly(α-hydroxyesters) during thermal processing of Bioglass®-containing composites. Acta Biomater 6:756–762
144. Blaker JJ, Maquet V, Jerome R, Boccaccini AR, Nazhat SN (2005) Mechanical properties of highly porous PDLLA/Bioglass®composite foams as scaffolds for bone tissue engineering. Acta Biomater 1:643–652

145. Meretoja VV, Helminen AO, Korventausta JJ, Haapa-aho V, Seppala JV et al (2006) Crosslinked poly(ε-caprolactone/D,L-lactide)/bioactive glass composite scaffolds for bone tissue engineering. J Biomed Mater Res A 77:261–268
146. Meretoja VV, Malin M, Seppala JV, Narhi TO (2009) Osteoblast response to continuous phase macroporous scaffolds under static and dynamic culture conditions. J Biomed Mater Res A 89:317–325
147. Yao J, Radin S, SL P, Ducheyne P (2005) The effect of bioactive glass content on synthesis and bioactivity of composite poly (lactic-co-glycolic acid)/bioactive glass substrate for tissue engineering. Biomaterials 26:1935–1943
148. Keshaw H, Georgiou G, Blaker JJ, Forbes A, Knowles JC et al (2009) Assessment of polymer/bioactive glass-composite microporous spheres for tissue regeneration applications. Tissue Eng A 15:1451–1461
149. El-Kady AM, Saad EA, El-Hady BMA, Farag MM (2010) Synthesis of silicate glass/poly (L-lactide) composite scaffolds by freeze-extraction technique: characterization and in vitro bioactivity evaluation. Ceram Int 36:995–1009
150. Tsigkou O, Hench LL, Boccaccini AR, Polak JM, Stevens MM (2007) Enhanced differentiation and mineralization of human fetal osteoblasts on PDLLA containing Bioglass®composite films in the absence of osteogenic supplements. J Biomed Mater Res A 80A:837–851
151. Maquet V, Boccaccini AR, Pravata L, Notingher I, Jerome R (2004) Porous poly(α-hydroxyacid)/Bioglass composite scaffolds for bone tissue engineering. I: preparation and in vitro characterisation. Biomaterials 25:4185–4194
152. Maquet V, Boccaccini AR, Pravata L, Notingher I, Jerome R (2003) Preparation, characterization, and in vitro degradation of bioresorbable and bioactive composites based on Bioglass-filled polylactide foams. J Biomed Mater Res A 66:335–346
153. Verrier S, Blaker JJ, Maquet V, Hench LL, Boccaccini AR (2004) PDLLA/Bioglass®composites for soft-tissue and hard-tissue engineering: an in vitro cell biology assessment. Biomaterials 25:3013–3021
154. Misra SK, Ohashi F, Valappil SP, Knowles JC, Roy I et al (2010) Characterization of carbon nanotube (MWCNT) containing P(3HB)/bioactive glass composites for tissue engineering applications. Acta Biomater 6:735–742
155. Misra SK, Philip SE, Chrzanowski W, Nazhat SN, Roy I et al (2009) Incorporation of vitamin E in poly(3-hydroxybutyrate)/Bioglass composite films: effect on surface properties and cell attachment. J R Soc Interface 6:401–409
156. Helen W, Gough JE (2008) Cell viability, proliferation and extracellular matrix production of human annulus fibrosus cells cultured within PDLLA/Bioglass®composite foam scaffolds in vitro. Acta Biomater 4:230–243
157. Helen W, Merry CLR, Blaker JJ, Gough JE (2007) Three-dimensional culture of annulus fibrosus cells within PDLLA/Bioglass®composite foam scaffolds: assessment of cell attachment, proliferation and extracellular matrix production. Biomaterials 28:2010–2020
158. Yang XB, Webb D, Blaker J, Boccaccini AR, Maquet V et al (2006) Evaluation of human bone marrow stromal cell growth on biodegradable polymer/Bioglass®composites. Biochem Biophys Res Commun 342:1098–1107
159. Barroca N, Daniel-da-Silva AL, Vilarinho PM, Fernandes MHV (2010) Tailoring the morphology of high molecular weight PLLA scaffolds through bioglass additions. Acta Biomater 6:3611–3620
160. Ishai O, Cohen LJ (1967) Elastic properties of filled and porous epoxy composites. Int J Mech Sci 9:539–546
161. Gibson LJ, Ashby MF (1997) Cellular solids: structure and properties. Cambridge University Press, Cambridge
162. Lu HH, Tang A, Oh SC, Spalazzi JP, Dionisio K (2005) Compositional effects on the formation of a calcium phosphate layer and the response of osteoblast-like cells on polymer-bioactive glass composites. Biomaterials 26:6323–6334

163. Widdows K, Kingdom JCP, Ansari T (2009) Double immuno-labelling of proliferating villous cytotrophoblasts in thick paraffin sections: integrating immuno-histochemistry and stereology in the human placenta. Placenta 30:735–738
164. Gubler M, Brunner TJ, Zehnder M, Waltimo T, Sener B et al (2008) Do bioactive glasses convey a disinfecting mechanism beyond a mere increase in pH? Int Endodont J 41:670–678
165. Waltimo T, Brunner TJ, Vollenweider M, Stark WJ, Zehnder M (2007) Antimicrobial effect of nanometric bioactive glass 45S5. J Dent Res 86:754–757
166. Waltimo T, Mohn D, Paque F, Brunner TJ, Stark WJ et al (2009) Fine-tuning of bioactive glass for root canal disinfection. J Dent Res 88:235–238
167. Keshaw H, Forbes A, Day RM (2005) Release of angiogenic growth factors from cells encapsulated in alginate beads with bioactive glass. Biomaterials 26:4171–4179
168. Moosvi SR, Day RM (2009) Bioactive glass modulation of intestinal epithelial cell restitution. Acta Biomater 5:76–83
169. Day RM, Boccaccini AR, Shurey S, Roether JA, Forbes A et al (2004) Assessment of polyglycolic acid mesh and bioactive glass for soft-tissue engineering scaffolds. Biomaterials 25:5857–5866
170. Leu A, Stieger SM, Dayton P, Ferrara KW, Leach JK (2009) Angiogenic response to bioactive glass promotes bone healing in an irradiated calvarial defect. Tissue Eng A 15:877–885
171. Andrade AL, Andrade SP, Domingues RZ (2006) In vivo performance of a sol–gel glass-coated collagen. J Biomed Mater Res Part B Appl Biomater 79B:122–128
172. Ross EA, Batich CD, Clapp WL, Sallustio JE, Lee NC (2003) Tissue adhesion to bioactive glass-coated silicone tubing in a rat model of peritoneal dialysis catheters and catheter tunnels. Kidney Int 63:702–708
173. Choi HY, Lee JE, Park HJ, Oum BS (2006) Effect of synthetic bone glass particulate on the fibrovascularization of porous polyethylene orbital implants. Ophthal Plast Reconstr Surg 22:121–125
174. Seed MP, Walsh DA (2008) Angiogenesis in inflammation: mechanism and clinical correlates. Progress in Inflammation Research. Birkhauser, Basel. 180 pp
175. Santos MI, Reis RL (2010) Vascularization in bone tissue engineering: Physiology, current strategies, major hurdles and future challenges. Macromol Biosci 10:12–27
176. Chapanian R, Amsden BG (2010) Combined and s^{eq}uential delivery of bioactive VEGF165 and HGF from poly(trimethylene carbonate) based photo-cross-linked elastomers. J Control Release 143:53–63
177. Chiu LLY, Radisic M (2010) Scaffolds with covalently immobilized VEGF and Angiopoietin-1 for vascularization of engineered tissues. Biomaterials 31:226–241
178. Leach JK, Kaigler D, Wang Z, Krebsbach PH, Mooney DJ (2006) Coating of VEGF-releasing scaffolds with bioactive glass for angiogenesis and bone regeneration. Biomaterials 27:3249–3255
179. Patel ZS, Young S, Tabata Y, Jansen JA, Wong MEK et al (2008) Dual delivery of an angiogenic and an osteogenic growth factor for bone regeneration in a critical size defect model. Bone 43:931–940
180. Briganti E, Spiller D, Mirtelli C, Kull S, Counoupas C et al (2010) A composite fibrin-based scaffold for controlled delivery of bioactive pro-angiogenetic growth factors. J Control Release 142:14–21
181. Kaigler D, Wang Z, Horger K, Mooney DJ, Krebsbach PH (2006) VEGF scaffolds enhance angiogenesis and bone regeneration in irradiated osseous defects. J Bone Miner Res 21:735–744
182. Rosengren Å, Oscarsson S, Mazzocchi M, Krajewski A, Ravaglioli A (2003) Protein adsorption onto two bioactive glass–ceramics. Biomaterials 24:147–155
183. Goller G (2004) The effect of bond coat on mechanical properties of plasma sprayed bioglass–titanium coatings. Ceram Int 30:351–355

184. Guo HB, Miao X, Chen Y, Cheang P, Khor KA (2004) Characterization of hydroxyapatite- and bioglass-316L fibre composites prepared by spark plasma sintering. Mater Lett 58:304–307
185. Verné E, Ferraris S, Vitale-Brovarone C, Spriano S, Bianchi CL et al (2010) Alkaline phosphatase grafting on bioactive glasses and glass ceramics. Acta Biomater 6:229–240
186. Annala T, Kellomäki M (2009) Sterilization of biodegradable polymers. In: Wuisman PIJM, Smit TH (eds) Degradable polymers for skeletal implants, 1st edn. Nova Science Publishers, Hauppauge, pp 123–138
187. Cordewener FW, van Geffen MF, Joziasse CAP, Schmitz JP, Bos RRM et al (2000) Cytotoxicity of poly(96L/4D-lactide): the influence of degradation and sterilization. Biomaterials 21:2433–2442
188. Janorkar AV, Metters AT, Hirt DE (2007) Degradation of Poly(L-Lactide) films under ultraviolet-induced photografting and sterilization conditions. J Appl Poly Sci 106:1042–1047
189. Jukes JM, van Blitterswijk CA, de Boer J (2010) Skeletal tissue engineering using embryonic stem cells. J Tissue Eng Regen Med 4:165–180
190. Oh CH, Hong SJ, Jeong I, Yu HS, Jegal SH, et al (2009) Development of robotic dispensed bioactive scaffolds and human adipose-derived stem cell culturing for bone tissue engineering. Tissue Eng C Methods
191. Zhang H, Ye XJ, Li JS (2009) Preparation and biocompatibility evaluation of apatite/wollastonite-derived porous bioactive glass ceramic scaffolds. Biomed Mater 4:45007
192. Harrison BS, Atala A (2007) Carbon nanotube applications for tissue engineering. Biomaterials 28:344–353
193. Boccaccini AR, Gerhardt LC (2010) Carbon nanotube composite scaffolds and coatings for tissue engineering applications. Trans Tech Publications, Zurich, pp 31–52.(Key Eng Mat Adv Bioceram Nanomed Tissue Eng)
194. Krug HF, Wick P (2011) Nanotoxikologie: Eine interdisziplinäre Herausforderung. Angewandte Chemie 123:1294–1314

Adv Biochem Engin/Biotechnol (2012) 126: 227–262
DOI: 10.1007/10_2011_118
© Springer-Verlag Berlin Heidelberg 2011
Published Online: 6 October 2011

Microenvironment Design for Stem Cell Fate Determination

Tali Re'em and Smadar Cohen

Abstract Stem cells are characterized by their dual ability for self-renewal and differentiation, potentially yielding large numbers of cells that can be used in cell therapy and tissue engineering for repairing devastating diseases. Attaining control over stem cell fate decision in culture is a great challenge since these cells integrate a complex array of "niche" signals, which regulate their fate. Given this, the recent findings that synthetic microenvironments can be designed to gain some level of control over stem cell fate are encouraging. This chapter provides an overview of the current state and knowledge of the design of synthetic microenvironments bio-inspired by the adult stem cell niche. We describe the biomaterials used for reconstituting the niche, highlighting the bioengineering principles used in the process. Such synthetic microenvironments constitute powerful tools for elucidating stem cell regulatory mechanisms that should fuel the development of advanced culture systems with accurate regulation of stem cell fate.

Keywords Stem cell · Microenvironment · ECM

Contents

1 Introduction..	228
1.1 Stem Cell Types and Their Microenvironments	228
2 Microenvironmental Soluble Factors...	232
2.1 Natural Factors..	232
2.2 Synthetic Factors	233
3 Synthetic Extracellular Matrix..	234
3.1 First Generation Materials as Artificial ECM	235
3.2 Bio-Inspired Materials and Scaffolds	237
4 Bioreactors for Dynamic Stem Cell Microenvironment...	251
5 Conclusions and Future Directions..	252
References	253

T. Re'em · S. Cohen (✉)
The Avram and Stella Goldstein-Goren Department of Biotechnology Engineering,
Ben-Gurion University of the Negev, 84105, Beer-Sheva Israel
e-mail: scohen@bgu.ac.il

1 Introduction

Stem cells possess tremendous clinical potential for use in cell-based therapeutic strategies, due to their capacity for expansion in culture and differentiation to specific cell lineages. Stem cells derived from bone marrow have been proven to be safe for use in human patients and have exhibited promising therapeutic effects in several diseases. Human embryonic stem cells are currently being investigated in phase I clinical studies for treating patients after spinal cord injury. Considering the enormous clinical potential of stem cells, extensive research efforts have been directed to find methods of better controlling cell fate and differentiation. One attractive approach involves microenvironmental design for stem cell cultivation.

The stem cell microenvironment can be broadly defined as the collection of all the surrounding external signals to which stem cells are exposed (Fig. 1). This would include the architectural space, physical engagement of the cell membrane with tethering molecules on neighboring cells or surfaces, signaling interactions at the interface of stem cells and other microenvironmental cells, paracrine and endocrine signals from local or distant sources, neural input and metabolic products of tissue activity [1, 2]. The interplay between all these components defines the complex and interactive microenvironment and determines stem cell fate, toward self-renewal or lineage-specific differentiation.

Effective replication of the stem cell microenvironment has been hypothesized to allow stem cells to reach their regenerative potential [3]. However, it is still a great challenge to engineer the multi-dimensional environment with the right context of all cues converging to lead the cells into one specific fate. Moreover, there is much we do not know about the natural cell microenvironment. Nevertheless, recent studies have demonstrated that artificial microenvironments, designed by implementing some bio-inspired components of the natural stem cell niche, are capable of controlling stem cell behavior.

Our goal in this review is to provide an overview of the current state and knowledge of the design of synthetic microenvironments, aimed to control stem cell fate determination. Specifically, we will describe the research conducted on three major components of the microenvironment: (1) the soluble factors (growth factors, cytokines and chemokines), (2) the insoluble matrix (extracellular matrix [ECM]) and (3) the mechanical stimuli to which the cells are exposed.

1.1 Stem Cell Types and Their Microenvironments

1.1.1 Bone-Marrow-Derived Stem Cells

The bone marrow (BM) environment, also termed "niche", has been studied since the late 1970s and is the best characterized stem cell microenvironment.

Microenvironment Design for Stem Cell Fate Determination 229

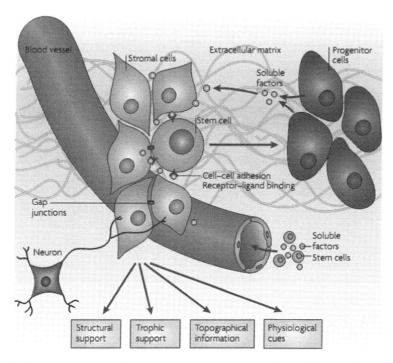

Fig. 1 Elements of the local microenvironment that participate in regulating the function of a stem cell in its niche, including soluble factors, extracellular matrix, architectural space, biophysical forces and intercellular interactions (reprinted by permission from [5])

Two major adult stem cell populations reside within the BM niche: the hematopoietic stem cells (HSCs) and the mesenchymal stem cells (MSCs).

HSCs, which regenerate the entire blood and immune system, have been used in BM transplantation to treat blood diseases (e.g., aplastic anemia, thalassemia and Gaucher's disease) and to reconstitute the patient's hematopoietic system after undergoing chemotherapy or radiotherapy to treat cancers. These stem cells seem to have little function outside their specific anatomic locations. It is the specific cues from neighboring cells, i.e., the osteoblasts and the stromal cells, that allow stem cells to persist and to change in number and fate.

The clinical need for ex vivo expansion of HSCs has motivated research into identifying the main BM niche regulators and unraveling the molecular pathways that mediate the HSC niche function (Fig. 2). Osteoblasts were shown to be of great importance for controlling the HSC niche size [4], but other cell types, like osteoclasts, endothelial cells and mesenchymal progenitors, have also been involved in modulating HSC fate (reviewed in [5]). In addition to the cellular elements, several signaling molecules, including osteopontin [6], trombopoietin [7] and N-cadherin [8], as well as other factors, like increased extracellular calcium-ion concentration [9] and low oxygen concentration [10, 11], have been shown to affect HSC behavior and to support the establishment of the niche

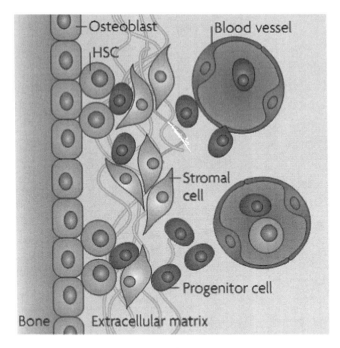

Fig. 2 Schematic model of the bone marrow stem cell microenvironment, also termed "bone marrow niche" (reprinted by permission from [5])

environment. Experimental evidence suggests that specific signaling pathways, such as Notch and Wnt, are activated in the BM and may have a role in HSC regulation [12].

Despite the success in identifying some of the components and signals of the HSC niche, a complete understanding of the mechanisms involved in HSC regulation still remains elusive, also due to the limited availability of appropriate ex vivo models which could mimic the complex niche organization. Cell culture systems developed so far do not entirely reproduce the physiological signals required to establish a functional niche structure, resulting in a limited efficiency in maintaining long-term repopulation of HSCs in vitro. Nevertheless, recent studies indicate the potential of three-dimensional (3D) scaffold-based perfusion systems as a suitable cultivation model for reconstructing ex vivo the BM stem cell niche, supporting HSC long-term expansion [13].

The bone marrow microenvironment houses an additional stem cell population, the BM-derived MSCs. These multipotent cells retain differentiation capabilities along a number of different mesenchymal lineages, including bone, cartilage, fat, tendon and muscle [14, 15]. Their differentiation has been shown to be manipulated exclusively in response to specific culture conditions; hence, in theory, they may serve as an ideal alternative to fully differentiated mesodermal cells [14]. Moreover, recent studies have indicated that these cells not only differentiate into mesodermal cells, but can also adopt the fate of endodermal and ectodermal cell types [16]. Some striking examples of the therapeutic use of BM-derived MSCs have demonstrated their great potential in a broad spectrum of clinical applications, including cardiovascular repair [17], spinal cord injury [18] and bone [19] and cartilage repair [20–23].

In addition to their wide differentiation potential, MSC utilization for cell therapy possesses several more advantages. First, these cells are easily obtained from bone marrow aspirates and can be simply isolated from HSCs due to their tendency to adhere to tissue culture plastic [15, 23]. Second, although these cells represent a very minor fraction of the total nucleated cell population in marrow, a very large cell population may be achieved in a very short term due to their relatively high expansion rate using standard cell culture techniques [22]. Third, MSCs are available throughout an entire human lifetime and their utilization would help to alleviate the problems of immune rejection and disease transmission associated with the use of allografts [20].

MSCs are also known to reside in other adult tissues, such as adipose tissue [24], yet they are most abundant in the BM [14]. Although stem cell niches should by concept exist in all organs and tissues, little information exists on the nature and the mechanisms that control these niches; and the existence of an adult stem cell pool in some tissues (such as in pancreatic islets) is still under debate [25, 26].

1.1.2 Embryonic and Induced Pluripotent Stem Cells

Whereas adult stem cells are mainly restricted to certain lineages, embryonic stem cells (ESCs) can differentiate into almost any cell type. The derivation of human ESCs (hESCs) from the inner cell mass of developing blastocysts in 1998 [27, 28] has opened visionary possibilities of cell therapy for almost any diseased organ.

Advances in stem cell biology have also enabled the generation of induced pluripotent stem cells (iPSCs), following reprogramming of adult somatic cells [29, 30]. This has opened the possibility of generating cells from the intended recipient of the cell therapy [31], supposedly solving the immune-rejection issue related to ESC use, although this concept is now under debate [32]. Moreover, the reprogrammed cells are not subjected to the same constraints of senescence that a more mature cell population encounters, such as the "Hayflick limit", and the generation of large numbers of pluripotent cells is therefore made possible. Since iPSCs mimic the ESC cell population and do not have their own defined microenvironment, we will refer only to the ESC microenvironment.

The microenvironment surrounding the developing embryo presents a number of spatially and temporally instructive biochemical cues within a complex and interactive milieu that guides and governs the sequential development and cell fate decisions during embryogenesis. However, unlike the relatively known adult stem cell microenvironment, the one surrounding the developing embryo is relatively obscure. This has not prevented researchers from testing the use of synthetic microenvironment analogs.

Gerecht et al. [34] were the first to demonstrate the generation of human embryoid bodies (hEBs) directly from hESC suspensions within 3D porous alginate scaffolds. The confining environments of the alginate scaffold pores enabled efficient formation of hEBs with a relatively high degree of cell proliferation and differentiation; encouraged round, small-sized hEBs; and induced vasculogenesis in the forming hEBs to a greater extent than the conventional EB formation investigated in cell suspensions in static or in rotating cultures [33]. In another study, Gerecht et al. [34] used hydrogels of hyaluronic acid (HA), a constituent of the ECM present during early stages of embryogenesis, to develop a synthetic microenvironment that maintained hESCs in their undifferentiated state, preserved their normal karyotype, and maintained their full differentiation capacity as indicated by embryoid body formation.

Since the pluripotent stem cell microenvironment has no definitive aspect, most of the research in the field is still focused on the design of adult stem cell microenvironments, subsequently applying the principal elements of these microenvironments to pluripotent stem cell research as well.

2 Microenvironmental Soluble Factors

2.1 Natural Factors

Soluble factors and cytokines were the first microenvironmental elements shown to be essential for the support of stem cell proliferation, differentiation and survival of distinct cell populations [35]. Due to their ease of manipulation in culture, most of the pioneer studies on stem cell expansion and differentiation used secreted factors as the principal induction factors. Thus, soluble factors are the best-characterized environmental signals impacting stem cell behavior. The factors exert their effect through activation of specific signaling pathways. In some cases, the role of the factors is mainly in specifying stem cell self-renewal and mediating cell survival, as Wnt proteins function in the HSC BM niche [5]. In other cases, the secreted factors play an instructive role in cell differentiation [19, 36–39]. It should be noticed that one factor may have an effect on a variety of signaling pathways, and the eventual cellular response is context-dependent. For example, the bone morphogenetic protein (BMP) signaling pathway as well as Wnt/beta-catenin signaling were shown to be essential for ESC self-renewal [40, 41]. However, various BMP proteins are also useful for chondrogenic and osteogenic differentiation of MSCs [19, 36, 37, 42, 43]. Specifically, the BMPs are known to activate the osteo-inductive Smad1/5/8 pathway. In our study, BMP 2 and 4 were shown for the first time to also activate the Janus-activated kinase (JAK)-STAT pathway in human MSCs (hMSCs). The activated STAT3 pathway in turn was shown to negatively modulate hMSC osteogenic differentiation (Fig. 3).

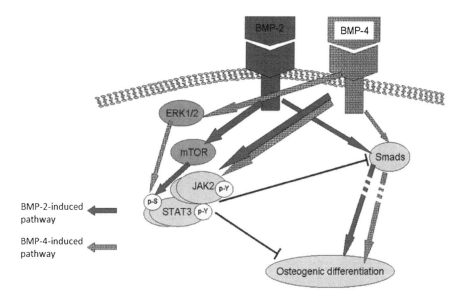

Fig. 3 Schematic model of the BMP functions in hMSC osteogenic differentiation. In addition to the BMP-activated osteo-inductive Smad pathway, BMP2 and BMP4 also activate the JAK-STAT pathway in hMSC through mTOR and ERK1/2 cascades, respectively. The STAT pathway serves as a negative modulator for BMP-induced hMSC osteogenic differentiation [42]

Compelling evidence now reveals that stem cell behavior can be fine-tuned by varying the concentration and combination of the factors of interest. Moreover, the spatio-temporal context is of great importance, influencing stem cell fate determination, as further discussed below.

2.2 Synthetic Factors

Synthetic small molecules targeting specific signaling pathways were shown to be useful in manipulating stem cell fate, by sustaining their pluripotency or inducing differentiation [44]. For instance, a small molecule named *stauprimide* was shown to increase the efficiency of the directed differentiation of mouse and human ESCs into endodermal lineages, as pancreatic or hepatic, in synergy with defined extracellular signaling cues [45]. In another study, the molecules *IDE1 and IDE 2* were also shown to direct endodermal differentiation [46] of mouse and human ESCs. Additionally, *(–)-indolactam V* is a small molecule that directs differentiation of hESCs into the pancreatic lineage [47].

Small molecules have been shown to replace transcription factors and enhance efficiency during somatic cell reprogramming (reviewed in [44]). Small molecules clearly offer some distinct advantages over genetic manipulation. For example, in contrast to genetic manipulation, the effects of small molecules are typically fast and reversible, providing more precise temporal regulation of protein function.

Synthetic molecules have been playing increasingly important roles in both elucidating the fundamental biology of stem cells and facilitating the development of therapeutic approaches to regenerative medicine. Such approaches will involve therapies using homogenous functional cells produced under chemically-defined conditions in vitro. The almost unlimited structural and functional diversity endowed by synthetic chemistry provides small molecules with unbounded potential for precisely controlling molecular interactions and/or recognition, a feature that can be extensively explored by design and screening.

In most cases, the signaling molecule, either natural or synthetic, is simply added to the growth medium of the cells. However, current stem cell research is increasingly oriented towards spatio-temporal presentation of signaling molecules, in similar fashion to that of the native ECM.

3 Synthetic Extracellular Matrix

The insoluble matrix, i.e., the ECM, is an important constituent of the microenvironment, affecting stem cell fate. Much attention has been given in previous years to the design of matrices able to promote the appropriate spatial cell arrangement. In addition to the obvious advantage of a high surface area per volume ratio compared to monolayer systems, culturing in 3D matrices also enhances cell–cell and cell–matrix interactions and allows better cell distribution [48]. Specifically, the cultivation of hESC in scaffolds was shown to result in an appropriate cell differentiation and neo-tissue formation, including blood vessels, and an integration with the host upon implantation [49], in contrast to the incomplete differentiation observed in two-dimensional (2D) studies.

Choosing the optimal scaffold requires studying stem cell interactions on both the molecular and cellular levels, methods of scaffold fabrication and optimization of scaffold properties to the specific needs of the stem cell population. The scaffold is required to be biocompatible, biodegradable over an appropriate time scale, and highly porous with large interconnected pores to provide efficient mass transport, cell permeation and interstitial fluid flow [20]. The scaffold temporarily provides the physical support for the seeded cells in culture until they produce their own ECM. Therefore, the matrix should be mechanically stable and suitable for cultivation either in a bioreactor or at the implant site [50]. However, the matrix should be also flexible enough to allow cell reorganization into a 3D tissue and its subsequent integration with the host tissues [51].

The ideal scaffold for stem cell applications is also required to introduce the stem cell to the right biochemical cues in a spatio-temporal fashion, similar to that of the native ECM. Preferably, the scaffold should dynamically interact with the cells and be adaptable to various cellular changes in culture.

3.1 First Generation Materials as Artificial ECM

The first generation materials used as artificial ECM consisted of previously known materials, used for applications such as drug or cell delivery systems. The natural materials extensively in use include ECM components and derivatives, such as collagen [20, 52, 53], fibrin [54, 55], gelatin [56–59], Matrigel [53, 60], hyaluronic acid (HA) [34, 61–63] and materials derived from plants and seaweeds, such as agarose [21, 64] and alginate [33, 51, 64–89]. Since collagen is a major component in native ECM and cells interact with collagen through integrin binding-mediated interactions, 3D collagen gels have been widely used for stem cell encapsulation, including MSCs [20] and ESCs [52, 53]. Gelatin is a product of collagen denaturation and as a porous scaffold it has been used for stem-cell-based tissue engineering applications due to its biocompatibility and lack of antigenicity [56–59]. Matrigel, a basement membrane matrix, extracted from Engelbreth–Holm–Swarm (EHS) mouse sarcoma cells, is rich in ECM-derived molecules and has been investigated extensively for the culture of stem cells. The complexity and derivation from natural tissues has motivated its use in cultures, particularly for ESCs, due to its mimicking of natural structures [53, 60].

Polysaccharides have been also used to form matrices. Hyaluronic acid (HA) is a polysaccharide found in many tissues and has been modified to form photo-polymerized hydrogels with controlled properties that allow for the encapsulation of viable cells. Cells may interact with HA through receptor binding, primarily CD44, and HA is degraded by hyaluronidases. Therefore, HA-based biomaterials have been utilized to regulate stem cell chondrogenic and osteogenic differentiation [62, 63]. Since it is one constituent of the ECM present during the early stages of embryogenesis, HA was used for the development of microenvironments that inhibited hESC differentiation [34].

Alginate is a seaweed-derived polysaccharide that forms hydrogels through ionic cross-linking. Although there are no direct cellular interactions, alginate forms stable hydrogels that become soluble through the dissociation of the crosslinks in the network due to the exchange of calcium by sodium ions. These hydrogels have been extensively used the encapsulation of stem cells for a variety of applications both in vitro and in vivo [67, 90, 91]. Moreover, alginate has also been used in the form of macro-porous scaffolds for the cultivation and differentiation of both ESCs [33] and MSCs [81].

Natural materials present some challenges. Since they usually are not well-defined and have a lot-to-lot variability, control over matrix mechanical properties and degradation rates is limited [92]. Moreover, naturally derived materials may provoke immune responses or harbor microbes or viruses [93]. Furthermore, such materials are often difficult to process without disrupting a potentially important hierarchical structure. In addition, hydrogels formed from natural materials generally have poor mechanical properties.

Synthetic materials are being widely investigated for stem cell culture. The wide diversity of matrix properties is tailored with respect to mechanics, chemistry and degradation. Both non-degradable and materials that degrade through either hydrolytic or enzymatic mechanisms have been synthesized [94]. Additionally, the processing of synthetic materials into desired structures may be much simpler than with natural materials. However, potential limitations in the use of synthetic materials include toxicity and a limited repertoire of cellular interactions, unless they are modified with adhesion peptides or designed to release biological molecules.

Due to their biocompatibility and use in medicine, poly(α-hydroxyesters) have been used extensively in the field of tissue engineering and for stem cell cultivation. Among the most popular are scaffolds made from poly(glycolic acid) (PGA), poly(lactic-acid) (PLA) and their co-polymer, poly(lactic-co-glycolic acid) (PLGA) [36, 95, 96]. For example, a scaffold made of a blend of PLGA/PLLA was seeded with hESCs and differentiation was induced through incorporation of the appropriate growth factors in culture media [98].

Poly(ethylene glycol) (PEG) hydrogels are popular matrices for encapsulating stem cells [96, 98–102]. PEG hydrogels are elastic, biocompatible and can be tailored to possess mechanical properties similar to various natural tissue types. But these nonionic and covalently cross-linked networks are very different from the self-assembled polyelectrolytes that comprise the bulk of natural ECMs. The PEG hydrogels are inherently hydrophilic, so protein adsorption and cell interactions are minimal. Thus, PEG hydrogels are often modified with tethered groups, such as adhesion peptides, to enable cellular interactions [102–105].

In addition to the variation existing in chemistry of the synthetic ECM, materials have been processed in different formats and porosity; the most common are:

- Macro-porous scaffolds—prepared by cross-linking (chemical or physical) of a biomaterial solution into the desired shape, with a subsequent solidification step and/or drying/freeze-drying. The porous form with its interconnected pores provides efficient mass transport, cell permeation and interstitial fluid flow. Such porous structures allow tissue growth in vitro or in-growth in situ [33, 70, 72, 74, 76, 77, 81, 84, 88, 105–109].
- Hydrogels—3D networks of hydrophilic polymers that absorb a large quantity of water as well as biological fluids, prepared by physical or covalent cross-linking of the polymer. Due to their aqueous nature, hydrogels simulate the hydrated structural aspect of native ECMs. 3D cell encapsulation can be achieved through in-situ formation of materials around stem cells [82, 83, 86, 104, 110, 111].
- Nanofibrous scaffolds—prepared by self-assembly of amphiphilic peptides or by electro-spinning. The nanofibrous structure mimics the cell environment, comprised of complex network of ECM molecules with nano–micro scale dimensions [112–115].

3.2 Bio-Inspired Materials and Scaffolds

Moving from the classical application of biomaterials as cell delivery vehicles, to attempts at replicating the microenvironmental cell niche in order to regulate stem cell fate, has required the re-design of biomaterials and scaffolds. The re-design was bio-inspired by the roles of the ECM, including adherence, migration, mechano-signals, growth factor presentation and cell differentiation [116]. The cues from the ECM, a complex network of collagen fibers, multi-adhesive matrix proteins and proteoglycans, have significant impacts on development during embryonic, fetal and neonatal stages as well as in adult tissues. The ECM network organizes cells into tissues and regulates cell growth, proliferation and function by signal transduction processes triggered by binding ECM ligands to cell receptors [117–122].

Attempts to gain cell recognition have been achieved by surface and bulk modification of the materials via chemical modification or adsorption of bioactive molecules, such as native long chains of ECM proteins as well as short peptide sequences derived from ECM proteins, interacting with cell receptors [123]. Advanced scaffold designs are now being developed to implement patterning, binding of ligands, sustained presentation and release of cytokines, and the structural and mechanical properties of specific tissues [124–126].

3.2.1 Peptide-Modified Scaffolds

Since the finding of signaling domains that are composed of several amino acids along the long chain of ECM proteins, and their interactions with cell membrane receptors, the short peptide fragments have been used for surface modification of materials in numerous studies. Though lacking the complete specificity and function of native ECM, the synthetic peptides can easily be covalently attached to the matrix, allowing control over ligand presentation, and thus the elucidation of cellular behavior on the biomaterial, in a more defined manner.

Moreover, short peptide sequences are relatively more stable during the modification process than the natural proteins, and nearly all modified peptides are available for cell binding. In contrast, the native ECM protein tends to be randomly-folded upon adsorption to the biomaterial surface, and thus the receptor binding domains are not always available. In addition, short peptides may be synthesized in large quantities and in relatively less expensive processes than the native proteins. The modification of materials with individual peptide sequences and epitopes derived from ECM molecules has proved to be a beneficial approach to controlling stem cell fate.

Introducing Adhesion Peptides

One of the most studied sequences for biomaterial modifications is the arginine–glycine–aspartic acid (RGD) sequence. It is a common adhesion motif in ECM proteins, such as fibronectin, collagen and laminin [127, 128], and various integrin receptors bind directly to it [129]. Synthetic, RGD-containing peptides have been incorporated into various materials and tissue engineering scaffolds to promote cell adhesion and improve their cell function [130]. The interaction of these peptides with cells has been shown to influence many processes within the cell, among them proliferation, migration and differentiation [131]. Activation of these processes is mediated by integrin-binding signal transduction to the cell interior, which regulates organization of the cytoskeleton, activates kinase signaling cascades, and modulates cell cycle and gene expression.

Hwang and colleagues showed that RGD-specific cell–matrix interactions could promote the differentiation of hESCs into chondrocytes. Moreover, their results suggested that RGD modification was even more efficient at inducing chondrogenic differentiation than when exogenous type I collagen (which contains RGD) was incorporated into the matrix [132]. Several studies indicated that RGD-modified PLLA and hydroxyapatite scaffolds improved hMSCs adherence to the matrix [133, 134]. Additional studies indicated that RGD-decoration of PEG hydrogels improved MSC viability [135] and promoted osteogenesis of MSCs in a dosage-dependent manner, as judged by early and late bone marker expression. It is suggested that rendering cell adhesion sites via introducing RGD into the scaffold can help the maintenance of MSC osteogenic potential in the 3D PEG hydrogels [103].

In another report, Li et al. [136] used a completely synthetic hydrogel system presenting adhesive RGD motifs as a culture environment for self-renewal of hESCs, replacing the conventional culture system on feeder layers of mouse embryonic fibroblasts (MEFs). The researchers fabricated hydrogel composed of poly(N-iso-propyl-acrylamide-co-acrylic acid) incorporating a semi-interpenetrating network of cell adhesion RGD peptides. This culture system allowed the simple and independent manipulation of cell adhesion ligand presentation and matrix stiffness. ESCs that were cultured on the substrates adhered to the surface, remained viable, maintained their morphology, and expressed markers of undifferentiated hESCs [136].

In our group, RGD peptide immobilized onto alginate macro-porous scaffolds was shown to enhance transforming growth factor-β1 (TGFβ1)-induced hMSC chondrogenic differentiation. The cell–matrix interactions facilitated by the immobilized RGD peptide were shown to be an essential feature of the cell microenvironment, allowing better cell accessibility to the chondrogenic-inducing molecule TGFβ1 (Fig. 4) [81]. In another study conducted by our group, alginate scaffolds were modified with an additional adhesive peptide, heparin binding peptide (HBP), which is a target for cell syndecan interactions. The HBP was immobilized onto alginate scaffolds, alone or in combination with RGD peptide. The integration of these multiple cell–matrix interactions was shown to promote

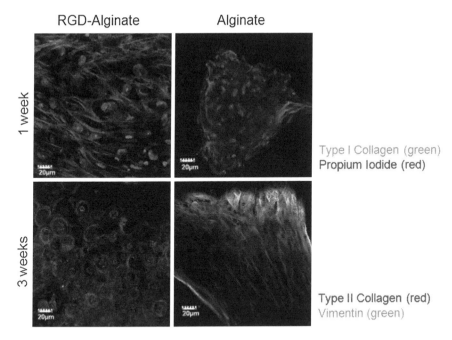

Fig. 4 *Upper panel* Alginate scaffolds containing short motifs of ECM adhesion proteins such as RGD encouraged hMSCs to spread and attach to the matrix (*left*), whereas on unmodified scaffolds (*right*) only cell–cell interactions were seen in 1-week hMSC constructs (collagen fibers, *green*; nuclei, *red*). *Lower panel* In the RGD-immobilized cell constructs, the production/secretion of collagen type II, a major component in the cartilaginous ECM, was more pronounced and the cells had the round morphology of committed chondrocytes (*left*), compared to unmodified alginate (type II collagen, *red*; vimentin, *green*) (*bar*: 20μm) (reprinted by permission from [81])

cardiac tissue regeneration, as judged by expression of cardiac cell markers, the arrangement of well-developed myocardial fibers and the formation of contractile tissue [137].

Other adhesive peptide fragments were identified and shown to have inductive properties. For instance, Martino et al. [55] were able to isolate peptide fragments from fibronectin that presented different specificities for the integrin $\alpha 5\beta 1$. Few of them were found to have an increased specificity for $\alpha 5\beta 1$ in contrast to the relatively promiscuous integrin binding of fibronectin. Interestingly, the presence of those peptides enhanced osteogenic differentiation of hMSCs in 2D and 3D fibrin matrix system, compared with the full-length protein, which provides both pro- and anti-differentiative cues in the natural MSC niche. In contrast, the attachment, spreading and proliferation were comparable with that on full-length fibronectin [55].

Introducing Matrix Metalloproteinase-Sensitive Peptides

The overall complexity of cell regulation is further increased by the dynamic nature of regulatory signal presentation, which changes in space and time. Also, the interactions between the cells and their environment occur in both directions. Cells both respond to and actively modify the properties of their environment by synthesizing or degrading the ECM, secreting cytokines, and communicating with other cells and matrix via molecular and physical signals.

Thus, bio-inspired scaffolds have been developed to incorporate enzyme-sensitive peptide sequences [100, 138–141]. Such enzymatically cleavable materials exploit the upregulation and downregulation of cell-secreted enzymes, such as matrix-metalloproteases (MMPs), to dictate material degradation on a local and cellular timescale. For example, Kraehenbuehl et al. [99] fabricated cross-linked PEG hydrogels incorporating adhesive (RGDSP) and enzymatically degradable MMP-sensitive peptide as cross-linkers of the matrix. The matrices promoted ESC differentiation into the cardiac lineage, probably via ligation of cell integrins relevant in early cardiac development ($\alpha 5\beta 1$, $\alpha v\beta 3$) to the matrix-immobilized RGDSP peptide. Further cardiac maturation was promoted by the MMP-sensitivity of the matrix, allowing cell-triggered matrix remodeling [99].

Moreover, in an attempt to mimic the dynamic nature of the stem cell microenvironment, where signaling events vary in time, the MMP-sensitive peptides can be also exploited as a linker for other functional peptides in order to control their temporal presentation in culture. For example, natively, hMSCs differentiating into chondrocytes initially produce the adhesion protein fibronectin, which is subsequently downregulated between days 7 and 10, while the excreted ECM is remodeled through enzyme production. Therefore, Salinas et al. [104] designed an adhesive sequence, RGD, with an MMP-13 specific cleavable linker to release RGD, mimicking the native differentiation timeline. The researchers showed that active MMP-13 production of encapsulated hMSCs increased from days 9–14 specifically in the chondrogenic differentiating cultures [104]. Their results pointed out the beneficial effect of the MMP-regulated presentation of RGD in the hydrogel on encapsulated hMSC chondrogenic differentiation.

Introducing Growth Factors-Derived Peptides

Instead of supplementing the signals as soluble factors, short peptides derived from growth factors and covalently bound to matrix were shown to be valuable in affecting stem cell fate. The potential advantages of such a strategy are better control over ligand presentation and the relatively prolonged stability of the short peptides compared to the full growth factor. For example, Saito et al. [142] could identify a BMP2-derived peptide (KIPKASSVPTELSAISTLYL), shown to enhance the osteogenic differentiation of a murine multipotent mesenchymal cell line, as indicated by alkaline phosphatase activity and gene expression profile. Moreover, the researchers demonstrated peptide binding to the BMP2-specific receptor [142]. The peptide was shown to induce prolonged ectopic calcification when immobilized onto a covalently cross-linked alginate gel and implanted into rat calf muscle, as compared to the full protein. These results suggest that the peptide remained active at the implanted site, continuously inducing the differentiation of osteoblast precursor cells into osteoblasts and activating osteoblasts to promote ectopic calcification [144]. These results were the basis for several additional experiments, indicating the potential of this strategy in the induction of MSC osteogenic differentiation for bone repair [144–146].

Other inductive BMP-derived peptides were identified and shown to be effective for osteogenic differentiation, including peptides from BMP4 [147], BMP7 [148] and BMP9 [149]. Recent studies indicate the beneficial effect of material modification with a combination of the above-mentioned BMP-derived peptide with the RGD adhesive peptide on the enhancement of MSC proliferation and osteogenic differentiation [150, 151].

Epitope Presentation Via Self-Assembling Peptide Scaffolds

Self-assembling peptides scaffolds have also been developed for the controlled differentiation of stem cells through careful selection of the constituent peptides. For example, Gelain et al. [112] attached several functional motifs including cell adhesion, differentiation and bone marrow (BM) homing motifs to a self-assembling peptide, and the functionalized peptides self-assembled into a nanofiber structure for neural stem cell (NSC) cultivation. Specifically, peptide scaffolds containing BM homing motifs significantly enhanced the neural cell survival without supplementation of extra soluble growth and neurotrophic factors to the routine cell culture media. Moreover, the gene expression profile showed selective and more uniform gene expression towards this cell lineage, compared with the control, Matrigel scaffold [112]. In another study, Silva et al. [113] demonstrated that laminin-derived isoleucine–lysine–valine–alanine–valine (IKVAV) peptide sequences could self-assemble to form gels that selectively direct mouse neural progenitor cells toward a neuronal fate [113].

3.2.2 Spatio-Temporal Presentation and Release of Signaling Molecules

Signaling molecules such as growth factors and cytokines are important for gaining control over stem cell fate decision. These molecules can be added to the culture media or secreted by the cells themselves. When exogenously supplied, their activity may be affected by diffusion limitations within the 3D cultivation system. Thus, current research is increasingly oriented towards presentation of these signaling molecules in the context of the matrix, in a spatio-temporal manner, similar to their presentation by the native ECM. The signaling

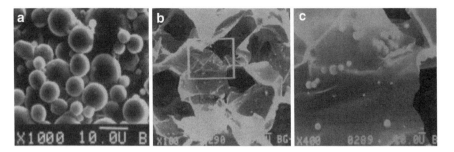

Fig. 5 Morphology of porous alginate scaffold incorporating PLGA microspheres capable of controlling the release of basic fibroblast growth factor. SEM picture of (**a**) microspheres, (**b–c**) the alginate composite scaffold (reprinted by permission from [80])

molecule may be delivered in a sustained release profile from matrix-embedded micro-particles, be covalently immobilized to the matrix, or be presented via affinity interactions.

Controlled Release Delivery of Signaling Molecules From Micro-Particles

In an attempt to design a dynamic cell microenvironment, where the regulatory signals often vary in space and time, microparticles for controlled delivery of signaling molecules have been developed and incorporated into the scaffolds (Fig. 5) [80]. In this way, molecule release is typically controlled through diffusion, particle degradation, or their combinations, leading to a wide range of factor delivery profiles. Moreover, the wide versatility of these systems is associated with other variables that can be modeled to obtain an optimal system for a specific application, such as particle composition, size and shape.

The incorporation of microparticles into scaffolds has been widely used to control stem cell fate decision. For example, gelatin microparticles loaded with TGFβ1 were co-encapsulated with rabbit MSCs in an injectable biodegradable hydrogel composites of oligo(poly(ethylene glycol) fumarate). The localized delivery of TGFβ1 resulted in an efficient chondrogenic differentiation, revealed by the increase in glycosaminoglycan content per DNA and the upregulation of chondrocyte-specific gene expression of type II collagen and aggrecan [152].

One recent study involved the combined application of immobilized RGD dextran-based hydrogel and microencapsulated vascular endothelial growth factor (VEGF) to induce vascular differentiation of hESCs [153]. HESCs within the bioactive hydrogels expressed higher levels of vascular markers, compared to spontaneously differentiated embryoid bodies (EBs). Carpenedo et al. [154] showed that using polymer microspheres for the delivery of morphogenic factors directly within EB microenvironments in a spatio-temporally controlled manner yielded homogeneous, synchronous and organized ESC differentiation. Degradable PLGA microspheres releasing retinoic acid were incorporated directly within

EBs and induced the formation of cystic spheroids uniquely resembling the phenotype and structure of early mouse embryos [154].

Covalent Binding of Signaling Molecules onto the Matrix

Covalent binding of signaling molecules onto the matrix facilitates their prolonged presentation and biological activity, in case long-term persistence of the signal is needed [155, 156]. The covalent binding of the factor to the scaffold overcomes the natural internalization process of the factor–receptor complex, which includes its endocytosis and degradation inside the cell. Thus, tethering the signaling molecule to the matrix may prevent ligand depletion from the environment. Moreover, tethered ligands represent a mimicry of ligands that are naturally insoluble, matrix-"tethered" i.e., a ligand which acts from the matrix, presented continuously to cell surface receptors [157]. However, as opposed to the temporal presentation of factors by natural ECM, covalent binding presents an unnatural permanent presentation.

Using this strategy, Fan et al. [156] showed that biomaterial surfaces covalently modified with epidermal growth factor (EGF) promoted both MSC spreading and survival, more strongly than saturating concentrations of soluble EGF added to the medium. By sustaining mitogen-activated protein kinase kinase-extracellular-regulated kinase signaling, tethered EGF increased the contact of MSCs and conferred resistance to cell death induced by the pro-inflammatory cytokine, Fas ligand. Thus, the authors suggested that tethered EGF may offer a protective advantage to MSCs in vivo during acute inflammatory reaction to tissue engineering scaffolds [156].

Immobilization of growth factors to biomaterial surfaces in specific spatially-defined patterns was carried out by bio-printing technologies [158, 159]. For example, inkjet bio-printing technology was used for the creation of an immobilized BMP2 printed pattern on fibrin substrate. The results indicated that the differentiation of primary muscle-derived stem cells towards the osteogenic lineage was confined to the printed patterns of BMP2, whereas cells off-pattern differentiated towards the myogenic lineage [158]. This approach may also be useful for understanding cell behavior towards immobilized biological patterns. For instance, recent work has focused on the effect of spatially-controlled gradients on MSC migration [160].

Affinity Binding and Presentation of Signaling Molecules

Despite the fact that growth factors and other signaling molecules were originally described as soluble molecules, evidence shows that the binding of signaling factors to ECM is a major mechanism regulating their activity. A bio-inspired approach for presentation and release of these factors exploits their natural interactions with native ECM. For instance, in the native cellular microenvironment, the proteoglycans heparin and heparan sulfate bind many growth factors, chemokines and cell adhesion molecules, collectively known as heparin-binding peptides, via high-affinity, specific electrostatic interactions [161, 162]. Such interactions are mediated by low- and high-sulfated sequences in these glycosaminoglycan (GAG) chains [163, 164]. In addition, binding affinity values were found to depend on the degree of polysaccharide sulfation [165, 166]. The proteoglycan interactions play a critical role in assembling protein–protein complexes, such as growth factor–receptor or enzyme–inhibitor, on the cell surface and in the ECM, which are directly involved in initiating cell signaling events or inhibiting biochemical pathways [167]. For instance, the interaction of heparin/heparan sulfate with fibroblast growth factor (FGF) has been shown to act as a template that bridges the factor and its receptor, thereby effectively lowering the concentration of FGF needed to initiate the signaling through its receptor and extending the response duration [168]. Moreover, recent works have shown that the presentation of VEGF *in trans* in association with heparan sulfate leads to enhanced signal transduction by facilitating the formation of receptor–ligand complexes and trapping of the active signaling complex [169].

Heparin-Based Biomaterials

Growing knowledge on the important roles of heparin/heparan sulfate has led to numerous attempts to incorporate these materials in stem cell microenvironmental design. For example, heparin functionalized PEG gels were shown to modulate protein adsorption and promote hMSC adhesion and osteogenic differentiation [170, 172]. In these experiments, the heparin was modified with methacrylate groups, copolymerized with dimethacrylated PEG to form a hydrogel, which enabled a localized delivery vehicle for basic-FGF and served as a synthetic ECM for the differentiation of hMSCs.

In another study, Willerth et al. [172] used a heparin-based delivery system for mouse ESC differentiation inside fibrin scaffolds. The delivery system presented three components: a bi-domain peptide, heparin and growth factor. The peptide contained a Factor XIIIa substrate derived from $\alpha 2$-plasmin inhibitor, allowing it to be covalently cross-linked into the fibrin scaffold, and a heparin-binding domain derived from antithrombin III, which bound heparin non-covalently and retained it inside the fibrin scaffold. The heparin could in turn bind growth factors and retain them inside the scaffold. This delivery system was used to deliver different molecules, such as platelet-derived growth factor (PDGF), at various doses. The controlled delivery of these molecules simultaneously increased the fraction of neural progenitors, neurons, and oligodendrocytes while decreasing the fraction of astrocytes obtained, compared to control cultures seeded inside unmodified fibrin scaffolds with no growth factors present in the medium [172].

Recently, Webber et al. [173] used a heparin-presenting nanofiber network to bind and deliver paracrine factors derived from hypoxic conditioned stem cell media to mimic the stem cell paracrine effect, for cardiovascular disease treatment.

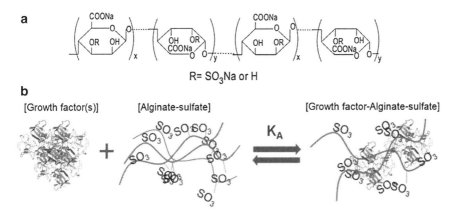

Fig. 6 Alginate–sulfate (**a**) and the model of reversible binding (**b**) (reprinted by permission from [203])

The self-assembling peptide nanofibers presenting heparin were capable of binding paracrine factors from a medium phase, primarily heparin-binding factors. When these factor-loaded materials were injected into the heart following coronary artery ligation in a mouse ischaemia–reperfusion model of acute myocardial infarction, significant preservation of hemodynamic function was observed [173].

Alginate–Sulfate Biomaterial

To mimic the natural interactions of heparin-binding proteins with heparin/heparan sulfate and obtain prolonged release of growth factors, our group has developed an innovative alginate biomaterial with affinity-binding sites for heparin-binding proteins, by sulfation of the uronic acid monomers in alginate (Fig. 6) [77]. Surface plasmon resonance (SPR) analysis revealed strong binding of various heparin-binding proteins to alginate–sulfate, but not to alginate [76, 77]. The equilibrium binding constants to alginate–sulfate were comparable to those obtained between the peptides and heparin. We found the release rate from bioconjugates of alginate–sulfate/growth factors to be dependent on the growth factor binding constants (K_A) and the initial concentrations of individual components forming the bioconjugate (Table 1). Additionally, the interactions of these molecules with alginate–sulfate were found to enhance their stability against enzymatic proteolysis induced by trypsin [82, 83]. This is of great importance if the delivered proteins are to remain active for prolonged periods of time in harsh environments, where extensive protein degradation takes place.

Injectable and implantable scaffolds composed of combinations of alginate–sulfate with unmodified alginate provided a unique type of a novel affinity-binding system, which has been capable of the sequential delivery of multiple signaling molecules, while retaining the already mentioned properties and characteristics of

Table 1 Equilibrium binding constants (K_A) calculated from the interactions of alginate–sulfate with proteins (SPR analysis) [76, 77]

$K_A (M^{-1})$	Protein
2.80×10^7	Acidic fibroblast growth factor
2.57×10^6	Basic fibroblast growth factor
9.93×10^6	Epidermal growth factor
5.36×10^7	Hepatocyte growth factor
1.01×10^8	Insulin-like growth factor-1
1.38×10^7	Interleukin-6
3.53×10^7	Platelet-derived growth factor-BB
2.06×10^8	Stromal cell derived factor-1
6.63×10^7	Transforming growth factor-$\beta 1$
1.81×10^6	Thrombopoietin
6.98×10^6	Vascular endothelial growth factor

the alginate as cell vehicle [76, 83]. For example, affinity-binding alginate scaffolds were shown to control the release of three known angiogenic factors, PDGF-BB, TGFβ1 and VEGF. In vitro release studies revealed a sequential order of protein release from the scaffold, as predicted by the values of the equilibrium binding constants to alginate–sulfate (see Table 1) and their initial concentrations: VEGF was released first, followed by PDGF-BB and TGFβ1. The sequential delivery of factors from the affinity-binding scaffold mimicked the signal cascade acting in angiogenesis. By contrast, factor release from the scaffolds lacking alginate–sulfate was rapid and was governed mainly by a burst effect. Subcutaneous implantation of triple-factor bound scaffolds resulted in superior vascularization, compared to factor-adsorbed or untreated scaffolds [76].

The affinity-binding alginate scaffolds were also used for the creation of a vascularized cardiac patch [70]. Such scaffolds containing a cocktail of pro-survival and angiogenic factors [insulin growth factor-1 (IGF-1), stromal cell-derived factor-1 (SDF-1) and VEGF] were seeded with rat neonatal cardiomyocytes, and then transplanted onto the omentum to achieve host-induced vascularization of the patch. Seven days later, the resultant vascularized patches were harvested and re-transplanted to replace the scar tissues of the infarcted rat hearts. The pre-vascularization within the affinity-binding patch was proven to significantly improve the therapeutic outcome, as judged by the structural and electrical integration into the scar tissue four weeks post-transplantation onto an infarcted heart. Moreover, the pre-vascularized cardiac patch was able to attenuate the deterioration of cardiac function, indicated by echocardiography and electrophysiology analyses [70].

In another study, the sequential delivery of IGF-1 and hepatocyte growth factor (HGF) from injectable affinity-binding alginate hydrogel was shown to promote myocardial repair in a similar rat model of acute myocardial infarction. The dual delivery of IGF-1 and HGF from affinity-binding alginate biomaterial was able to preserve scar thickness, attenuate infarct expansion and reduce scar fibrosis after 4 weeks, concomitantly with increased angiogenesis and mature blood vessel formation at the infarct. Furthermore, this treatment prevented cell apoptosis and induced cardiomyocyte cell cycle re-entry, an indication of endogenous regeneration of cardiac muscle [83].

Alginate–sulfate has several advantages over the use of heparin, as it forms a stable non-immunogenic hydrogel when combined with unmodified alginate, and its mass production may be achieved at relatively low cost compared to that of heparin. In contrast to alginate, exogenous delivered heparin and heparan sulfate molecules are rapidly degraded in the human body and thus they have limited use as a prolonged and efficient drug delivery system. Moreover, heparin and heparan sulfate are very heterogeneous macromolecules with multiple biological functions beyond being a depot for growth factors, and thus controlling their behavior and effect in vivo is limited. For example, heparin, commonly used as an anticoagulant drug, was recently suggested also to play a key role in inflammatory processes [174].

Thus, we believe that the affinity-binding platform of alginate–sulfate will be a powerful tool in the design of stem cell microenvironments that better mimic the natural presentation and release patterns of the inductive cues which control stem cell fate decision.

3.2.3 Matrix Chemistry

A different strategy utilizing simple matrix chemistry has recently been introduced to fine-tune the subtle design of the stem cell microenvironment, or, in other cases, even to create inductive microenvironments without the use of the conventional natural regulatory molecules.

A good example of microenvironment fine-tuning is the photo-degradable hydrogels, developed for "on-demand" degradation of the physical network structure [175]. These hydrogels are prepared by cross-linking via photo-labile moieties. Thus, the degradation rate and extent as well as the resulting material properties, such as stiffness, can be predictably manipulated with light intensity and wavelength. This precise control of gel cross-linking density was used to examine the influence of gel structure on hMSC spreading. These cells exhibited rounded cell morphology within a densely cross-linked gel; however, when the gel cross-linking density was reduced through photo-degradation, cell spreading was observed. Thus, cell morphology can be manipulated by irradiation and degradation of these hydrogels at any time in culture. Moreover, photo-degradable linkages can also be exploited to locally modify the chemical environment within a hydrogel by incorporation of tethered, but photo-labile, biologically active functionalities. For instance, a photo-labile group was coupled to the adhesive peptide RGD, and incorporated into PEG-based hydrogel, producing a hydrogel with photolytic control over its chemical interaction with encapsulated cells. The temporal presentation of the adhesive RGD peptide was shown to be beneficial for hMSC chondrogenic differentiation [125].

A recent study by Benoit and co-workers pointed out the possibility of controlling stem cell decision by using simple chemistry alone. The researchers screened immobilized small molecules for their ability to induce hMSCs to differentiate down several pathways. The small molecules were chosen to incorporate functionalities found in the extracellular environment of the target cell types. Specifically, PEG hydrogels were modified with small amounts of carboxyl groups to mimic glycosaminoglycans in cartilage, phosphate groups for their role in bone mineralization, or *t*-butyl groups to mimic the lipid-rich environment in adipose tissue. HMSCs were encapsulated in the tethered PEG hydrogels and cultured in standard hMSC media, without the added cytokines or steroids typically used in differentiation media. Most interestingly, these synthetic matrices were successful in inducing the differentiation of hMSCs into the chondrogenic, osteogenic or adipogenic pathways respectively [111]. This study was the first example where synthetic matrices were shown to control induction of multiple hMSC lineages purely through interactions with small-molecule chemical functional groups tethered to the hydrogel material. Strategies using simple chemistry to control complex biological processes would be particularly powerful as they could make production of therapeutic materials simpler, cheaper and more easily controlled.

3.2.4 Mechanical Features of the Matrix

Stem cells sense and respond to the mechanical properties of the extracellular matrix. These properties are determined primarily by the matrix chemical composition, water content and structure. Previous studies using differentiated cell types have demonstrated that the mechanical properties of a material affect cell behaviors such as growth and migration [176–182]. In particular, adhesion ligands, which bind to integrins and other cell surface receptors, serve as mechanical transducers between the external material and the internal cytoskeleton of the cell, allowing cells to sense and respond to the stiffness of their substrates. Tensional homeostasis with the microenvironment thereby induces cellular cytoskeletal organization and alters gene regulatory pathways.

The extent of matrix mechanical effects on stem cell fate in 3D microenvironments and the underlying biophysical mechanisms are still under investigation. Recent studies have demonstrated the integral role of mechanical cues in the commitment of stem cell fate. For example, McBeath et al. [183] demonstrated that hMSC lineage commitment could be largely regulated by cell shape and size via related changes in the cytoskeleton tension. In their study, they used spatial micro-patterning of adhesion molecules to control cell shape and degree of spreading with single cell precision. HMSCs patterned on a small island tended to undergo adipogenic differentiation, whereas cells patterned on a larger island were able to spread and develop high cytoskeleton tension, and tended to undergo osteogenic differentiation [183]. The researchers suggested an explanatory mechanism for their results, in which the commitment signal for the osteogenic differentiation required actin–myosin-generated tension.

In an additional study, the correlation between tension forces and cell differentiation in monolayer cultures of hMSCs was nicely demonstrated [184]. Ruiz et al. [184] cultured hMSC sheets on micro-patterns of controlled shape and exposed them to a mixture of pro-osteogenic and pro-adipogenic morphogens. They showed that gradients of mechanical forces can drive a pattern of differentiation dictated by cell spatial arrangement and corresponding cytoskeletal stress. HMSCs at the edge of multicellular islands differentiated into the osteogenic lineage, whereas those in the center became adipocytes. Interestingly, changing the shape of the multicellular sheet modulated the locations of osteogenic versus adipogenic differentiation. Measuring traction forces revealed gradients of stress that preceded and mirrored the patterns of differentiation, where regions of high stress resulted in osteogenesis, whereas stem cells in regions of low stress differentiated to adipocytes. These findings demonstrate a role for mechanical forces created within multicellular organization in spatial cell differentiation. Such geometric control is also useful for controlling cell–cell interactions, as in the aggregation of ESCs in vitro into embryoid bodies (EBs), a preliminary step toward their differentiation. The conventional methods of culturing EBs are poorly controlled and result in the formation of heterogeneous structures with a wide range of sizes and shapes. Spatial control over the EB formation can lead to a more homogeneous, and thereby more efficiently controlled differentiation. For instance, Karp et al. [185] developed micro-fabricated cell-repellent PEG hydrogel in micro-wells as templates to initiate the controlled formation of homogeneous EBs. Their approach resulted in synthetic microenvironments that enhanced the differentiation of ESCs and significantly reduced variability in the expression of differentiation markers. They were also able to pattern EBs into shapes that do not naturally occur, such a triangles and curves; nevertheless the biological implications for stem cell fate are not clear [185].

The effects of matrix physical attributes such as matrix stiffness on stem cell fate were first examined by Engelr et al. [186]. In their study, the elasticity of the matrix was identified as a key factor of the stem cell micro-environment, specifying stem cell commitment. MSCs were cultivated on 2D polyacrylamide (PA) gels with varying matrix elasticity, set by degree of cross-linking, and adhesion was provided by coating the gels with collagen I. The researchers showed that MSCs differentiate into tissues that most closely match the mechanical properties of the PA substrate upon which they were cultured. MSCs that were cultured on stiff (bone-like) gels differentiated into osteoblasts, those that were cultured on medium stiffness (muscle-like) gels differentiated into muscle cells, and those that were cultured on compliant gels (neural-like) differentiated into neural cells—all in identical serum conditions. These findings were attributed to the cytoskeleton tension forces that activated molecular pathways of cell differentiation in a manner dependent on substrate stiffness [186].

In another study, the commitment of MSC populations changed in response to the scaffold rigidity; however, cell fate was not correlated with cell morphology. Instead, matrix stiffness regulated integrin binding as well as reorganization of adhesion ligands on the nanoscale, both of which were traction-dependent and correlated with osteogenic commitment of MSC populations. These findings suggested that cells interpret changes in the physical properties of adhesion substrates as changes in adhesion-ligand presentation [187].

Whereas all the above-mentioned studies utilize static polymer systems, their native counterparts reside in a dynamic environment in which elasticity may change spatially and/or temporally. In a recent study, Tse et al. [188] tried to explore the potential signal of physiological stiffness gradients to MSC differentiation. To that end, they cultured MSCs on a photo-polymerized PA hydrogel of varying stiffness and provided the first evidence that MSCs indeed appear to undergo directed migration, or durotaxis, up stiffness gradients rather than remain stationary. Temporal assessment of morphology and differentiation markers indicated that MSCs migrated to stiffer matrix and then differentiated into a more contractile myogenic phenotype. This study may indicate that stiffness variation, not just stiffness alone, can be an important regulator of MSC behavior [188].

3.2.5 Complex Hierarchical Matrices

New generations of interactive biomaterial scaffolds are now being developed for the construction of hierarchically complex tissues, such as vascular networks, interfaces, structural hierarchy, and other complex functional features. These scaffolds need to establish compositional gradients or sub-compartments, temporal changes, and control over cell-driven tissue and organ morphogenesis. For example, in an attempt to design scaffolds for the appropriate layered structure of various native tissues (as skin and cartilage–bone interface), layered systems have been developed, to allow the creation of optimized, tissue-specific biological environments in each respective layer via variations in mechanical, structural and chemical properties. The bi-layered system designs were shown to be of great potential for organized tissue growth, when implanted in vivo, either seeded with mature cells [189], or acellular [190]. Furthermore, recent studies have pointed out that the use of bi-layered system may be a powerful tool for the spatially-controlled simultaneous induction of stem cells into distinct lineages.

One example is the design of a composite bi-layered hydrogel of PEGylated fibrinogen and type I collagen, aimed at approximating the layered structure of skin. The bi-layered matrix was able to control the bidirectional differentiation of adipose-derived stem cells into endothelial cells and pericytes. Specifically, matrix-driven phenotypic changes into a fibroblast-like morphology were observed in the collagen layer, whereas a tube-like morphology was simultaneously detected in the PEGylated fibrin layer. The matrix composition dictated the lineage specification and was not driven by soluble factors [191].

In another study, injectable, biodegradable hydrogel composites of cross-linked oligo(poly(ethylene glycol) fumarate) (OPF) and gelatin microparticles (MPs) were utilized to fabricate a bi-layered osteochondral construct, consisting of a chondrogenic layer and an osteogenic layer. Rabbit MSCs were encapsulated in both layers, and were able to undergo chondrogenic differentiation, in the presence of TGFβ1-loaded MPs. Although simultaneous differentiation of MSCs into osteoblasts was limited in the osteogenic layer, there was significantly enhanced chondrogenic differentiation of MSCs in the chondrogenic layer. This effect was further enhanced when MSCs pre-differentiated into osteoblastic cells were encapsulated in the osteogenic layer [192].

In addition to the suggested layered composite approach, other studies utilized delivery of controlled factor gradients via MPs in polymer scaffolds for the induction of stem cells into a complex tissue [193, 194]. For example, BMP2- and TGFβ1-loaded PLGA MPs were utilized with a gradient scaffold fabrication technology to produce MP-based scaffolds containing opposing gradients of these signals. The results indicated that hMSC-seeded gradient scaffolds produced regionalized ECM, similar to that of the osteochondral tissue. Overall, these studies demonstrate the fabrication of layered hydrogel composites that mimic the structure and function of osteochondral tissue, along with the application of these composites as cell and growth factor carriers [193].

4 Bioreactors for Dynamic Stem Cell Microenvironment

The microenvironment consists of additional cues, such as local blood perfusion and hypoxia in the bone marrow niche, known to be key players in the determination of HSC fate. Recently, Winkler et al. [195] established a positional hierarchy between HSCs and lineage-restricted hematopoietic progenitor cells (HPCs) relative to the blood flow rate. The most potent HSCs were exposed to negligible blood flow, whereas lineage-restricted HPCs as well as stromal cells such as endothelial cells, MSCs and osteoblast cells were located in vascular niches, perfused by rapid blood flow. This study along with many others suggests that stem cell microenvironment design should consider the dynamic nature of the native niche, i.e., including the parameters of fluid flow kinetics and mass transport (especially oxygen diffusion) in the 3D cell cultures, as well as other time-constant-related factors and metabolite diffusion. Ex vivo, in static 3D cell cultures with the lack of vasculature, the transport of nutrients and dissolved oxygen from the bulk medium to the surface of the cell constructs is limited, as is the internal mass transfer rate from the surface of the construct into its core.

In an attempt to overcome mass transport limitations and to enable controlled biophysical and mechanical stimuli, bioreactors have been implemented for creating the 3D stem cell microenvironment. The bioreactor systems, from simple conventional spinner flasks, the rotary wall vessels, up to the latest perfusion vessels, were found to improve cell viability, proliferation and differentiation. For example, 3D cultivation in spinner flasks improved, to some extent, tissue homogeneity and viability of hMSC [196]. In another study, rotary cell culture systems (RCCS), developed by NASA, were shown to enable 3D cell cultivation under medium mixing with a minimal shear stress on the cultivated cells. It was shown that cultivation of hESCs within a rotating bioreactor increased the cell proliferation rate and maintained cell viability in the culture [197].

Although rotary vessels have improved mass transfer to the construct surface, they did not solve the problem of the limiting internal mass transfer within the construct. To address this challenge, perfusion bioreactors, designed to force the culture medium into the cell constructs, have been developed [69, 88] and shown to be superior to conventional bioreactors. For example, 3D cultivation in perfusion bioreactors was shown to encourage MSC expansion and their subsequent osteogenic differentiation in vivo [198]. Sikavitsas et al. [199] have shown that fluid flow not only mitigates nutrient transport limitations in 3D perfusion cultures of MSCs, but also provides mechanical stimulation to seeded cells in the form of fluid shear stress, resulting in increased deposition of mineralized matrix. Cellular constructs cultured in a flow perfusion bioreactor yielded a significant increase in matrix mineralization after 16 days compared with those cultured statically, when cultured in the presence of osteogenic supplements [200].

Our group has developed a novel perfusion bioreactor that is capable of cultivating multiple 3D cellular constructs in one flow chamber. Its unique features provided a homogeneous fluid flow along the bioreactor cross-section and maximal exposure of the cellular constructs to the perfusing medium. By employing this advanced perfusion bioreactor, cardiac cell constructs were shown to maintain almost 100% of the seeded cells viable, while less than 60% of the cells in static cultures were viable after 7-day cultivation [69]. Moreover, a thick (>500 μm) cardiac tissue was generated, composed of elongated and aligned cells with a massive striation, resembling the native adult heart [88].

In addition to the great advantage of superior mass transport, bioreactors can also provide control over the cellular environment in terms of biochemical and physical regulatory signals, such as mechanical and electrical stimuli, and enable on-line monitoring and response [87, 201]. For additional information on the potential utilization of advanced bioreactor systems for stem cell expansion and differentiation, the reader is referred to the reviews in [13, 202].

5 Conclusions and Future Directions

Biomaterials with controllable physical, chemical and biological properties are potentially a powerful tool for directing stem cells towards expansion or specific-lineage differentiation. Clearly, the studies presented point out the tremendous potential in using bio-inspired biomaterials for stem cell applications. However, this field is still in its earliest stages. We still need to better understand the full, complex and dynamic interplay between the key parameters of the microenvironment controlling stem cell fate. This knowledge will enable the creation of novel intelligent biomaterials that combine the advantages of native and synthetic materials. These biomaterials should be sophisticated enough to elicit in vivo like cell responses yet simple and practical enough for use in biology and medicine. These advanced materials may incorporate the spatial presentation of multiple regulatory molecules (such as peptides and growth factors), with temporally controlled sequential delivery. Novel biomaterials such as affinity-binding alginate may serve as a platform for such devices. The utilization of such controllable biomaterials may also enable the creation of innovative complex hierarchical matrices for the potential regeneration of organized tissues such as the osteochondral tissue. New synthetic small molecules with regulatory functions, as well as new synthetic polymers for control over stem cell fate decisions, are yet to be found, possibly via high-throughput screening technologies. Moreover, the design and application of "smart" bioreactors will facilitate better understanding of the dynamic nature of stem cell microenvironment, and subsequently precise control over biochemical and physical signals for stem cell fate decisions. Finally, scale-up technologies will be needed for realizing stem cell potential in clinical research. To conclude, given the complexity of stem cell fate control, much has still to be learned, yet our growing knowledge has already made major breakthroughs in manipulating stem cell fate, raising more high expectations from future research.

References

1. Scadden DT (2006) The stem-cell niche as an entity of action. Nature 441:1075–1079
2. Metallo CM, Mohr JC, Detzel CJ et al (2007) Engineering the stem cell microenvironment. Biotechnol Prog 23:18–23
3. Barrilleaux B, Phinney DG, Prockop DJ, O'Connor KC (2006) Review: ex vivo engineering of living tissues with adult stem cells. Tissue Eng 12:3007–3019
4. Calvi LM, Adams GB, Weibrecht KW et al (2003) Osteoblastic cells regulate the haematopoietic stem cell niche. Nature 425:841–846
5. Jones DL, Wagers AJ (2008) No place like home: anatomy and function of the stem cell niche. Nat Rev Mol Cell Biol 9:11–21
6. Nilsson SK, Johnston HM, Whitty GA et al (2005) Osteopontin, a key component of the hematopoietic stem cell niche and regulator of primitive hematopoietic progenitor cells. Blood 106:1232–1239
7. Yoshihara H, Arai F, Hosokawa K et al (2007) Thrombopoietin/MPL signaling regulates hematopoietic stem cell quiescence and interaction with the osteoblastic niche. Cell Stem Cell 1:685–697
8. Arai F, Hirao A, Ohmura M et al (2004) Tie2/angiopoietin-1 signaling regulates hematopoietic stem cell quiescence in the bone marrow niche. Cell 118:149–161
9. Adams GB, Chabner KT, Alley IR et al (2006) Stem cell engraftment at the endosteal niche is specified by the calcium-sensing receptor. Nature 439:599–603
10. Parmar K, Mauch P, Vergilio JA, Sackstein R, Down JD (2007) Distribution of hematopoietic stem cells in the bone marrow according to regional hypoxia. Proc Natl Acad Sci USA 104:5431–5436
11. Jang YY, Sharkis SJ (2007) A low level of reactive oxygen species selects for primitive hematopoietic stem cells that may reside in the low-oxygenic niche. Blood 110:3056–3063
12. Duncan AW, Rattis FM, DiMascio LN et al (2005) Integration of Notch and Wnt signaling in hematopoietic stem cell maintenance. Nat Immunol 6:314–322
13. Di Maggio N, Piccinini E, Jaworski M et al (2011) Toward modeling the bone marrow niche using scaffold-based 3D culture systems. Biomaterials 32:321–329
14. Pittenger MF, Mackay AM, Beck SC et al (1999) Multilineage potential of adult human mesenchymal stem cells. Science 284:143–147

15. Prockop DJ (1997) Marrow stromal cells as stem cells for nonhematopoietic tissues. Science 276:71–74
16. Zhao LR, Duan WM, Reyes M et al (2002) Human bone marrow stem cells exhibit neural phenotypes and ameliorate neurological deficits after grafting into the ischemic brain of rats. Exp Neurol 174:11–20
17. Miyahara Y, Nagaya N, Kataoka M et al (2006) Monolayered mesenchymal stem cells repair scarred myocardium after myocardial infarction. Nat Med 12:459–465
18. Hofstetter CP, Schwarz EJ, Hess D et al (2002) Marrow stromal cells form guiding strands in the injured spinal cord and promote recovery. Proc Natl Acad Sci USA 99:2199–2204
19. Maegawa N, Kawamura K, Hirose M et al (2007) Enhancement of osteoblastic differentiation of mesenchymal stromal cells cultured by selective combination of bone morphogenetic protein-2 (BMP-2) and fibroblast growth factor-2 (FGF-2). J Tissue Eng Regen Med 1:306–313
20. Meinel L, Hofmann S, Karageorgiou V et al. (2004) Engineering cartilage-like tissue using human mesenchymal stem cells and silk protein scaffolds. Biotechnol Bioeng 88:379–391
21. Mauck RL, Yuan X, Tuan RS (2006) Chondrogenic differentiation and functional maturation of bovine mesenchymal stem cells in long-term agarose culture. Osteoarthritis Cartilage 14:179–189
22. Martin I, Obradovic B, Treppo S et al (2000) Modulation of the mechanical properties of tissue engineered cartilage. Biorheology 37:141–147
23. Johnstone B, Hering TM, Caplan AI, Goldberg VM, Yoo JU (1998) In vitro chondrogenesis of bone marrow-derived mesenchymal progenitor cells. Exp Cell Res 238:265–272
24. Zuk PA, Zhu M, Mizuno H et al (2001) Multilineage cells from human adipose tissue: implications for cell-based therapies. Tissue Eng 7:211–228
25. Dor Y, Brown J, Martinez OI, Melton DA (2004) Adult pancreatic beta-cells are formed by self-duplication rather than stem-cell differentiation. Nature 429:41–46
26. Smukler SR, Arntfield ME, Razavi R et al (2011) The adult mouse and human pancreas contain rare multipotent stem cells that express insulin. Cell Stem Cell 8:281–293
27. Shamblott MJ, Axelman J, Wang S et al (1998) Derivation of pluripotent stem cells from cultured human primordial germ cells. Proc Natl Acad Sci USA 95:13726–13731
28. Thomson JA, Itskovitz-Eldor J, Shapiro SS et al (1998) Embryonic stem cell lines derived from human blastocysts. Science 282:1145–1147
29. Takahashi K, Tanabe K, Ohnuki M et al (2007) Induction of pluripotent stem cells from adult human fibroblasts by defined factors. Cell 131:861–872
30. Yu J, Vodyanik MA, Smuga-Otto K et al. (2007) Induced pluripotent stem cell lines derived from human somatic cells. Science 318:1917–1920
31. Takahashi K, Yamanaka S (2006) Induction of pluripotent stem cells from mouse embryonic and adult fibroblast cultures by defined factors. Cell 126:663–676
32. Zhao T, Zhang ZN, Rong Z, Xu Y (2011) Immunogenicity of induced pluripotent stem cells. Nature 474:212–215
33. Gerecht-Nir S, Cohen S, Ziskind A, Itskovitz-Eldor J (2004) Three-dimensional porous alginate scaffolds provide a conducive environment for generation of well-vascularized embryoid bodies from human embryonic stem cells. Biotechnol Bioeng 88:313–320
34. Gerecht S, Burdick JA, Ferreira LS et al (2007) Hyaluronic acid hydrogel for controlled self-renewal and differentiation of human embryonic stem cells. Proc Natl Acad Sci USA 104:11298–11303
35. Watt FM, Hogan BL (2000) Out of Eden: stem cells and their niches. Science 287:1427–1430
36. Kang SW, La WG, Kang JM, Park JH, Kim BS (2008) Bone morphogenetic protein-2 enhances bone regeneration mediated by transplantation of osteogenically undifferentiated bone marrow-derived mesenchymal stem cells. Biotechnol Lett 30:1163–1168
37. Karamboulas K, Dranse HJ, Underhill TM (2010) Regulation of BMP-dependent chondrogenesis in early limb mesenchyme by TGFbeta signals. J Cell Sci 123:2068–2076

38. Nostro MC, Farida SSO, Audrey H et al (2011) Stage-specific signaling through TGFb family members and WNT regulates patterning and pancreatic specification of human pluripotent stem cells. Development 138:861–871
39. Tuli R, Tuli S, Nandi S et al (2003) Transforming growth factor-beta-mediated chondrogenesis of human mesenchymal progenitor cells involves N-cadherin and mitogen-activated protein kinase and Wnt signaling cross-talk. J Biol Chem 278:41227–41236
40. Ying QL, Nichols J, Chambers I, Smith A (2003) BMP induction of Id proteins suppresses differentiation and sustains embryonic stem cell self-renewal in collaboration with STAT3. Cell 115:281–292
41. Sato N, Meijer L, Skaltsounis L, Greengard P, Brivanlou AH (2004) Maintenance of pluripotency in human and mouse embryonic stem cells through activation of Wnt signaling by a pharmacological GSK-3-specific inhibitor. Nat Med 10:55–63
42. Levy O, Ruvinov E, Reem T, Granot Y, Cohen S (2010) Highly efficient osteogenic differentiation of human mesenchymal stem cells by eradication of STAT3 signaling. Int J Biochem Cell Biol 42:1823–1830
43. Levy O, Dvir T, Tsur-Gang O, Granot Y, Cohen S (2008) Signal transducer and activator of transcription 3-A key molecular switch for human mesenchymal stem cell proliferation. Int J Biochem Cell Biol 40:2606–2618
44. Li W, Ding S (2009) Small molecules that modulate embryonic stem cell fate and somatic cell reprogramming. Trends Pharmacol Sci 31:36–45
45. Zhu S, Wurdak H, Wang J et al (2009) A small molecule primes embryonic stem cells for differentiation. Cell Stem Cell 4:416–426
46. Borowiak M, Maehr R, Chen S et al (2009) Small molecules efficiently direct endodermal differentiation of mouse and human embryonic stem cells. Cell Stem Cell 4:348–358
47. Chen S, Borowiak M, Fox JL et al (2009) A small molecule that directs differentiation of human ESCs into the pancreatic lineage. Nat Chem Biol 5:258–265
48. Tun T, Miyoshi H, Aung T et al (2002) Effect of growth factors on ex vivo bone marrow cell expansion using three-dimensional matrix support. Artif Organs 26:333–339
49. Levenberg S, Golub JS, Amit M, Itskovitz-Eldor J, Langer R (2002) Endothelial cells derived from human embryonic stem cells. Proc Natl Acad Sci USA 99:4391–4396
50. Kong HJ, Smith MK, Mooney DJ (2003) Designing alginate hydrogels to maintain viability of immobilized cells. Biomaterials 24:4023–4029
51. Dar A, Shachar M, Leor J, Cohen S (2002) Optimization of cardiac cell seeding and distribution in 3D porous alginate scaffolds. Biotechnol Bioeng 80:305–312
52. McCloskey KE, Gilroy ME, Nerem RM (2005) Use of embryonic stem cell-derived endothelial cells as a cell source to generate vessel structures in vitro. Tissue Eng 11: 497–505
53. Chen SS, Fitzgerald W, Zimmerberg J, Kleinman HK, Margolis L (2007) Cell–cell and cell-extracellular matrix interactions regulate embryonic stem cell differentiation. Stem Cells 25:553–561
54. Bhang SH, Lee YE, Cho SW et al (2007) Basic fibroblast growth factor promotes bone marrow stromal cell transplantation-mediated neural regeneration in traumatic brain injury. Biochem Biophys Res Commun 359:40–45
55. Martino MM, Mochizuki M, Rothenfluh DA et al (2009) Controlling integrin specificity and stem cell differentiation in 2D and 3D environments through regulation of fibronectin domain stability. Biomaterials 30:1089–1097
56. Rosenthal A, Macdonald A, Voldman J (2007) Cell patterning chip for controlling the stem cell microenvironment. Biomaterials 28:3208–3216
57. Tielens S, Declercq H, Gorski T et al (2007) Gelatin-based microcarriers as embryonic stem cell delivery system in bone tissue engineering: an in vitro study. Biomacromolecules 8:825–832
58. Rohanizadeh R, Swain MV, Mason RS (2008) Gelatin sponges (Gelfoam) as a scaffold for osteoblasts. J Mater Sci Mater Med 19:1173–1182

59. Zeng X, Zeng YS, Ma YH et al (2011) Bone marrow mesenchymal stem cells in a three dimensional gelatin sponge scaffold attenuate inflammation, promote angiogenesis and reduce cavity formation in experimental spinal cord injury. Cell Transplant 20. doi: 10.3727/096368911X566181
60. Ruhnke M, Ungefroren H, Zehle G et al (2003) Long-term culture and differentiation of rat embryonic stem cell-like cells into neuronal, glial, endothelial, and hepatic lineages. Stem Cells 21:428–436
61. Radice M, Brun P, Cortivo R et al (2000) Hyaluronan-based biopolymers as delivery vehicles for bone-marrow-derived mesenchymal progenitors. J Biomed Mater Res 50:101–109
62. Lisignoli G, Cristino S, Piacentini A et al (2005) Cellular and molecular events during chondrogenesis of human mesenchymal stromal cells grown in a three-dimensional hyaluronan based scaffold. Biomaterials 26:5677–5686
63. Kim J, Kim IS, Cho TH et al (2007) Bone regeneration using hyaluronic acid-based hydrogel with bone morphogenic protein-2 and human mesenchymal stem cells. Biomaterials 28:1830–1837
64. Awad HA, Wickham MQ, Leddy HA, Gimble JM, Guilak F (2004) Chondrogenic differentiation of adipose-derived adult stem cells in agarose, alginate, and gelatin scaffolds. Biomaterials 25:3211–3222
65. Alsberg E, Anderson KW, Albeiruti A, Franceschi RT, Mooney DJ (2001) Cell-interactive alginate hydrogels for bone tissue engineering. J Dent Res 80:2025–2029
66. Alsberg E, Anderson KW, Albeiruti A, Rowley JA, Mooney DJ (2002) Engineering growing tissues. Proc Natl Acad Sci USA 99:12025–12030
67. Chang JC, Hsu SH, Chen DC (2009) The promotion of chondrogenesis in adipose-derived adult stem cells by an RGD-chimeric protein in 3D alginate culture. Biomaterials 30:6265–6275
68. Connelly JT, Garcia AJ, Levenston ME (2007) Inhibition of in vitro chondrogenesis in RGD-modified three-dimensional alginate gels. Biomaterials 28:1071–1083
69. Dvir T, Benishti N, Shachar M, Cohen S (2006) A novel perfusion bioreactor providing a homogenous milieu for tissue regeneration. Tissue Eng 12:2843–2852
70. Dvir T, Kedem A, Ruvinov E et al (2009) Prevascularization of cardiac patch on the omentum improves its therapeutic outcome. Proc Natl Acad Sci USA 106:14990–14995
71. Dvir-Ginzberg M, Elkayam T, Aflalo ED, Agbaria R, Cohen S (2004) Ultrastructural and functional investigations of adult hepatocyte spheroids during in vitro cultivation. Tissue Eng 10:1806–1817
72. Dvir-Ginzberg M, Elkayam T, Cohen S (2008) Induced differentiation and maturation of newborn liver cells into functional hepatic tissue in macroporous alginate scaffolds. FASEB J 22:1440–1449
73. Dvir-Ginzberg M, Gamlieli-Bonshtein I, Agbaria R, Cohen S (2003) Liver tissue engineering within alginate scaffolds: effects of cell-seeding density on hepatocyte viability, morphology, and function. Tissue Eng 9:757–766
74. Dvir-Ginzberg M, Konson A, Cohen S, Agbaria R (2007) Entrapment of retroviral vector producer cells in three-dimensional alginate scaffolds for potential use in cancer gene therapy. J Biomed Mater Res B Appl Biomater 80:59–66
75. Elkayam T, Amitay-Shaprut S, Dvir-Ginzberg M, Harel M, Cohen S (2006) Enhancing the drug metabolism activities of C3A–a human hepatocyte cell line–by tissue engineering within alginate scaffolds. Tissue Eng 12:1357–1368
76. Freeman I, Cohen S (2009) The influence of the sequential delivery of angiogenic factors from affinity-binding alginate scaffolds on vascularization. Biomaterials 30:2122–2131
77. Freeman I, Kedem A, Cohen S (2008) The effect of sulfation of alginate hydrogels on the specific binding and controlled release of heparin-binding proteins. Biomaterials 29:3260–3268
78. Glicklis R, Shapiro L, Agbaria R, Merchuk JC, Cohen S (2000) Hepatocyte behavior within three-dimensional porous alginate scaffolds. Biotechnol Bioeng 67:344–353

79. Kedem A, Perets A, Gamlieli-Bonshtein I et al (2005) Vascular endothelial growth factor-releasing scaffolds enhance vascularization and engraftment of hepatocytes transplanted on liver lobes. Tissue Eng 11:715–722
80. Perets A, Baruch Y, Weisbuch F et al (2003) Enhancing the vascularization of three-dimensional porous alginate scaffolds by incorporating controlled release basic fibroblast growth factor microspheres. J Biomed Mater Res A 65:489–497
81. Re'em T, Tsur-Gang O, Cohen S (2010) The effect of immobilized RGD peptide in macroporous alginate scaffolds on TGFbeta1-induced chondrogenesis of human mesenchymal stem cells. Biomaterials 31:6746–6755
82. Ruvinov E, Leor J, Cohen S (2010) The effects of controlled HGF delivery from an affinity-binding alginate biomaterial on angiogenesis and blood perfusion in a hindlimb ischemia model. Biomaterials 31:4573–4582
83. Ruvinov E, Leor J, Cohen S (2011) The promotion of myocardial repair by the sequential delivery of IGF-1 and HGF from an injectable alginate biomaterial in a model of acute myocardial infarction. Biomaterials 32:565–578
84. Shachar M, Tsur-Gang O, Dvir T, Leor J, Cohen S (2011) The effect of immobilized RGD peptide in alginate scaffolds on cardiac tissue engineering. Acta Biomater 7:152–162
85. Stevens MM, Qanadilo HF, Langer R, Prasad Shastri V (2004) A rapid-curing alginate gel system: utility in periosteum-derived cartilage tissue engineering. Biomaterials 25:887–894
86. Levenberg S, Huang NF, Tsur-Gang O, Ruvinov E, Landa N et al (2009) The effects of peptide-based modification of alginate on left ventricular remodeling and function after myocardial infarction. Biomaterials 30:189–195
87. Barash Y, Dvir T, Tandeitnik P et al (2010) Electric field stimulation integrated into perfusion bioreactor for cardiac tissue engineering. Tissue Eng Part C Methods 16:1417–1426
88. Dvir T, Levy O, Shachar M, Granot Y, Cohen S (2007) Activation of the ERK1/2 cascade via pulsatile interstitial fluid flow promotes cardiac tissue assembly. Tissue Eng 13:2185–2193
89. Wang L, Shelton RM, Cooper PR et al (2003) Evaluation of sodium alginate for bone marrow cell tissue engineering. Biomaterials 24:3475–3481
90. Ma HL, Hung SC, Lin SY, Chen YL, Lo WH (2003) Chondrogenesis of human mesenchymal stem cells encapsulated in alginate beads. J Biomed Mater Res A 64:273–281
91. Dean SK, Yulyana Y, Williams G, Sidhu KS, Tuch BE (2006) Differentiation of encapsulated embryonic stem cells after transplantation. Transplantation 82:1175–1184
92. Lee CH, Singla A, Lee Y (2001) Biomedical applications of collagen. Int J Pharm 221:1–22
93. Schmidt CE, Baier JM (2000) Acellular vascular tissues: natural biomaterials for tissue repair and tissue engineering. Biomaterials 21:2215–2231
94. Ifkovits JL, Burdick JA (2007) Review: photopolymerizable and degradable biomaterials for tissue engineering applications. Tissue Eng 13:2369–2385
95. Chastain SR, Kundu AK, Dhar S, Calvert JW, Putnam AJ (2006) Adhesion of mesenchymal stem cells to polymer scaffolds occurs via distinct ECM ligands and controls their osteogenic differentiation. J Biomed Mater Res A 78:73–85
96. Martin I, Shastri VP, Padera RF et al (2001) Selective differentiation of mammalian bone marrow stromal cells cultured on three-dimensional polymer foams. J Biomed Mater Res 55:229–235
97. Levenberg S, Huang NF, Lavik E et al (2003) Differentiation of human embryonic stem cells on three-dimensional polymer scaffolds. Proc Natl Acad Sci USA 100:12741–12746
98. Hwang NS, Kim MS, Sampattavanich S et al (2006) Effects of three-dimensional culture and growth factors on the chondrogenic differentiation of murine embryonic stem cells. Stem Cells 24:284–291
99. Kraehenbuehl TP, Zammaretti P, Van der Vlies AJ et al (2008) Three-dimensional extracellular matrix-directed cardioprogenitor differentiation: systematic modulation of a synthetic cell-responsive PEG-hydrogel. Biomaterials 29:2757–2766

100. Salinas CN, Cole BB, Kasko AM, Anseth KS (2007) Chondrogenic differentiation potential of human mesenchymal stem cells photoencapsulated within poly(ethylene glycol)-arginine-glycine-aspartic acid-serine thiol-methacrylate mixed-mode networks. Tissue Eng 13:1025–1034
101. Salinas CN, Anseth KS (2009) Decorin moieties tethered into PEG networks induce chondrogenesis of human mesenchymal stem cells. J Biomed Mater Res A 90:456–464
102. Salinas CN, Anseth KS (2008) The influence of the RGD peptide motif and its contextual presentation in PEG gels on human mesenchymal stem cell viability. J Tissue Eng Regen Med 2:296–304
103. Yang F, Williams CG, Wang DA et al (2005) The effect of incorporating RGD adhesive peptide in polyethylene glycol diacrylate hydrogel on osteogenesis of bone marrow stromal cells. Biomaterials 26:5991–5998
104. Salinas CN, Anseth KS (2008) The enhancement of chondrogenic differentiation of human mesenchymal stem cells by enzymatically regulated RGD functionalities. Biomaterials 29:2370–2377
105. Callegari A, Bollini S, Iop L et al (2007) Neovascularization induced by porous collagen scaffold implanted on intact and cryoinjured rat hearts. Biomaterials 28:5449–5461
106. Fujimoto KL, Tobita K, Merryman WD et al (2007) An elastic, biodegradable cardiac patch induces contractile smooth muscle and improves cardiac remodeling and function in subacute myocardial infarction. J Am Coll Cardiol 49:2292–2300
107. Gaballa MA, Sunkomat JN, Thai H et al (2006) Grafting an acellular 3-dimensional collagen scaffold onto a non-transmural infarcted myocardium induces neo-angiogenesis and reduces cardiac remodeling. J Heart Lung Transplant 25:946–954
108. Kochupura PV, Azeloglu EU, Kelly DJ et al (2005) Tissue-engineered myocardial patch derived from extracellular matrix provides regional mechanical function. Circulation 112:I-144–I-149
109. Glicklis R, Merchuk JC, Cohen S (2004) Modeling mass transfer in hepatocyte spheroids via cell viability, spheroid size, and hepatocellular functions. Biotechnol Bioeng 86:672–680
110. Kloxin AM, Kloxin CJ, Bowman CN, Anseth KS (2010) Mechanical properties of cellularly responsive hydrogels and their experimental determination. Adv Mater 22:3484–3494
111. Benoit DS, Schwartz MP, Durney AR, Anseth KS (2008) Small functional groups for controlled differentiation of hydrogel-encapsulated human mesenchymal stem cells. Nat Mater 7:816–823
112. Gelain F, Bottai D, Vescovi A, Zhang S (2006) Designer self-assembling peptide nanofiber scaffolds for adult mouse neural stem cell 3-dimensional cultures. PLoS One 1:e119
113. Silva GA, Czeisler C, Niece KL et al (2004) Selective differentiation of neural progenitor cells by high-epitope density nanofibers. Science 303:1352–1355
114. Zhao X, Zhang S (2007) Designer self-assembling peptide materials. Macromol Biosci 7:13–22
115. Ki CS, Park SY, Kim HJ et al (2008) Development of 3-D nanofibrous fibroin scaffold with high porosity by electrospinning: implications for bone regeneration. Biotechnol Lett 30:405–410
116. Dvir T, Tsur-Gang O, Cohen S (2005) "Designer" scaffolds for tissue engineering and regeneration. Israel J Chem 45:487–494
116. Jones FS, Jones PL (2000) The tenascin family of ECM glycoproteins: structure, function, and regulation during embryonic development and tissue remodeling. Dev Dyn 218:235–259
118. Loeser RF (2002) Integrins and cell signaling in chondrocytes. Biorheology 39:119–124
119. Makino H, Sugiyama H, Kashihara N (2000) Apoptosis and extracellular matrix-cell interactions in kidney disease. Kidney Int Suppl 77:S67–75
120. Boudreau NJ, Jones PL (1999) Extracellular matrix and integrin signalling: the shape of things to come. Biochem J 339 (Pt 3):481–488

121. Lukashev ME, Werb Z (1998) ECM signalling: orchestrating cell behaviour and misbehaviour. Trends in Cell Biol 8:437–441
122. Gumbiner BM (1996) Cell adhesion: the molecular basis of tissue architecture and morphogenesis. Cell 84:345–357
123. Shin H, Jo S, Mikos AG (2003) Biomimetic materials for tissue engineering. Biomaterials 24:4353–4364
124. Engelmayr GC, Jr., Cheng M, Bettinger CJ et al (2008) Accordion-like honeycombs for tissue engineering of cardiac anisotropy. Nat Mater 7:1003–1010
125. Kloxin AM, Kasko AM, Salinas CN, Anseth KS (2009) Photodegradable hydrogels for dynamic tuning of physical and chemical properties. Science 324:59–63
126. Zhao X, Kim J, Cezar CA et al (2011) Active scaffolds for on-demand drug and cell delivery. Proc Natl Acad Sci USA 108:67–72
127. Pierschbacher MD, Ruoslahti E (1984) Variants of the cell recognition site of fibronectin that retain attachment-promoting activity. Proc Natl Acad Sci USA 81:5985–5988
128. Ruoslahti E, Pierschbacher MD (1986) Arg-Gly-Asp: a versatile cell recognition signal. Cell 44:517–518
129. Ruoslahti E, Pierschbacher MD (1987) New perspectives in cell adhesion: RGD and integrins. Science 238:491–497
130. Hersel U, Dahmen C, Kessler H (2003) RGD modified polymers: biomaterials for stimulated cell adhesion and beyond. Biomaterials 24:4385–4415
131. Rowley JA, Mooney DJ (2002) Alginate type and RGD density control myoblast phenotype. J Biomed Mater Res 60:217–223
132. Hwang NS, Varghese S, Zhang Z, Elisseeff J (2006) Chondrogenic differentiation of human embryonic stem cell-derived cells in arginine-glycine-aspartate-modified hydrogels. Tissue Eng 12:2695–2706
133. Alvarez-Barreto JF, Shreve MC, Deangelis PL, Sikavitsas VI (2007) Preparation of a functionally flexible, three-dimensional, biomimetic poly(L-lactic acid) scaffold with improved cell adhesion. Tissue Eng 13:1205–1217
134. Sawyer AA, Weeks DM, Kelpke SS, McCracken MS, Bellis SL (2005) The effect of the addition of a polyglutamate motif to RGD on peptide tethering to hydroxyapatite and the promotion of mesenchymal stem cell adhesion. Biomaterials 26:7046–7056
135. Nuttelman CR, Tripodi MC, Anseth KS (2005) Synthetic hydrogel niches that promote hMSC viability. Matrix Biol 24:208–218
136. Li YJ, Chung EH, Rodriguez RT, Firpo MT, Healy KE (2006) Hydrogels as artificial matrices for human embryonic stem cell self-renewal. J Biomed Mater Res A 79:1–5
137. Sapir Y, Kryukov O, Cohen S (2011) Integration of multiple cell–matrix interactions into alginate scaffolds for promoting cardiac tissue regeneration. Biomaterials 32:1838–1847
138. Lutolf MP, Lauer-Fields JL, Schmoekel HG et al (2003) Synthetic matrix metalloproteinase-sensitive hydrogels for the conduction of tissue regeneration: engineering cell-invasion characteristics. Proc Natl Acad Sci USA 100:5413–5418
139. Raeber GP, Lutolf MP, Hubbell JA (2005) Molecularly engineered PEG hydrogels: a novel model system for proteolytically mediated cell migration. Biophys J 89:1374–1388
140. Lutolf MP, Weber FE, Schmoekel HG et al (2003) Repair of bone defects using synthetic mimetics of collagenous extracellular matrices. Nat Biotechnol 21:513–518
141. Lutolf MP, Hubbell JA (2005) Synthetic biomaterials as instructive extracellular microenvironments for morphogenesis in tissue engineering. Nat Biotechnol 23:47–55
142. Saito A, Suzuki Y, Ogata S, Ohtsuki C, Tanihara M (2003) Activation of osteo-progenitor cells by a novel synthetic peptide derived from the bone morphogenetic protein-2 knuckle epitope. Biochim Biophys Acta 1651:60–67
143. Saito A, Suzuki Y, Ogata S, Ohtsuki C, Tanihara M (2004) Prolonged ectopic calcification induced by BMP-2-derived synthetic peptide. J Biomed Mater Res A 70:115–121
144. Saito A, Suzuki Y, Ogata S, Ohtsuki C, Tanihara M (2005) Accelerated bone repair with the use of a synthetic BMP-2-derived peptide and bone-marrow stromal cells. J Biomed Mater Res A 72:77–82

145. Saito A, Suzuki Y, Kitamura M et al (2006) Repair of 20-mm long rabbit radial bone defects using BMP-derived peptide combined with an alpha-tricalcium phosphate scaffold. J Biomed Mater Res A 77:700–706
146. Lee JS, Murphy WL (2010) Modular peptides promote human mesenchymal stem cell differentiation on biomaterial surfaces. Acta Biomater 6:21–28
147. Choi YJ, Lee JY, Park JH et al (2010) The identification of a heparin binding domain peptide from bone morphogenetic protein-4 and its role on osteogenesis. Biomaterials 31:7226–7238
148. Chen Y, Webster TJ (2009) Increased osteoblast functions in the presence of BMP-7 short peptides for nanostructured biomaterial applications. J Biomed Mater Res A 91:296–304
149. Bergeron E, Senta H, Mailloux A et al (2009) Murine preosteoblast differentiation induced by a peptide derived from bone morphogenetic proteins-9. Tissue Eng Part A 15:3341–3349
150. Moore NM, Lin NJ, Gallant ND, Becker ML (2011) Synergistic enhancement of human bone marrow stromal cell proliferation and osteogenic differentiation on BMP-2-derived and RGD peptide concentration gradients. Acta Biomater 7:2091–2100
151. Zouani OF, Chollet C, Guillotin B, Durrieu MC (2010) Differentiation of pre-osteoblast cells on poly(ethylene terephthalate) grafted with RGD and/or BMPs mimetic peptides. Biomaterials 31:8245–8253
152. Park H, Temenoff JS, Tabata Y, Caplan AI, Mikos AG (2007) Injectable biodegradable hydrogel composites for rabbit marrow mesenchymal stem cell and growth factor delivery for cartilage tissue engineering. Biomaterials 28:3217–3227
153. Ferreira LS, Gerecht S, Fuller J et al (2007) Bioactive hydrogel scaffolds for controllable vascular differentiation of human embryonic stem cells. Biomaterials 28:2706–2717
154. Carpenedo RL, Bratt-Leal AM, Marklein RA et al (2009) Homogeneous and organized differentiation within embryoid bodies induced by microsphere-mediated delivery of small molecules. Biomaterials 30:2507–2515
155. Zisch AH, Schenk U, Schense JC, Sakiyama-Elbert SE, Hubbell JA (2001) Covalently conjugated VEGF–fibrin matrices for endothelialization. J Control Release 72:101–113
156. Fan VH, Tamama K, Au A et al (2007) Tethered epidermal growth factor provides a survival advantage to mesenchymal stem cells. Stem Cells 25:1241–1251
157. Swindle CS, Tran KT, Johnson TD et al (2001) Epidermal growth factor (EGF)-like repeats of human tenascin-C as ligands for EGF receptor. J Cell Biol 154:459–468
158. Phillippi JA, Miller E, Weiss L et al (2008) Microenvironments engineered by inkjet bioprinting spatially direct adult stem cells toward muscle- and bone-like subpopulations. Stem Cell 26:127–134
159. Miller ED, Fisher GW, Weiss LE, Walker LM, Campbell PG (2006) Dose-dependent cell growth in response to concentration modulated patterns of FGF-2 printed on fibrin. Biomaterials 27:2213–2221
160. Miller ED, Li K, Kanade T et al (2011) Spatially directed guidance of stem cell population migration by immobilized patterns of growth factors. Biomaterials 32:2775–2785
161. Bishop JR, Schuksz M, Esko JD (2007) Heparan sulphate proteoglycans fine-tune mammalian physiology. Nature 446:1030–1037
162. Chen BL, Arakawa T, Hsu E et al (1994) Strategies to suppress aggregation of recombinant keratinocyte growth factor during liquid formulation development. J Pharm Sci 83:1657–1661
163. Shriver Z, Liu D, Sasisekharan R (2002) Emerging views of heparan sulfate glycosaminoglycan structure/activity relationships modulating dynamic biological functions. Trends Cardiovasc Med 12:71–77
164. Casu B, Lindahl U (2001) Structure and biological interactions of heparin and heparan sulfate. Adv Carbohydr Chem Biochem 57:159–206
165. Wu ZL, Zhang L, Yabe T et al (2003) The involvement of heparan sulfate (HS) in FGF1/HS/FGFR1 signaling complex. J Biol Chem 278:17121–17129

166. Pye DA, Vives RR, Turnbulli JE, Hyde P, Gallagher JT (1998) Heparan sulfate oligosaccharides require 6-O-sulfation for promotion of basic fibroblast growth factor mitogenic activity. J Biol Chem 273:22936–22942
167. Raman R, Sasisekharan V, Sasisekharan R (2005) Structural insights into biological roles of protein-glycosaminoglycan interactions. Chem Biol 12:267–277
168. Forsten-Williams K, Chua CC, Nugent MA (2005) The kinetics of FGF-2 binding to heparan sulfate proteoglycans and MAP kinase signaling. J Theor Biol 233:483–499
169. Jakobsson L, Kreuger J, Holmborn K et al (2006) Heparan sulfate in trans potentiates VEGFR-mediated angiogenesis. Dev Cell 10:625–634
170. Benoit DS, Anseth KS (2005) Heparin functionalized PEG gels that modulate protein adsorption for hMSC adhesion and differentiation. Acta Biomater 1:461–470
171. Benoit DS, Collins SD, Anseth KS (2007) Multifunctional hydrogels that promote osteogenic hMSC differentiation through stimulation and sequestering of BMP2. Adv Funct Mater 17:2085–2093
172. Willerth SM, Rader A, Sakiyama-Elbert SE (2008) The effect of controlled growth factor delivery on embryonic stem cell differentiation inside fibrin scaffolds. Stem Cell Res 1:205–218
173. Webber MJ, Han X, Murthy SN et al (2010) Capturing the stem cell paracrine effect using heparin-presenting nanofibres to treat cardiovascular diseases. J Tissue Eng Regen Med 4:600–610
174. Oschatz C, Maas C, Lecher B et al (2011) Mast cells increase vascular permeability by heparin-initiated bradykinin formation in vivo. Immunity 34:258–268
175. Kloxin AM, Tibbitt MW, Anseth KS (2010) Synthesis of photodegradable hydrogels as dynamically tunable cell culture platforms. Nat Protoc 5:1867–1887
176. Guo WH, Frey MT, Burnham NA, Wang YL (2006) Substrate rigidity regulates the formation and maintenance of tissues. Biophys J 90:2213–2220
177. Engler AJ, Griffin MA, Sen S et al (2004) Myotubes differentiate optimally on substrates with tissue-like stiffness: pathological implications for soft or stiff microenvironments. J Cell Biol 166:877–887
178. Boontheekul T, Hill EE, Kong HJ, Mooney DJ (2007) Regulating myoblast phenotype through controlled gel stiffness and degradation. Tissue Eng 13:1431–1442
179. Collin O, Tracqui P, Stephanou A et al (2006) Spatiotemporal dynamics of actin-rich adhesion microdomains: influence of substrate flexibility. J Cell Sci 119:1914–1925
180. Engler AJ, Carag-Krieger C, Johnson CP et al (2008) Embryonic cardiomyocytes beat best on a matrix with heart-like elasticity: scar-like rigidity inhibits beating. J Cell Sci 121:3794–3802
181. Saez A, Ghibaudo M, Buguin A, Silberzan P, Ladoux B (2007) Rigidity-driven growth and migration of epithelial cells on microstructured anisotropic substrates. Proc Natl Acad Sci USA 104:8281–8286
182. Discher DE, Janmey P, Wang YL (2005) Tissue cells feel and respond to the stiffness of their substrate. Science 310:1139–1143
183. McBeath R, Pirone DM, Nelson CM, Bhadriraju K, Chen CS (2004) Cell shape, cytoskeletal tension, and RhoA regulate stem cell lineage commitment. Dev Cell 6:483–495
184. Ruiz SA, Chen CS (2008) Emergence of patterned stem cell differentiation within multicellular structures. Stem Cells 26:2921–2927
185. Karp JM, Yeh J, Eng G et al (2007) Controlling size, shape and homogeneity of embryoid bodies using poly(ethylene glycol) microwells. Lab Chip 7:786–794
186. Engler AJ, Sen S, Sweeney HL, Discher DE (2006) Matrix elasticity directs stem cell lineage specification. Cell 126:677–689
187. Huebsch N, Arany PR, Mao AS et al (2010) Harnessing traction-mediated manipulation of the cell/matrix interface to control stem-cell fate. Nat Mater 9:518–526
188. Tse JR, Engler AJ () Stiffness gradients mimicking in vivo tissue variation regulate mesenchymal stem cell fate. PLoS One 6:e15978

189. Schek RM, Taboas JM, Segvich SJ, Hollister SJ, Krebsbach PH (2004) Engineered osteochondral grafts using biphasic composite solid free-form fabricated scaffolds. Tissue Eng 10:1376–1385
190. Holland TA, Bodde EW, Baggett LS et al (2005) Osteochondral repair in the rabbit model utilizing bilayered, degradable oligo(poly(ethylene glycol) fumarate) hydrogel scaffolds. J Biomed Mater Res A 75:156–167
191. Natesan S, Zhang G, Baer DG et al (2011) A bilayer construct controls adipose-derived stem cell differentiation into endothelial cells and pericytes without growth factor stimulation. Tissue Eng Part A 17:941–953
192. Guo X, Park H, Liu G et al (2009) In vitro generation of an osteochondral construct using injectable hydrogel composites encapsulating rabbit marrow mesenchymal stem cells. Biomaterials 30:2741–2752
193. Dormer NH, Singh M, Wang L, Berkland CJ, Detamore MS (2010) Osteochondral interface tissue engineering using macroscopic gradients of bioactive signals. Ann Biomed Eng 38:2167–2182
194. Wang X, Wenk E, Zhang X et al (2009) Growth factor gradients via microsphere delivery in biopolymer scaffolds for osteochondral tissue engineering. J Control Release 134:81–90
195. Winkler IG, Barbier V, Wadley R et al (2010) Positioning of bone marrow hematopoietic and stromal cells relative to blood flow in vivo: serially reconstituting hematopoietic stem cells reside in distinct nonperfused niches. Blood 116:375–385
196. Meinel L, Karageorgiou V, Fajardo R et al (2004) Bone tissue engineering using human mesenchymal stem cells: effects of scaffold material and medium flow. Ann Biomed Eng 32:112–122
197. Gerecht-Nir S, Cohen S, Itskovitz-Eldor J (2004) Bioreactor cultivation enhances the efficiency of human embryoid body (hEB) formation and differentiation. Biotechnol Bioeng 86:493–502
198. Braccini A, Wendt D, Jaquiery C et al (2005) Three-dimensional perfusion culture of human bone marrow cells and generation of osteoinductive grafts. Stem Cell 23:1066–1072
199. Sikavitsas VI, Bancroft GN, Holtorf HL, Jansen JA, Mikos AG (2003) Mineralized matrix deposition by marrow stromal osteoblasts in 3D perfusion culture increases with increasing fluid shear forces. Proc Natl Acad Sci USA 100:14683–14688
200. van den Dolder J, Bancroft GN, Sikavitsas VI et al (2003) Flow perfusion culture of marrow stromal osteoblasts in titanium fiber mesh. J Biomed Mater Res A 64:235–241
201. Terraciano V, Hwang N, Moroni L et al (2007) Differential response of adult and embryonic mesenchymal progenitor cells to mechanical compression in hydrogels. Stem Cells 25:2730–2738
202. Rodrigues CA, Fernandes TG, Diogo MM, da Silva CL,Cabral JM (2011) Stem cell cultivation in bioreactors. Biotechnol Adv. doi: 10.1016/j.biotechadv.2011.06.009
203. Ruvinov E, Cohen S (2011) Instructive biomaterials for myocardial regeneration and repair. In: Zilberman M (ed) Active implants and scaffolds for tissue regeneration. Springer, Berlin.

Stem Cell Differentiation Depending on Different Surfaces

Sonja Kress, Anne Neumann, Birgit Weyand and Cornelia Kasper

Abstract Mesenchymal stem cells and 3D biomaterials are a potent assembly in tissue engineering. Today, a sizable number of biomaterials has been characterized for special tissue engineering applications. However, diverse material properties, such as soft or hard biomaterials, have a specific influence on cell behavior. Not only the cell attachment and proliferation, but also differentiation is controlled by the microenvironment. Material characteristics such as pore size, stiffness, roughness, and geometry affect not only the cell attachment and proliferation, but also the differentiation behavior of mesenchymal stem cells. Optimization of these features might enable direct differentiation without adjustment of the culture medium by applying expensive growth or differentiation factors. Future aspects include the design of multilayered biomaterials, where each zone fulfills a distinct function. Moreover, the embedding of growth and differentiation factors into the matrix with a controlled release rate might be advantageous to direct differentiation.

Keywords 3D biomaterials · Focal adhesion · Geometry · Mesenchymal stem cells · Roughness · Stiffness

---

S. Kress · A. Neumann
Institute for Technical Chemistry, Leibniz University Hannover,
Callinstraße 3, 30167 Hanover, Germany
e-mail: kress@iftc.uni-hannover.de

A. Neumann
e-mail: neumann@iftc.uni-hannover.de

B. Weyand
Department of Plastic, Hand and Reconstructive Surgery,
Hannover Medical School, OE 6260, Carl-Neubergstr. 1,
30625 Hanover, Germany
e-mail: weyand.birgit@mh-hannover.de

C. Kasper (✉)
Department of Biotechnology, Institute for Applied Microbiology,
University of Material Resource and Life Science, Muthgasse 18,
1180 Vienna, Austria
e-mail: cornelia.kasper@boku.ac.at

Abbreviations

3D	Three-dimensional
AD	adipose
BM	Bone marrow
ECM	Extracellular matrix
ERK	Extracellular-signal-regulated kinase
HAP	Hydroxyapatite
hMSC	Human mesenchymal stem cells
MSC	Mesenchymal stem cells
UC	Umbilical cord

Contents

1	Introduction	264
2	Bone-Marrow-Derived Mesenchymal Stem Cells and Their Alternatives	265
3	Biomaterials as an Artificial Extracellular Matrix	265
4	The Influence of Biomaterial Properties on Stem Cell Differentiation	267
	4.1 Porosity and Pore Size	268
	4.2 Biomaterial Stiffness	270
	4.3 Microscale and Nanoscale Topography	270
	4.4 Geometry	272
	4.5 Composite Materials	273
5	Sensing the Microenvironment	275
6	Further Perspectives	277
References		278

1 Introduction

Tissue engineering is a young and interdisciplinary field of biotechnology which aims at regeneration or replacement of defective tissues. To achieve full regeneration of a tissue, many scientific and technical questions have to be answered and the optimal parameters for each tissue have to be defined.

Generally, there are three approaches in tissue engineering [1]:

1. The isolation of cells to inject them into the damaged region, in order to support the needed function.
2. The introduction of growth and differentiation factors into the target location.
3. The use of three-dimensional (3D) biomaterials in combination with cells.

The first two approaches can only be considered for small defects; therefore, the seeding of cells on a 3D matrix has become the method of choice. Within this approach, the selection of the cell type and the composition of the culture medium play an important role. The influence of soluble or immobilized growth factors on the differentiation of mesenchymal stem cells (MSC) has been studied in detail [2–4]. Moreover, the application of human MSC (hMSC) has become more and more attractive for tissue engineering [5], but also for clinical trails [6]. However, the key parameter of the third approach seems to be the selection of the proper biomaterial [7]. Presently, the physical characteristics which might control stem cell fate have rarely been explored. This chapter will highlight an important control mechanism of stem cell differentiation: the interaction of MSC with their surrounding microenvironment.

2 Bone-Marrow-Derived Mesenchymal Stem Cells and Their Alternatives

Today, hMSC isolated from bone marrow bm are considered as the "gold standard" [8]. Among other criteria, they have the ability to differentiate into several lineages, such as adipocytes, chondrocytes, and osteoblasts. Nevertheless, bm as a source of MSC has several disadvantages, such as an invasive and painful collection procedure with a high risk of infection, a low frequency of MSC, and a differing quality depending on the donor's age [8]. Thus, MSC derived from other sources, for example, adipose tissue [9], umbilical cord uc tissue [10, 11], and uc blood [12], represent an alternative and have shown promising results. Especially MSC derived from postnatal tissues have gained the attention of several research groups, since these tissues are easily accessible and can be processed directly after birth. Furthermore, the frequency of MSC is much higher in postnatal tissues, and the cells show a higher proliferation capacity compared with bm-derived MSC. Summing up the increasing research activities over the last decade, uc MSC seem to be a valuable cell source with a great potential for cell-based therapies which initiate tissue repair [8, 13].

3 Biomaterials as an Artificial Extracellular Matrix

Although it is well known that MSC and tissue cells grow best in a defined 3D surrounding of macromolecules, most tissue engineering experiments are still conducted on flat-bottom surfaces of culture vessels made of glass or plastic [14]. The first experiments to culture adherent cells on appropriate biomaterials were performed during the mid-1980s [15]. Since then, research activity in the area of biomaterials for tissue engineering has increased continuously (Fig. 1). Today, it is known that an appropriate biomaterial mimics the extracellular matrix (ECM) in vivo and a culture substrate in vitro. For tissue engineering applications the 3D biomaterial should mimic the ECM properties of the specific tissue that needs to be regenerated. In vitro these artificial matrices support cell proliferation, migration,

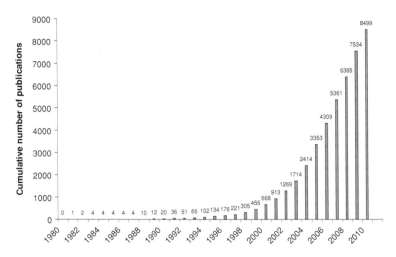

Fig. 1 Cumulative number of publications over the last two decades dealing with biomaterial and tissue engineering (entries in PubMed with the terms "biomaterial" and "tissue engineering" until December 2010)

orientation, adhesion, contraction, maturation, and the synthesis of ECM components. After transplantation, the main function of the matrix in vivo is to provide mechanical support until the newly formed tissue assumes this role [16]. Therefore, it is important to adjust the biomaterial properties as precisely as possible [17]. The biomaterial features which have a huge impact on cell differentiation are the surface charge, the substructure, the mechanical properties of the material, and the degradation kinetics [17].

In a healthy tissue, cell differentiation is controlled by various environmental factors of the cell surroundings, such as interstitial body fluids, fibrous tissue, muscles, and tendons as well as the skeleton, which serve together as a natural matrix [18]. The selection of an artificial biomaterial for in vitro cultivation and differentiation is often a critical parameter. For tissue engineering applications, four groups of biomaterials have been studied intensively [19] (Table 1).

However, naturally derived biomaterials often lack the mechanical strength required by certain tissues, such as bone [7]. Other disadvantages are the relatively low availability of these materials and their batch-to-batch variation [16]. Synthetic inorganic materials, such as β-tricalcium phosphate and hydroxyapatite (HAP), have often been used to engineer hard tissue matrices [20]. These materials are considered to be osteoinductive as their surface supports osteoblastic adhesion, growth, and differentiation [7]. But the rather brittle behavior of pure synthetic inorganic materials led to the fabrication of composite materials [21], which combine synthetic inorganic materials with naturally derived components [22] (Sect. 4.5).

Table 1 Biomaterials for tissue engineering applications

Origin	Biomaterial	References
Synthetic organic materials	Aliphatic polyester	[89]
	Poly(ethylene glycol)	[90]
	Polycaprolactone	[91–93]
	Poly(lactic acid)	[94, 95]
	Poly(lactic-*co*-glycolic acid)	[96, 97]
Synthetic inorganic materials	β-Tricalcium phosphate	[98]
	Hydroxyapatite	[98–102]
	Glass ceramics	[103, 104]
Organic materials of natural origin	Collagen	[105–109]
	Gelatine	[110–112]
	Fibrin	[113, 114]
	Chitin/chitosan	[115–118]
	Agarose	[111, 112]
	Alginate	[119, 120]
Inorganic materials of natural origin	Corraline hydroxyapatite	[121, 122]
	Magnesium (alloys)	[123–125]

Generally, the ideal biomaterial should fulfill the following demands [17, 19, 22]. It should be:

1. Biocompatible.
2. Biodegradable or bioresorbable with a controllable degradation rate. Moreover, the degradation products should be nontoxic and should not influence the tissue regeneration.
3. Three-dimensional (3D) and highly porous with an interconnected architecture to allow an even cell distribution and to facilitate the transportation of oxygen, nutrients, and waste products.
4. Mechanically robust to withstand forces in the area of transplantation.
5. Equipped with a suitable surface for cell attachment, proliferation, and differentiation.
6. Easily processed into various shapes and sizes to facilitate incorporation in the damaged region.
7. Sterilizable to avoid bacterial, fungal, and viral contamination.

4 The Influence of Biomaterial Properties on Stem Cell Differentiation

Controlling stem cell differentiation is one of the key issues in tissue engineering. It is well known that differentiation is influenced by a complex interplay of crucial parameters, for example, the combination of growth factors. But recent studies demonstrated that not only soluble growth factors but also the local microenvironment plays an important role in initiating and controlling MSC differentiation [23]. The number of reports demonstrating that cellular behavior is modulated by biomaterial properties such as pore size, topography, geometry, and stiffness is increasing steadily. These properties are described to control cell distribution, adhesion, migration, cell shape, proliferation, and finally differentiation. Within this chapter we will focus on the optimal biomaterial parameters for MSC differentiation into the osteogenic lineage.

4.1 Porosity and Pore Size

The porosity (percentage of void space in a solid) and the pore size of biomaterials play a critical role in tissue formation. As stated earlier, a certain interconnected porous network is necessary to distribute oxygen as well as nutrients and to eliminate waste material, but optimal pore sizes for the regeneration of different tissues have not yet been reported [24]. For bone tissue engineering applications, pore sizes between 10 and 2250 µm have been used, resulting in various degrees of tissue formation and ingrowth [24–26]. Nevertheless, early studies of Hulbert et al. [27] showed the minimum pore size required to generate mineralized bone to be 100 µm. Matrices with a smaller pore size resulted in either ingrowth of unmineralized tissue or fibrous tissue. Supposedly, too small pore sizes limit the transport of nutrients and the cell migration [24]. It was demonstrated that only pores sizes above 300 µm result in vascularized tissue formation [28]. Additionally, increased porosity and pore sizes facilitate bone ingrowth [29], even though this depends on the biomaterial [28]. In contrast, if the pores are too large, the decrease in surface area limits cell adhesion [30] and increasing porosity affects the load-bearing capacity [31]. Therefore, porosity and pore size have to be within a specific range to maintain the balance between the optimal pore size for cell migration, mass transportation, vascularization, and specific surface area for cell attachment as well as mechanical stability.

A variety of optimal pore sizes for different tissues have been proposed (Table 2). These diverse numbers for the optimal pore size and porosity might be caused by results obtained from various cell types or MSC derived from different origins. Another reason could be the difference in biomaterial fabrication, which results in differing pore architectures [32].

In many cases it has been reported that for bone tissue engineering applications pores greater than 300 µm facilitate capillary formation and therefore direct osteogenesis [25, 28, 33]. To achieve osteoconduction, pore sizes between 100 and 400 µm are generally preferred [34].

Furthermore, differing porosities and pore sizes have described for bone tissue in vivo. Human and mammalian bone is classified into two types:

1. Cortical bone, which is compact and usually located in the shaft of long bones (diaphysis) and in the outer bone shell

Table 2 Proposed optimal porosities and pore sizes for the regeneration of different tissues

Differentiation	Cells	Biomaterial	Optimal pore size (μm)	Optimal porosity (%)	References
Osteogenic	Primary fetal bovine osteoblasts	Polycaprolactone–hydroxyapatite	450–750	60–70	[26]
Osteogenic	bm-hMSC	β-Tricalcium phosphate	200–600	65	[29]
Osteogenic	Rabbit bm-derived osteoblasts	Natural coral	200	36	[126]
Osteogenic	bm-hMSC	Coralline hydroxyapatite	200	75	[127]
Osteogenic	Rat bm MSC	Sintered titanium fiber mesh	250	86	[128]
Osteogenic	ad-hMSC	ZrO$_2$ ceramic	600	80–89	[129, 130]
Osteogenic	Canine bm-MSC	β-Tricalcium phosphate	400–500	70	[131]
Chondrogenic	ad-hMSC	Polycaprolactone	370–400	80–97	[32]
Chondrogenic	bm-hMSC	Polycaprolactone	100–150	–	[132]
Chondrogenic	Rabbit chondrocytes	Chitosan-based hyaluronic acid hybrid polymer fibers	400	–	[133]
Chondrogenic	Human chondrocytes	Polycarbonate	8	–	[134]
Chondrogenic	Bovine chondrocytes	Polyurethane	<5–60	–	[135]
Myogenic	Canine bm MSC	50:50 poly(lactic-co-glycolic acid)	50–200	–	[136]
Hepatic	ad-hMSC	Poly(lactic-co-glycolic acid)	120–200	50	[137]

ad adipose-derived, *bm* bone marrow, *hMSC* human mesenchymal stem cells, *MSC* mesenchymal stem cells

2. Trabecular bone, also known as spongy bone and which is located in the end of long bones (epiphysis), in vertebrae, and in flat bones such as the pelvis

Cortical bone is much denser, with a porosity of only 5–10% [35], whereas the porosity of trabecular bone ranges between 78 and 92% [36]. Because of the low porosity of cortical bone, there are no descriptions of pore sizes. However, various canals such as the Haversian canals cross the cortical bone. The Haversian canals contain blood vessels and nerve cells and are therefore responsible for nutrient supply and the conduction of nervous stimuli. These longitudinal channels have an average diameter of 50 μm [37]. The trabecular spacing was measured by Hakulinen et al. [38] via a quantitative ultrasound technique. A mean pore size of 719 ± 110 μm was calculated and compared with the data obtained by micro-computed tomography. The micro-computed tomography data showed a much higher variation of trabecular spacing, with pore sizes ranging from 482 to

947 μm. The difference in bone architecture indicates that the selection of the biomaterial with a certain pore size depends on the location to where it is going to be transplanted.

4.2 Biomaterial Stiffness

Even though most cells in vivo attach to and proliferate on rather soft matrices, in vitro research today is mainly performed on stiff surfaces such as glass and plastic [14]. However, the influence of matrix stiffness on many cellular processes, for example, migration [39, 40], adhesion [41, 42], cell shape [40, 42, 43], and proliferation [44], has already been verified. Furthermore, many recent publications confirm that matrix stiffness determines the fate of MSC and consequently their differentiation [45, 46]. The results of these experiments provide the evidence that MSC are able to sense the stiffness of their local microenvironment [47]. Discher et al. stated that cells respond to the stiffness of their surrounding environment by adjusting their adhesion and their cytoskeleton [138]. Their feedback to matrix stiffness is probably a change in the expression of integrins, cadherins, and cytoskeletal proteins [43].

It has been found that hMSC respond in vitro differently to elasticities similar to the in vivo tissue stiffness of brain (0.1–1 kPa), muscle (8–17 kPa), and nascent bone (more than 34 kPa) [45]. In vitro hMSC expressed neurogenic, myogenic, and osteogenic key markers when cultivated in the same medium, but on biomaterials with the stated elasticities. Directed differentiation was demonstrated without the addition of expensive growth factors [46]. Furthermore, Engler et al. [45] stated that soluble growth factors tend to be less selective than matrix stiffness regarding cell differentiation. MSC were cultured on a soft matrix (0.1–1 kPa) in growth medium for either 1 or 3 weeks. Then either myogenic or osteogenic differentiation medium was added. A decrease in gene expression of myogenic or osteogenic markers was only observed during the initial week of cultivation. After one week, MSC committed to the lineage specified by the substrate stiffness. These results emphasize that the surface stiffness of biomaterials has a significant influence on the determination of cell fate [46]. But by which mechanism do MSC sense the biomaterial stiffness and how do they transmit the information into differentiation signals? A short answer to this question will be given in Sect. 5.

4.3 Microscale and Nanoscale Topography

It is widely accepted that the topography of a biomaterials surface determines the biologic reaction. The topography of matrices influences the cell adhesion [48], orientation, and differentiation. However, these effects seem to be cell-dependent. Different cell types have been observed to show a preference for either a smooth or a rough topography [49]. According to their origin, osteoblasts prefer rougher surfaces, whereas fibroblasts favor smooth surfaces. In bone tissue engineering, it has been shown that an increased surface roughness enhances osteogenic differentiation [18, 50, 51]. However, the results are conflicting. Some publications report no change in response to rough surfaces [52] or even a reduction in cellular response [53]. One significant problem might be the low reproducibility of roughness. Even though two surfaces exhibit the same roughness, their topography can appear very different. Furthermore, intrabatch and interbatch variations of biomaterials may cause difficulties. Other reports suggest the spacing between ligated integrins, which are essential for adhesion and signal transmission, to be less than 70 nm. Thus, larger nanoscale spacing fails to trigger differentiation signals [54–56].

However, the behavior and functionality of cells may be influenced by different topographic sizes, ranging from macroscale to microscale and nanoscale features [57]. The smallest feature size shown to affect cell behavior is 10 nm [58]. A sizable number of publications have described the positive influence of nanoscale patterns on the differentiation of MSC. Osteogenic differentiation, for example, has been observed on nanopattern surfaces of 17–25 nm [59], 50 nm [60], and 100 nm [61, 62]. Myogenic differentiation has been demonstrated by cultivating bovine aortic endothelial cells on nanotubes of 1 μm length and 30 nm average pore diameter [63]. Even differentiation of hMSC into the neuronal lineage has been monitored by using nanogratings of 350 nm as a culture substrate [64].

In several studies, cells grown on microscale rough surfaces demonstrated a lower proliferation rate compared with those grown on smooth surfaces. Furthermore, it was found that cells could not cross over large grooves, glens, holes, and craters [18]. Cells formed a confluent layer within these irregular surface areas much faster than in regions of even surfaces owing to the limited space. Consequently, cells in grooves, glens, holes, and craters showed the pileup phenomenon, which results in bone nodule formation and finally in osteogenic differentiation. Thus, cells grown on surfaces with high roughness displayed a higher alkaline phosphatase activity and bone morphogenic protein production, which supports osteogenic differentiation. Graziano et al. also proved this effect by cultivating hMSC on convex and concave surfaces [139]. Cells cultured on concave surfaces showed better cell–matrix interactions, and after 30 days of cultivation, the production of specific bone proteins, such as osteonectin and bone sialoprotein, was demonstrated.

Similar effects have been observed in our research group by cultivating adMSC and ucMSC for 35 days on a macroporous, interconnected zirconium dioxide (ZrO_2) biomaterial (Sponceram, Zellwerk) with an average pore size of 600 μm. Scanning electron microscopy images illustrated a confluent cell layer in the cavities of the matrix (Fig. 2). Within these areas an increased bone nodule formation was observed. Moreover, the presence of specific bone proteins such as collagen 1, osteopontin, and bone morphogenic protein was verified using polymerase chain reaction analysis.

The influence of microscale surface topography on adipogenic differentiation has also been observed [65]. The results indicated an advanced adipogenic

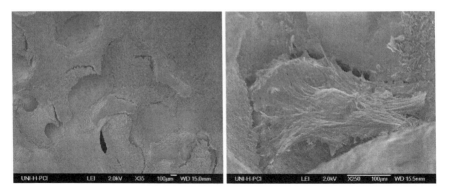

Fig. 2 Scanning electron microscopy images from adipose-derived mesenchymal stem cells (*left*) and umbilical cord mesenchymal stem cells (*right*) cultivated for 5 weeks on a macroporous ZrO_2 biomaterial (Sponceram, Zellwerk)

differentiation by an increased rate of lipid production on patterned surfaces with grooves of 3 μm width, 1.5 μm depth, and 100 μm separation.

Summarizing the publications on the differentiation of MSC influenced by surface topography, we conclude that surface roughness may be used as an effective tool to modulate cell differentiation.

4.4 Geometry

Cells can sense not only the stiffness or roughness of their underlying substrate, but also the edges of their microenvironment. This so-called pattern edge phenomenon influences cell division, cell migration, and the cytoskeleton dynamics [66, 67]. Nevertheless, the influence of two-dimensional substrate geometries on stem cell differentiation has rarely been examined. Wan et al. [68] hypothesized that stem cell proliferation and differentiation is directed through position-defined mechanical stress and the resulting morphological cell changes. To prove their hypothesis they cultivated human adMSC on different two-dimensional patterns (rings with different diameters and rectangles with differing sizes). Osteogenic and adipogenic differentiation occurred in the same regions depending on the medium, containing cells with a small and elongated morphology. Within the ring patterns differentiation decreased with the radius from the inner to the outer ring, where mostly large spreading cells were observed. Differentiation within the rectangles occurred mainly close to the short-axis edge. Finally, it can be stated that the geometry-derived forces affect cell morphology and cytoskeletal organization, which direct the stem cell proliferation and differentiation. Proliferation may be inhibited and differentiation may be induced by low levels of cytoskeletal tension due to increased cell–cell contacts. Similar results have been obtained by [23].

They confirmed that the geometrical shape of the surface adhesion area affects the cell differentiation behavior. A pattern array with various geometries, including octagon, pentagon, right triangle, square, trapezoid, and triangle, has been designed to cultivate hMSC. Analytical analysis of adipogenic differentiation was performed on days 3, 5, 10, and 14. Striking differences were found in the rate of differentiation depending on the geometrical shape. To exclude the possibility that the surface area and therefore the population density influences the differentiation, the area of each pattern was calculated for comparison. In summary, it could be stated that only the pattern geometry has an influence on the adipogenic differentiation rate, but not the pattern area. However, a too low pattern area and therefore a decreasing cell population does not result in cell differentiation. As shown earlier, differentiation requires a minimum cell density [69]. The results of Luo et al. [23] presume that total differentiation within one pattern is elicited by the cells on the periphery of the pattern, which seem to be able to sense the edges of the pattern geometry [66, 67]. Future work will use high-throughput screening analysis tools to study the correlation between pattern geometry and various differentiation directions.

4.5 Composite Materials

Composite materials combine synthetic inorganic materials with naturally derived components [22] to construct a stable and bioactive matrix. The combination of two components allows a precise adjustment of biomaterial properties. Four different strategies to manufacture composite materials are well known (Fig. 3):

1. The mechanical mixture of a polymer and a ceramic component
2. The embedding of bioactive molecules
3. The application of microspheres
4. Surface functionalization

The most common method to fabricate composite materials is the mechanical mixture of a polymer and a ceramic component. The interaction of an organic and an inorganic component seems to be very attractive to regenerate hard tissues, such as bone, because these composite matrices mimic natural bone. HAP and β-tricalcium phosphate are often used as inorganic components since they are considered to be osteoinductive. Mostly, they are mixed with the polymer as nanoparticles or microparticles.

The embedding of bioactive molecules into a suitable matrix is another interesting approach for stem cell differentiation. The goal is to control the release of signal molecules, for example, growth and differentiation factors. The release should be dose-dependent and exact in time and location. Different natural and synthetic polymers, such as poly(lactic-*co*-glycolic acid) [70, 71], gelatin [72], alginate [73], and fibrin [74], have been investigated for this application. These modified polymers can be combined with synthetic inorganic materials such as

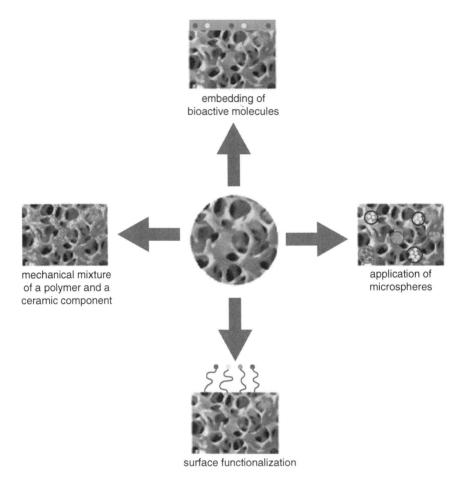

Fig. 3 Different fabrication methods for ceramic-based composite biomaterials

ceramics, resulting in a mechanically stable and bioactive matrix. The combination of these components can be realized by coating a porous ceramic with the modified polymer. However, the polymer coating can plug the pores of the ceramic and therefore negatively influence the transportation of nutrients and waste material. Thus, only biodegradable polymers such as polycaprolactone [75] or polyvinylpyrrolidone [76] in combination with highly porous ceramics can be applied. The release of the bioactive molecules is caused by the degradation of the matrix. Degradation signifies the cleavage of polymers, due to either an enzymatic or a hydrolytic process. Hence, the dose and the lifetime depend on the degradation rate. Finally, the result is a thin layer of bioactive molecules surrounding the matrix.

Another promising alternative to combine natural and synthetic components is the application of microspheres. Microspheres are small spherical polymer particles with a porous inner matrix and a variable surface. Bioactive molecules can be captured within the inner matrix. Generally, microspheres have a diameter of 1–500 μm. Smaller particles (10–1000 nm) are described as nanospheres. The biodegradable polymer microspheres are embedded in the pores of a ceramic matrix prior to cultivation. The major advantage of this method compared with the embedding of bioactive molecules into a polymer coating is the combination of different polymers with differing degradation profiles. Consequently, it is possible to create a time-controlled release of multiple growth and differentiation factors. Therefore, this method can be used to direct the differentiation behavior of MSC. The time-dependent release can be controlled by variation of the polymer properties. The most commonly used polymer for this application is poly(lactic-*co*-glycolic acid) [70, 71]. Moreover, alginates [77] and chitosans [78] are frequently applied. An in vivo microenvironment can be simulated in vitro by applying different bioactive molecules and polymers with defined release profiles.

Another possibility to create a bioactive surface on a ceramic matrix is its specific modification with biologic compounds. Early studies showed a functionalization of the surface with ECM proteins, such as fibronectin and laminin [79]. These proteins support cell adhesion and proliferation. It is well known that only specific amino acid sequences of these proteins interact with the integrins of the cells; therefore, only short amino acid sequences are used for surface modification. The sequence arginine–glycine–aspartic acid [80–82] supports cell adhesion and has been studied in detail. Moreover, the amino acid sequences tyrosine–isoleucine–glycine–serine–arginine [83], arginine–glutamine–aspartic acid–valine [84], and isoleucine–lysine–valine–alanine–valine [83] are used for tissue engineering surface modification. These sequences can be introduced into the 3D matrix network by physical, chemical, photochemical, and ionic cross-linking.

5 Sensing the Microenvironment

The modulating effect of matrix stiffness, topography, and geometry upon cellular responses to biomaterials has been studied in detail over the past decade. It is obvious that cells have the ability to sense their microenvironment and react to the properties of their surroundings in a different manner. But what mechanism do the cells use to identify their substrate and how do they process the information obtained? Prior to sensing the surface elasticity, roughness, or geometry, the cells need to adhere to the substrate. The adhesion procedure is modulated by integrins, which are located in the cell membrane. Three mechanochemical features are important during the adhesion process [85]: (1) the biomaterial–integrin binding forces have to pass a critical threshold, (2) the integrins must mechanically link the artificial matrix to the cytoskeleton in order to transmit extracellular forces to the cell interior, and (3) the transmitted forces have to be translated into biochemical signals (mechanotransduction), resulting in a cellular response. But not only integrins are involved in the procedure of cell–matrix adhesion and signal

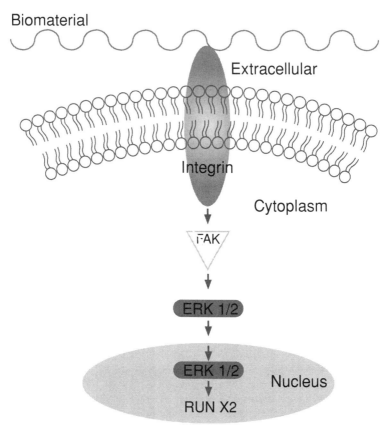

Fig. 4 The focal adhesion kinase pathway. The first step in the transmission of extracellular forces is the cell adhesion on the substrate modulated by integrins. The integrins activate the intracellular focal adhesion kinase and via multiple phosphorylation steps the extracellular-signal-regulated kinase is translocated into the nucleus, where it affects the expression of specific transcription factors, such as Runx2

transmission. Diverse protein networks dynamically link the artificial ECM to the intracellular actin cytoskeleton. These supramolecular structures are called focal adhesions. Proteins joining the focal adhesion complex have the ability to translate mechanical forces into biochemical signals and pass the information to the nucleus. However, the exact signaling pathways linking focal adhesion with the commitment of MSC are not completely understood. Nevertheless, several studies suggest that for osteogenic lineage commitment integrins activate the focal adhesion kinase, which influences cellular events through adhesion-dependent phosphorylation of downstreaming signaling molecules, especially the extracellular-signal-regulated kinase (ERK) [86]. The ERK is a member of the mitogen-activated protein kinase family, which acts as a mediator of cellular differentiation.

The final step in modulating differentiation is the translocation of ERK 1/2 to the nucleus, where it affects the expression of specific cellular transcription factors, such as Runx2, which regulates the expression of many osteoblast genes (Fig. 4) [46, 55, 85]. Even the timescale for this pathway has been identified rudimentarily. The ERK is activated 30 min after the adhesion, which leads to the phosphorylation of the osteogenic transcription factors within 8 days. After 16 days an increased level of osteogenic markers was observed and after 21 days mineralized matrix was formed. These results confirm the observation in the study of Engler et al. [45] that after 1 week a reprogramming of MSC by changing the differentiation medium is not possible.

It might be true that through focal adhesion and the focal adhesion kinase pathway cells can sense their surrounding stiffness, topography, and geometry. However, the complete differentiation process is influenced by a variety of signals [85].

6 Further Perspectives

The affect of local microenvironment on cell behavior has important implications for regenerating tissues or healing disease-affected tissues. Therefore, further investigations in optimizing 3D structures of biomaterials which imitate a specific in vivo ECM in detail could be the key to future success in clinical tissue engineering. The results of Engler et al. [45] demonstrated that the substrate features might even be more important than the composition of the medium. Hence, a combination of, for example, different topologies or pore sizes within one biomaterial could be used to differentially functionalize implants without adding growth or differentiation factors. Thus, after transplantation different "zones" within one implant fulfill distinct functions [57]. These materials could, for example, be advantageous to create bone–cartilage constructs. Kon et al. [87] achieved osteochondral regeneration in a sheep model by using a multilayered biomaterial. The region mimicking the cartilage features contained 100% collagen type 1. In contrast, the area mimicking the bone ECM exhibited only 30% collagen type 1 and 70% HAP, which is considered to be osteoinductive. The transition between these two regions is a layer of 60% collagen 1 and 40% HAP. However, the fabrication of matrices with different features still remains a technical challenge [31].

Another approach to optimize the directed differentiation of MSC on 3D matrices is the embedding of growth and differentiation factors into the biomaterial substrate. A controlled release of these factors ensures the optimal supply of the developing tissue. A future step in the optimization process could be a combination of tissue engineering and gene therapy. In 2002, Samuel et al. [88] reported the delivery of plasmid DNA to chondrocytes via a collagen–glycosaminoglycan matrix. The idea is to genetically modify the cells which have direct contact to the biomaterial in order for them to produce the required growth factors themselves and to modulate their own differentiation.

References

1. Langer R, Vacanti JP (1993) Tissue engineering. Science 260(5110):920–926
2. Kohara H, Tabata Y (2010) Review: tissue engineering technology to enhance cell recruitment for regeneration therapy. J Med Biol Eng 30(5):267–276
3. Calori GM et al (2009) Bone morphogenetic proteins and tissue engineering: future directions. Injury 40:67–76
4. Porter JR, Ruckh TT, Popat KC (2009) Bone tissue engineering: a review in bone biomimetics and drug delivery strategies. Biotechnol Prog 25(6):1539–1560
5. Mauney JR, Volloch V, Kaplan DL (2005) Role of adult mesenchymal stem cells in bone tissue-engineering applications: current status and future prospects. Tissue Eng 11(5–6):787–802
6. Quarto R et al (2001) Repair of large bone defects with the use of autologous bone marrow stromal cells. New Engl J Med 344(5):385–386
7. Ma PX (2004) Scaffolds for tissue fabrication. Mater Today 7(5):30–40
8. Moretti P, HatlapatkaT, Marten D, Lavrentieva A, Majore I, Hass R, Kasper C (eds) (2010) Mesenchymal stromal cells derived from the human umbilical cord tissues: primitive cells with potential for clinical and tissue engineering applications. Springer, Berlin, pp 29–54
9. Zuk PA et al (2002) Human adipose tissue is a source of multipotent stem cells. Mol Biol Cell 13(12):4279–4295
10. Majore I et al (2009) Characterization of mesenchymal stem cell-like cultures derived from human umbilical cord. Hum Gene Ther 20(11):1491–1491
11. Majore I et al (2009) Identification of subpopulations in mesenchymal stem cell-like cultures from human umbilical cord. Cell Commun Signal 7
12. Kogler G et al (2004) A new human somatic stem cell from placental cord blood with intrinsic pluripotent differentiation potential. J Exp Med 200(2):123–135
13. Hatlapatka T et al (2011) Optimization of culture conditions for the expansion of umbilical cord-derived mesenchymal stem or stromal cell-like cells using xeno-free culture conditions. Tissue Eng Part C Methods
14. Colley HE et al (2009) Plasma polymer coatings to support mesenchymal stem cell adhesion, growth and differentiation on variable stiffness silicone elastomers. Plasma Process Polym 6(12):831–839
15. Takezawa T (2003) A strategy for the development of tissue engineering scaffolds that regulate cell behavior. Biomaterials 24(13):2267–2275
16. Kim BS, Mooney DJ (1998) Development of biocompatible synthetic extracellular matrices for tissue engineering. Trends Biotechnol 16(5):224–230
17. Eisenbarth E (2007) Biomaterials for tissue engineering. Adv Eng Mater 9(12):1051–1060
18. Kawahara H et al (2004) In vitro study on bone formation and surface topography from the standpoint of biomechanics. J Mater Sci Mater Med 15(12):1297–1307
19. Hutmacher DW (2001) Scaffold design and fabrication technologies for engineering tissues—state of the art and future perspectives. J Biomater Sci Polym Ed 12(1):107–124
20. Arinzeh TL et al (2005) A comparative study of biphasic calcium phosphate ceramics for human mesenchymal stem-cell-induced bone formation. Biomaterials 26(17):3631–3638
21. Wang M (2006) Composite Scaffolds for bone tissue engineering. Am J Biochem Biotechnol 2(2):80–84
22. Cao Y et al (2005) Scaffolds, stem cells, and tissue engineering: a potent combination! Aust J Chem 58(10):691–703

23. Luo W, Jones SR, Yousaf MN (2008) Geometric control of stem cell differentiation rate on surfaces. Langmuir 24(21):12129–12133
24. Roosa SMM et al (2010) The pore size of polycaprolactone scaffolds has limited influence on bone regeneration in an in vivo model. J Biomed Mater Res Part A 92A(1):359–368
25. Tsuruga E et al (1997) Pore size of porous hydroxyapatite as the cell-substratum controls BMP-induced osteogenesis. J Biochem 121(2):317–324
26. Shor L et al (2007) Fabrication of three-dimensional polycaprolactone/hydroxyapatite tissue scaffolds and osteoblast-scaffold interactions in vitro. Biomaterials 28(35):5291–5297
27. Hulbert SF et al (1970) Potential of ceramic materials as permanently implantable skeletal prostheses. J Biomed Mater Res 4:433–456
28. Karageorgiou V, Kaplan D (2005) Porosity of 3D biomaterial scaffolds and osteogenesis. Biomaterials 26(27):5474–5491
29. Kasten P et al (2008) Porosity and pore size of beta-tricalcium phosphate scaffold can influence protein production and osteogenic differentiation of human mesenchymal stem cells: an in vitro and in vivo study. Acta Biomater 4(6):1904–1915
30. Yannas I (1992) Tissue regeneration by use of collagen-glycosaminoglycan copolymers. Clin Mater 9:179–187
31. Khoda AKMB, Ozbolat IT, Koc B (2011) Engineered tissue scaffolds with variational porous architecture. J Biomech Eng Trans ASME 133(1):011001–011012
32. Oh SH et al (2010) Investigation of pore size effect on chondrogenic differentiation of adipose stem cells using a pore size gradient scaffold. Biomacromolecules 11(8):1948–1955
33. Kuboki Y, Jin QM, Takita H (2001) Geometry of carriers controlling phenotypic expression in BMP-induced osteogenesis and chondrogenesis. J Bone Joint Surg Am 83A:S105–S115
34. Cyster LA et al (2005) The influence of dispersant concentration on the pore morphology of hydroxyapatite ceramics for bone tissue engineering. Biomaterials 26(7):697–702
35. Bone structure. http://www.engin.umich.edu/class/bme456/bonestructure/bonestructure.htm. Accessed 15 Mar 2011
36. Grimm MJ, Williams JL (1997) Measurements of permeability in human calcaneal trabecular bone. J Biomech 30(7):743–745
37. Benninghoff D, Drenckhahn D (eds) (2003) Anatomie, 16th edn. Fischer, Munich
38. Hakulinen MA et al (2006) Ultrasonic characterization of human trabecular bone microstructure. Phys Med Biol 51(6):1633–1648
39. Pelham RJ, Wang YL (1997) Cell locomotion and focal adhesions are regulated by substrate flexibility. Proc Natl Acad Sci USA 94(25):13661–13665
40. Ni Y, Chiang MYM (2007) Cell morphology and migration linked to substrate rigidity. Soft Matter 3(10):1285–1292
41. Schneider A et al (2006) Polyelectrolyte multilayers with a tunable young's modulus: influence of film stiffness on cell adhesion. Langmuir 22(3):1193–1200
42. Yeung T et al (2005) Effects of substrate stiffness on cell morphology, cytoskeletal structure, and adhesion. Cell Motil Cytoskelet 60(1):24–34
43. McBeath R et al (2004) Cell shape, cytoskeletal tension, and RhoA regulate stem cell lineage commitment. Dev Cell 6(4):483–495
44. Wang HB, Dembo M, Wang YL (2000) Substrate flexibility regulates growth and apoptosis of normal but not transformed cells. Am J Physiol Cell Physiol 279(5):C1345–C1350
45. Engler AJ et al (2006) Matrix elasticity directs stem cell lineage specification. Cell 126(4):677–689
46. Rowlands AS, George PA, Cooper-White JJ (2008) Directing osteogenic and myogenic differentiation of MSCs: interplay of stiffness and adhesive ligand presentation. Am J Physiol Cell Physiol 295(4):C1037–C1044
47. Wang LS et al (2010) The role of stiffness of gelatin-hydroxyphenylpropionic acid hydrogels formed by enzyme-mediated crosslinking on the differentiation of human mesenchymal stem cell. Biomaterials 31(33):8608–8616
48. Wan YQ et al (2005) Adhesion and proliferation of OCT-1 osteoblast-like cells on micro- and nano-scale topography structured poly(L-lactide). Biomaterials 26(21):4453–4459

49. Brunette DM (2001) Titanium in medicine. In: Tengvall P, Brunette DM, Textor M, Thomsen P (eds) Principles of cell behavior on titanium surfaces and their application to implanted devices. Springer, Berlin, pp 485–512
50. Boyan BD et al (2002) Osteoblast-mediated mineral deposition in culture is dependent on surface microtopography. Calcif Tissue Int 71(6):519–529
51. Perizzolo D, Lacefield WR, Brunette DM (2001) Interaction between topography and coating in the formation of bone nodules in culture for hydroxyapatite- and titanium-coated micro machined surfaces. J Biomed Mater Res 56(4):494–503
52. Castellani R et al (1999) Response of rat bone marrow cells to differently roughened titanium discs. Clin Oral Implant Res 10(5):369–378
53. ter Brugge PJ, Wolke JGC, Jansen JA (2002) Effect of calcium phosphate coating crystallinity and implant surface roughness on differentiation of rat bone marrow cells. J Biomed Mater Res 60(1):70–78
54. Cavalcanti-Adam EA et al (2007) Cell spreading and focal adhesion dynamics are regulated by spacing of integrin ligands. Biophys J 92(8):2964–2974
55. Zhao LZ et al (2011) Suppressed primary osteoblast functions on nanoporous titania surface. J Biomed Mater Res Part A 96A(1):100–107
56. Arnold M et al (2004) Activation of integrin function by nanopatterned adhesive interfaces. Chemphyschem 5(3):383–388
57. McNamara LE (2010) Nanotopographical control of stem cell differentiation. J Tissue Eng 13
58. Dalby MJ et al (2004) Investigating the limits of filopodial sensing: a brief report using SEM to image the interaction between 10 nm high nano-topography and fibroblast filopodia. Cell Biol Int 28(3):229–236
59. Li JJ et al (2009) Surface characterization and biocompatibility of micro- and nano-hydroxyapatite/chitosan-gelatin network films. Mater Sci Eng C Biomim Supramol Syst 29(4):1207–1215
60. Dalby MJ et al (2007) The control of human mesenchymal cell differentiation using nanoscale symmetry and disorder. Nat Mater 6(12):997–1003
61. Oh S et al (2009) Stem cell fate dictated solely by altered nanotube dimension. Proc Natl Acad Sci USA 106(7):2130–2135
62. Oh S et al (2006) Significantly accelerated osteoblast cell growth on aligned TiO_2 nanotubes. J Biomed Mater Res Part A 78A(1):97–103
63. Peng L et al (2009) The effect of TiO2 nanotubes on endothelial function and smooth muscle proliferation. Biomaterials 30(7):1268–1272
64. Yim EK, Pang SW, Leong KW (2007) Synthetic nanostructures inducing differentiation of human mesenchymal stem cells into neuronal lineage. Exp Cell Res 313(9):1820–1829
65. Chaubey A et al (2008) Surface patterning: tool to modulate stem cell differentiation in an adipose system. J Biomed Mater Res Part B Appl Biomater 84B(1):70–78
66. Jiang X et al (2005) Directing cell migration with asymmetric micropatterns. Proc Natl Acad Sci USA 102(4):975–978
67. Thery M et al (2006) Anisotropy of cell adhesive microenvironment governs cell internal organization and orientation of polarity. Proc Natl Acad Sci USA 103(52):19771–19776
68. Wan LQ et al (2010) Geometric control of human stem cell morphology and differentiation. Integr Biol 2(7–8):346–353
69. Gerber I, Gwynn I (2001) Infuence of cell isolation, cell culture density, and cell nutrition on differentiation of rat calvarial osteoblast-like cells in vitro. Eur Cells Mater 2:10–20
70. Cohen S et al (1991) Controlled delivery systems for proteins based on poly(lactic glycolic acid) microspheres. Pharm Res 8(6):713–720
71. Benoit JP et al (2000) Development of microspheres for neurological disorders: from basics to clinical applications. J Control Release 65(1–2):285–296
72. Young S et al (2005) Gelatin as a delivery vehicle for the controlled release of bioactive molecules. J Control Release 109(1–3):256–274

73. Gombotz WR, Pettit DK (1995) Biodegradable polymers for protein and peptide drug-delivery. Bioconjug Chem 6(4):332–351
74. Schense JC et al (2000) Enzymatic incorporation of bioactive peptides into fibrin matrices enhances neurite extension. Nat Biotechnol 18(4):415–419
75. Kim HW, Knowles JC, Kim HE (2004) Development of hydroxyapatite bone scaffold for controlled drug release via poly(epsilon-caprolactone) and hydroxyapatite hybrid coatings. J Biomed Mater Res B Appl Biomater 70(2):240–249
76. Arcos D et al (1997) Ibuprofen release from hydrophilic ceramic-polymer composites. Biomaterials 18(18):1235–1242
77. Chan LW, Lee HY, Heng PWS (2002) Production of alginate microspheres by internal gelation using an emulsification method. Int J Pharm 242(1–2):259–262
78. Zhang Y, Zhang MQ (2002) Three-dimensional macroporous calcium phosphate bioceramics with nested chitosan sponges for load-bearing bone implants. J Biomed Mater Res 61(1):1–8
79. Shin H, Jo S, Mikos AG (2003) Biomimetic materials for tissue engineering. Biomaterials 24(24):4353–4364
80. Craig WS et al (1995) Concept and progress in the development of RGD-containing peptide pharmaceuticals. Biopolymers 37(2):157–175
81. Massia SP, Hubbell JA (1990) Covalent surface immobilization of Arg-Gly-Asp- and Tyr-Ile-Gly-Ser-Arg-containing peptides to obtain well-defined cell-adhesive substrates. Anal Biochem 187(2):292–301
82. Yu JS et al (2010) The use of human mesenchymal stem cells encapsulated in RGD modified alginate microspheres in the repair of myocardial infarction in the rat. Biomaterials 31(27):7012–7020
83. Ranieri JP et al (1995) Neuronal cell attachment to fluorinated ethylene–propylene films with covalently immobilized laminin oligopeptides Yigsr and Ikvav.2. J Biomed Mater Res 29(6):779–785
84. Massia SP, Hubbell JA (1992) Vascular endothelial-cell adhesion and spreading promoted by the peptide REDV of the IIICS region of plasma fibronectin is mediated by integrin alpha-4-beta-1. J Biol Chem 267(20):14019–14026
85. Biggs MJP, Dalby MJ (2010) Focal adhesions in osteoneogenesis. Proc Inst Mech Eng Part H J Eng Med 224(H12):1441–1453
86. Klees RF et al (2005) Laminin-5 induces osteogenic gene expression in human mesenchymal stem cells through an ERK-dependent pathway. Mol Biol Cell 16(2):881–890
87. Kon E et al (2010) Orderly osteochondral regeneration in a sheep model using a novel nano-composite multilayered biomaterial. J Orthop Res 28(1):116–124
88. Samuel RE et al (2002) Delivery of plasmid DNA to articular chondrocytes via novel collagen-glycosaminoglycan matrices. Hum Gene Ther 13(7):791–802
89. Lovell CS et al (2010) Analysis and modeling of the mechanical properties of novel thermotropic polymer biomaterials. Polymer 51(9):2013–2020
90. Sung HJ et al (2010) Poly(ethylene glycol) as a sensitive regulator of cell survival fate on polymeric biomaterials: the interplay of cell adhesion and pro-oxidant signaling mechanisms. Soft Matter 6(20):5196–5205
91. Shalumon KT et al (2011) Preparation, characterization and cell attachment studies of electrospun multi-scale poly(caprolactone) fibrous scaffolds for tissue engineering. J Macromol Sci Part A Pure Appl Chem 48(1):21–30
92. da Silva MLA et al (2010) Cartilage tissue engineering using electrospun PCL nanofiber meshes and MSCs. Biomacromolecules 11(12):3228–3236
93. Woodruff MA, Hutmacher DW (2010) The return of a forgotten polymer-polycaprolactone in the 21st century. Prog Polym Sci 35(10):1217–1256
94. Koch MA et al (2010) Perfusion cell seeding on large porous PLA/calcium phosphate composite scaffolds in a perfusion bioreactor system under varying perfusion parameters. J Biomed Mater Res A 95(4):1011–1018

95. Gupta B, Revagade N, Hilborn J (2007) Poly(lactic acid) fiber: an overview. Prog Polym Sci 32(4):455–482
96. Stevanovic M, Uskokovic D (2009) Poly(lactide-*co*-glycolide)-based micro and nanoparticles for the controlled drug delivery of vitamins. Curr Nanosci 5(1):1–14
97. Lu JM et al (2009) Current advances in research and clinical applications of PLGA-based nanotechnology. Expert Rev Mol Diagn 9(4):325–341
98. Barrere F, van Blitterswijk CA, de Groot K (2006) Bone regeneration: molecular and cellular interactions with calcium phosphate ceramics. Int J Nanomed 1(3):317–332
99. Emadi R et al (2010) Nanostructured forsterite coating strengthens porous hydroxyapatite for bone tissue engineering. J Am Ceram Soc 93(9):2679–2683
100. Chen JG et al (2010) In situ grown fibrous composites of poly(DL-lactide) and hydroxyapatite as potential tissue engineering scaffolds. Polymer 51(26):6268–6277
101. Uskokovic V, Uskokovic DP (2011) Nanosized hydroxyapatite and other calcium phosphates: chemistry of formation and application as drug and gene delivery agents. J Biomed Mater Res Part B Appl Biomater 96B(1):152–191
102. Swetha M et al (2010) Biocomposites containing natural polymers and hydroxyapatite for bone tissue engineering. Int J Biol Macromol 47(1):1–4
103. Bellucci D et al (2010) Potassium based bioactive glass for bone tissue engineering. Ceram Int 36(8):2449–2453
104. Mourino V, Newby P, Boccaccini AR (2010) Preparation and characterization of gallium releasing 3D alginate coated 45S5 bioglass (R) based scaffolds for bone tissue engineering. Adv Eng Mater 12(7):B283–B291
105. Mullen LM et al (2010) Binding and release characteristics of insulin-like growth factor-1 from a collagen-glycosaminoglycan scaffold. Tissue Eng Part C Methods 16(6):1439–1448
106. Yuan T et al (2010) Chondrogenic differentiation and immunological properties of mesenchymal stem cells in collagen type I hydrogel. Biotechnol Prog 26(6):1749–1758
107. Zorlutuna P, Vadgama P, Hasirci V (2010) Both sides nanopatterned tubular collagen scaffolds as tissue-engineered vascular grafts. J Tissue Eng Regen Med 4(8):628–637
108. Zheng WF, Zhang W, Jiang XY (2010) Biomimetic collagen nanofibrous materials for bone tissue engineering. Adv Eng Mater 12(9):B451–B466
109. Lee CH, Singla A, Lee Y (2001) Biomedical applications of collagen. Int J Pharm 221(1–2):1–22
110. Kosmala JD, Henthorn DB, Brannon-Peppas L (2000) Preparation of interpenetrating networks of gelatin and dextran as degradable biomaterials. Biomaterials 21(20):2019–2023
111. Tripathi A, Kathuria N, Kumar A (2009) Elastic and macroporous agarose-gelatin cryogels with isotropic and anisotropic porosity for tissue engineering. J Biomed Mater Res Part A 90A(3):680–694
112. Leddy HA, Awad HA, Guilak F (2004) Molecular diffusion in tissue-engineered cartilage constructs: effects of scaffold material time, and culture conditions. J Biomed Mater Res Part B Appl Biomater 70B(2):397–406
113. Bensaid W et al (2003) A biodegradable fibrin scaffold for mesenchymal stem cell transplantation. Biomaterials 24(14):2497–2502
114. Kim SH et al (2009) Recent research trends of fibrin gels for the applications of regenerative medicine. Tissue Eng Regen Med 6(1–3):273–286
115. Oliveira JT et al (2011) Novel melt-processable chitosan-polybutylene succinate fibre scaffolds for cartilage tissue engineering. J Biomater Sci Polym Ed 22(4–6):773–788
116. Jayakumar R et al (2010) Biomedical applications of chitin and chitosan based nanomaterials—a short review. Carbohydr Polym 82(2):227–232
117. Venkatesan J, Kim SK (2010) Chitosan composites for bone tissue engineering—an overview. Mar Drugs 8(8):2252–2266
118. VandeVord PJ et al (2002) Evaluation of the biocompatibility of a chitosan scaffold in mice. J Biomed Mater Res 59(3):585–590
119. Eiselt P et al (2000) Porous carriers for biomedical applications based on alginate hydrogels. Biomaterials 21(19):1921–1927

120. Choi MS et al (2008) Chondrogenic differentiation of human adipose-derived stem cells in alginate sponge scaffolds. Tissue Eng Regen Med 5(4–6):842–848
121. Ebraheim NA, Mekhail AO, Darwich M (1997) Open reduction and internal fixation with bone grafting of clavicular nonunion. J Trauma Inj Infect Crit Care 42(4):701–704
122. Kretlow JD et al (2010) Uncultured marrow mononuclear cells delivered within fibrin glue hydrogels to porous scaffolds enhance bone regeneration within critical-sized rat cranial defects. Tissue Eng Part A 16(12):3555–3568
123. Staiger MP et al (2006) Magnesium and its alloys as orthopedic biomaterials: a review. Biomaterials 27(9):1728–1734
124. Brar HS et al (2009) Magnesium as a biodegradable and bioabsorbable material for medical implants. JOM 61(9):31–34
125. Zeng RC et al (2008) Progress and challenge for magnesium alloys as biomaterials. Adv Eng Mater 10(8):B3–B14
126. Chen FL et al (2002) Bone graft in the shape of human mandibular condyle reconstruction via seeding marrow-derived osteoblasts into porous coral in a nude mice model. J Oral Maxillofac Surg 60(10):1155–1159
127. Mygind T et al (2007) Mesenchymal stem cell ingrowth and differentiation on coralline hydroxyapatite scaffolds. Biomaterials 28(6):1036–1047
128. van den Dolder J et al (2003) Bone tissue reconstruction using titanium fiber mesh combined with rat bone marrow stromal cells. Biomaterials 24(10):1745–1750
129. Diederichs S et al (2009) Dynamic cultivation of human mesenchymal stem cells in a rotating bed bioreactor system based on the Z (R) RP platform. Biotechnol Prog 25(6):1762–1771
130. Roker S et al (2009) Novel 3D biomaterials for tissue engineering based on collagen and macroporous ceramics. Materialwiss Werkstofftech 40(3):54, 224–225
131. Yuan J et al (2007) Repair of canine mandibular bone defects with bone marrow stromal cells and porous beta-tricalcium phosphate. Biomaterials 28(6):1005–1013
132. Kim HJ, Lee JH, Im GI (2010) Chondrogenesis using mesenchymal stem cells and PCL scaffolds. J Biomed Mater Res Part A 92A(2):659–666
133. Yamane S et al (2007) Effect of pore size on in vitro cartilage formation using chitosan-based hyaluronic acid hybrid polymer fibers. J Biomed Mater Res Part A 81A(3):586–593
134. Lee SJ et al (2004) Response of human chondrocytes on polymer surfaces with different micropore sizes for tissue-engineered cartilage. J Appl Polym Sci 92(5):2784–2790
135. Chia SL et al (2006) Biodegradable elastomeric polyurethane membranes as chondrocyte carriers for cartilage repair. Tissue Eng 12(7):1945–1953
136. Cho SW et al (2004) Smooth muscle-like tissues engineered with bone marrow stromal cells. Biomaterials 25(15):2979–2986
137. Wang M et al (2010) Hepatogenesis of adipose-derived stem cells on poly-lactide-co-glycolide scaffolds: in vitro and in vivo studies. Tissue Eng Part C Methods 16(5):1041–1050
138. Discher DE et al (2005) Tissue cells feel and respond to the stiffness of their substrate. Science 310:1139–1143
139. Graziano A et al (2007) Scaffold's surface geometry significantly affects human stem cell bone tissue engineering. J Cell Physiol 214:166–172

Designing the Biocompatibility of Biohybrids

Frank Witte, Ivonne Bartsch and Elmar Willbold

Abstract Biohybrid has been used as a fashionable term in scientific publications during the past years to describe a functional unit consisting of a bioactive and a structural component. The bioactive part of the biohybrid could consist of cells or bioactive molecules, while the structural part is of biological or non-biological origin. Biohybrids are currently used as implants and transplants in regenerative medicine or in vitro applications such as assays, biosensors or bioreactors. However, a clear definition of a biohybrid has not been given yet. This chapter reviews the current applications of biohybrids and identifies the challenges of biohybrids in in vivo applications. A classification of biohybrids according to their functional use and application is provided.

Keywords Artificial organs · Biocompatibility · Biohybrid · Biomaterial · Drug delivery · Tissue engineering

Contents

1 Introduction 286
2 Compatibility of the Structural Compound 287
3 Compatibility of the Bioactive Compound 288
4 Classification of Biohybrids 289
    4.1 Classification of Biohybrids According to Their Function 290
    4.2 Classification of Biohybrids According to Their Therapeutical Application 291
5 General Advantages and Limitations of Biohybrids 293
6 Conclusions 293
References 293

F. Witte (✉) · I. Bartsch · E. Willbold
Laboratory for Biomechanics and Biomaterials, Orthopaedic Clinic,
Hannover Medical School, Anna-von-Borries-Straße 1-7,
30625 Hannover, Germany
e-mail: witte.frank@mh-hannover.de

F. Witte · I. Bartsch · E. Willbold
Center for Biocompatibility and Implant-Immunology, CrossBIT,
Hannover Medical School, Feodor-Lynen-Straße 31,
30625 Hannover, Germany

1 Introduction

During the last decades, great progress was achieved in many fields of biomedical research. Although classical clinical or pharmacological research remain the major scientific fields, technical aspects of many biomedical issues have become more and more important. Many different fields in biomedical research and also in clinical applications took advantage of ideas and concepts of neighbouring scientific areas, especially from engineering sciences. Although the clinical use of ortheses, prostheses or even implants as compensation or replacement of lost body parts is known since prehistoric times, the development of more complex technical solutions was the outcome of the past twenty years of research. The sustained emergence and the increasing importance of all aspects of regenerative medicine and stem cell based therapies opened many fascinating approaches for the treatment of various serious health problems. Amongst other merely technical oriented approaches, the development of biohybrid systems hereby faced a remarkable upturn.

In the literature, there is no specific definition for a biohybrid system, but generally, biohybrids are regarded as the functional combination of both a bioactive and a structural component, thus adding the benefits and advantages of both components. The structural component is normally referred to as a biomaterial and consists of metals, polymers, ceramics, decellularised extracellular matrices or composites. The bioactive component could either consist of active molecules or cells.

Based on this idea, cells are combined, for example, with carbon nanofibers to be used in nerve and spinal cord regeneration or cells are linked to hip prothesis or surfaces of cardiovascular devices to create so called biofunctionalized surfaces or implants. However, the combination of cells and materials have been always in the focus of tissue engineering approaches and in regenerative medicine, since surgeons and patients are demanding faster healing after surgery and sustained functionality of the implant. The combination of bioactive cells and a mechanical framework is necessary, since the treatment with sole cell constructs or suspensions is usually lacking sufficient mechanical support or guidance.

However, biohybrids are also used for innovative analytical tools in vitro. In this application the biocompatibility and immunogenicity of the bioactive and structural component is not important. However, in any case the cytocompatiblity of the structural component seems to be important in these applications, especially in bioreactors for the production of bioactive molecules.

In this review, we want to elaborate which applications are currently summarized under the term biohybrid and which challenges and design strategies exist for biohybrids in various clinical applications, especially in musculoskeletal approaches. However, since there is no classification for the term biohybrid, it can create some confusion if used in brief abstracts without any further explanation. Therefore, we would like to suggest a classification system of biohybrids in this chapter.

2 Compatibility of the Structural Compound

The structural component of a biohybrid is normally a biomaterial which needs to prove its biocompatibility in vivo. According to the definition of the Society for Biomaterials (also known as the "Williams' definition" [1]), biocompatibility is the ability of a material to perform with an appropriate host response in a specific application. According to Anderson et al. [2], placing a biomaterial in vivo always requires an injection, insertion or surgical intervention which is associated with damaging tissues and therefore is associated physiologically with inflammation, wound healing and a foreign body reaction (FBR).

The damage of an intact tissue leads to three episodes of wound healing: inflammation, proliferation and remodelling. The inflammation caused by a material inserted in a living tissue can be divided into an acute and a chronic form. The acute inflammation is identified by the presence of the fast invading cells as neutrophils, monocytes and lymphocytes. Due to vessel leakages caused by the implantation method platelets degranulate and will attract leucocytes. Depending on the age of injury, the dominating cell type differs. The first cells arriving at the location after implantation are polymorphnuclear cells (PMN), especially neutrophils. Their recruitment is initiated by cytokines, e.g. chemoattractants from injured cells or histamine by mast cells. The neutrophils will disappear within the first two days after the implantation of the biomaterial [2]. The chronic form of an inflammation is characterised by the presence of macrophages, the formation of foreign body giant cells and the encapsulation of the biomaterial with fibrous tissue. Monocytes invaded from the blood will arrive at the implantation site and will differentiate into macrophages. They produce several chemokines like TNF-α, IL-6, G-CSF and GM-CSF to attract additional monocytes and macrophages to the area of interest. Macrophages adherent to the material surface produce TNF-α, IL-1 and IL-6 to activate regional T-lymphocytes.

During the last decades, the biomaterial research concentrated on designing materials that are well tolerated by the living organism, e.g. a newly formed fibrous capsule that shields the implant from the tissue without initiating a severe FBR. But today it has been realised, that specific cell-implant interactions could be advantageous for the acceptance of the implanted foreign body to the host tissue [3]. Modulating the implant surface can modulate the immune reaction towards accepting or tolerating the biomaterial.

The human immune system is divided into two different types: the innate and the adaptive immune system. The innate immune system is non-specific and consists of natural barriers, e.g. skin and mucous membranes, phagocytes and the complement system. The adaptive immune system is mainly comprised of lymphocytes, and it can be distinguished between the humoral and the cell-mediated immunity. The innate and the adaptive immune system work together, hand in hand, and cannot be separated completely from each other.

Immediately after the insertion of a biomaterial into a living tissue, proteins from blood and interstitial fluids will adsorb to the biomaterial surface [3].

The Vroman-effect describes the changes in protein adsorption and release which is determined by the mobility of the protein and the affinity to the implant surface. At the moment of insertion, the proteins with the highest mobility will arrive at the implant site first and adsorb at the surfaces. Later, they will be replaced by slower proteins with a higher affinity to the implant material. The blood-material-interactions are described by the hematocompatibility of the biomaterial. Several surface modifications may improve the hematocompatibility and the biocompatibility of the biomaterial. Due to the adsorption of proteins at the implant surface directly after implantation, PMNs, monocytes and macrophages are attracted and will bind to that layer by specific protein receptors. A modulation of the surface protein deposition layer would restrict the binding of proteins and select required types of proteins.

Other types of modification are the alteration of the surface roughness and topography or the mimicking of the extracellular matrix. Also the loading with anti-inflammatory mediators could improve the acceptance of the biomaterial and down-regulate the foreign body reaction.

However, when these materials are used as scaffolds, matrices or substrates for anchoring of living cells or layers of biological active molecules, the host response to the bioactive composite of the biohybrid has to be taken into account.

3 Compatibility of the Bioactive Compound

A critical issue in the therapeutic use of any bioactive implant is its biosafety. The biological compound must be well tolerated by the host immune system, and it needs to prove therapeutic effect over a longer period of time. Especially when using differentiated or redifferentiated cells, absolutely no remaining proliferating cells causing any types of tumors or teratomas are tolerated and the biological compound has to integrate site-specifically.

The failure or lacking of one of these issues excludes the bioactive compound from any use in humans. Therefore, extended in vitro and in vivo testing has to be performed and the data have to be analyzed very carefully to avoid any health risk.

An always existing risk associated with cell or tissue transplantation is the graft-versus-host-reaction (GvHR). The GvHR is an immunological reaction between an organism (host) and inserted cells or tissues (graft). In the majority of cases, the origin of the transplanted cells or tissues is allogeneic, that means that the donor and the recipient are different individuals of the same species. In this case the specific major histocompatibility complex (MHC) of the donor maybe different from that one of the host which leads to the rejection of the transplant. Current therapy of that rejection is the suppression of the recipient's immune system to inhibit the defence of the grafted implant by administrating steroids.

A frequently investigated cell type in biohybrids for regenerative medicine is the mesenchymal stem cell (MSC). MSCs are pluripotent cells that can differentiate into a variety of cell types [4] and can be used in several therapies. Because of the expression of MHC class I the MSCs will escape the lysis by natural killer cells. Furthermore, MSCs inhibit immune cell proliferation and activation [5]. In this respect the MSCs have an immune modulating potential and are capable of producing an immunosuppressive environment expressing cytokines [6]. Placing the MSCs into a three-dimensional construct (structural compound) will allow for implantation at a specific location into the organism. The choice of the scaffold material can affect the MSC differentiation and has to be taken into account. As an example, the chondrogenic differentiation is favoured by the presence of carboxyl- or hydroxyl-groups, whereas the osteogenic differentiation is benefited by the attendance of amino- and sulfhydryl-groups in vitro [7]. The elasticity of the matrix influences the determination into neuronal, muscle or bone lineage specification, as shown in cell culture experiments [8].

Prior to seeding the scaffold material, the cells have to be harvested from bone marrow, muscle, adipose tissue, umbilical cord or lung. Following harvesting, the cells have to be isolated and cultured. Contrary to the knowledge of the unlimited potential of self-renewal of human MSCs, the cells cultured in vitro underlay senescence and showed reduced differentiation potential [9]. Furthermore, Miura et al. demonstrated that the continuous passaging of murine MSCs was leading to the formation of a fibrosarcoma after implantation in vivo [10]. Moreover, the effect of the reduced supply of cells with oxygen and nutrients in the center of any scaffolding material on the differentiation and excretory activity of the cells needs to be considered in biohybrids for regenerative medicine [5].

As an investigative tool, non-destructive in vivo fluorescence imaging is a valuable technique to enhance the design process of the structural component and investigating the GvHR in the same animal at different time points [11]. Repetitive measurements in the same animal provide an excellent option to follow the biocompatibility and GvHR using targeted optical fluorescent imaging probes. Moreover, the degradation pathway of fluorescent degradable scaffold components can be followed by tagging the structural components of the scaffold.

4 Classification of Biohybrids

The successful use of biohybrids is reported multifold with great variation in various biomedical applications. However, no systematic classification exists so far. Here we suggest a simple classification of biohybrids according to their functional unit and according to their possible therapeutical application.

4.1 Classification of Biohybrids According to Their Function

The functional unit of a biohybrid may be described by one of the following subgroups:

4.1.1 Three-Dimensional Spheroid Biohybrids

The use of three-dimensional spheroids has a longstanding tradition. First established the 1950s [12], this powerful culture system was used very successfully for developmental studies [13], in tissue engineering [14], as a basic step during differentiation of embryonic stem cells [15, 16] and as a monitoring system for drugs and other bioactive agents [17, 18]. More recently, three-dimensional spheroids were also further developed by introducing scaffolds of biological [19] or non-biological origin [20] as structural component. The structural component is of minor importance and quantity in three-dimensional spheroids because they consist mainly of cells and their extracellular matrix. However, because the three-dimensional spheroid technique is mainly an in vitro technique, the biocompatibility of the structural component is only of minor importance.

4.1.2 Biohybrid Biosensors

Biohybrid biosensors have many functional similarities with three-dimensional cultures. However in this use, the cells grow two-dimensionally and are combined with chiplike scaffolds or matrices for sensing [21, 22]. Since these biosensing approaches are applied mainly in vitro, biocompatibility is currently a minor issue.

4.1.3 Drug Delivery Biohybrids

One of the most interesting clinical fields of biohybrid applications is drug delivery. For patients who need a continuously available therapeutic, it would be a significant benefit, if a long-term drug depot could be implanted which releases the therapeutic(s) at a controlled and physiological rate. Inevitably, these constructs have to be deposited in the organism and biocompatibility is a major concern [23]. The structural part in these drug releasing biohybrids may consist of various biomaterials, e.g. polylactid acids [24], hydrogels [25] or specific combinations [26]. Naturally, the structural part must be highly biocompatible and non-toxic, independent from the fact whether the structural part consists of a permanent or a biodegradable substance.

4.1.4 Encapsulating Scaffold Biohybrids

One of the most important and promising applications for biohybrids are cells in scaffolds for tissue repair. The bioactive component of the biohybrid may consist of proteins [27] or cells, while the scaffold serves as necessary placeholder or mechanical framework for the cells in all aspects of tissue engineering. Tissue engineering is one of the most quickly expanding areas in biomedicine [28]. There is an overwhelming amount of very different approaches focusing on many different cell types, tissues and functions which cannot be reviewed in detail here. However, independent from its composition, these are the most challenging biohybrid constructs, since the host has not only to face a structural non-biological component, but moreover foreign proteins, cells or even a functionalized tissue with hardly predictable reactions of the host immune system.

4.2 Classification of Biohybrids According to Their Therapeutical Application

The most common use of the term biohybrid can be found in various fields of regenerative medicine. It is obvious that biohybrids can be classified according to their therapeutical use in different organ systems. Examples of therapeutical biohybrids and their use are given below:

4.2.1 Neuronal Biohybrids

The central nervous system is the most complex human organ system and its unique capacity is the foundation which distinguishes humans from their evolutionary relatives. Therefore, non- or malfunction of parts of the nervous system deeply impacts not only individual sensory or physiological functions, but also affects a human's self-image. Whereas the human peripheral nervous system has some capacity to regenerate, axon regeneration in the central nervous system is extremely limited [29] and the presence of adult stem cells has been discovered only recently [30, 31]. Nevertheless, neural prostheses for the electric stimulation of specific neural regions (e.g. cochlea implants, retina implants, deep brain stimulation) have a partly long tradition and were successfully applied in clinical settings [32]. Only recently, attempts were made to introduce real biohybrid constructs for the therapy of degenerative diseases [33]. The biocompatibility of neuronal prostheses is determined by the reaction of the central nervous system specific microglia cells, which are the keyplayers in the immune system beyond the blood–brain barrier [34].

4.2.2 Kidney and Liver Biohybrids

There are several approaches to develop artificial organs as an alternative in case of failure or malfunction of the kidney or the liver system. With respect to kidney, tubular epithelial cells growing on semi-permeable hollow fibres were developed [35, 36], mostly as an external device for hemopurification. Inevitably, these are mostly technical driven developments and most innovations occur with respect to the engineering of the membrane compartment [37]. While kidney biohybrid devices are already in clinical use, the development of an effective liver assistance technology remains challenging [38].

4.2.3 Pancreas Biohybrids

Worldwide, diabetes is one of the most widespread diseases, causing enormous costs burdening social health systems and affecting the quality of life of the affected people. It is not astonishing that the therapeutic treatment of diabetes is in the focus not only of pharmaceutical companies but also in basic science, especially because the aetiology of diabetes is well known. Although the reasons and consequences of diabetes may be complex and manifold, the main focus of biohybrid research lies on engineering cells which form the islets of Langerhans. Great success has been achieved with the transplantation of islets of Langerhans from donors; however, this method is limited by the availability of donor tissue and problems of transplant rejection, since current immunosuppressive regimes do not prevent graft rejection [39]. Therefore current research is focussing on producing islet cells from different stem cell sources and to develop an encapsulated cell therapy in the shape of a bioartificial pancreas, protecting the grafted cells from the host immune system [40].

4.2.4 Cardiovascular Biohybrids

In the cardiovascular system, regenerative medicine concentrates on the cure of cardiomyocyte loss using stem cells of different origins [41, 42], permanent or biodegradable stents [43] with or without drug-eluting capacities to prevent e.g. restenosis [44] and tissue engineered vessels [45]. Here, synthetic non-degradable polymers, partly with functionalized surfaces, are used as well as degradable polymeric scaffolds and biopolymers. In combination with bioactive molecules, these scaffolds are able to promote and guide vascular regeneration processes [46] which lead to cellular colonization and finally to the replacement of the artificial graft by autologous cells and extracellular matrix.

4.2.5 Musculoskeletal Biohybrids

The structural component is of main interest in musculoskeletal biohybrids, since these biohybrid systems are mechanically challenged implants with or without functionalized surfaces in orthopaedic and dental applications. These functionalized surfaces may contain elements of the extracellular matrix, ceramics, ions or synthetic polymers responsible for an enhanced osseointegration, especially for improving the adhesion of osteoblasts [47–49], bioactive proteins which serve as growth factors [50, 51], antiinflammatory proteins [52] or antimicrobial substances to prevent biofilm production [53, 54].

5 General Advantages and Limitations of Biohybrids

The general advantages of implantable biohybrids are the sustainable release of physiologically or therapeutically needed molecules (drugs) and the ability for lifelong replacement of a specific tissue function. However, therapeutic cell transplantation in combination with biomaterials usually conforms to the regulations of a regenerative medicine product (RMP) which requires a more complex process for certification than sole structural treatment strategies using only biomaterials [55]. If the biomaterial is biodegradable, it becomes even more complex since the degradation (by-)products need to prove general biocompatibility as well as cytocompatibility to the transplanted cells. Overall, these complex regulatory processes may prevent the interest of companies in biohybrid implants if the advantage for the clinical treatment and the market is not clearly obvious.

6 Conclusions

Biohybrids are designed functional units composed of a structural and a bioactive component which serves for a specific purpose in regenerative medicine, health care or environmental sciences. As implants, these highly complex systems need to prove the biocompatibility of the biomaterial and the immune tolerance of the transplanted cells or bioactive molecules. In this sense, biohybrid implants are clearly more challenging than the sole application of its components. However, great hope and therapeutical potential is currently associated with this sustainable long-term therapeutical approach. We have attempted to provide a classification to keep the term biohybrid in a clear context in biomedical research.

References

1. Williams DF (1999) The Williams dictionary of biomaterials. Liverpool University Press, Liverpool
2. Anderson JM, Rodriguez A, Chang DT (2008) Foreign body reaction to biomaterials. Semin Immunol 20:86–100
3. Franz S, Rammelt S, Scharnweber D, Simon JC (2011) Immune responses to implants—a review of the implications for the design of immunomodulatory biomaterials. Biomaterials 32:6692–6709

4. Beyer Nardi N, da Silva Meirelles L (2006) Mesenchymal stem cells: isolation, in vitro expansion and characterization. Handb Exp Pharmacol 174:249–282
5. Satija NK, Singh VK, Verma YK, Gupta P, Sharma S, Afrin F, Sharma M, Sharma P, Tripathi RP, Gurudutta GU (2009) Mesenchymal stem cell-based therapy: a new paradigm in regenerative medicine. J Cell Mol Med 13:4385–4402
6. Ryan JM, Barry FP, Murphy JM, Mahon BP (2005) Mesenchymal stem cells avoid allogeneic rejection. J Inflamm (Lond) 2:8
7. Curran JM, Chen R, Hunt JA (2006) The guidance of human mesenchymal stem cell differentiation in vitro by controlled modifications to the cell substrate. Biomaterials 27:4783–4793
8. Engler AJ, Sen S, Sweeney HL, Discher DE (2006) Matrix elasticity directs stem cell lineage specification. Cell 126:677–689
9. Wagner W, Horn P, Castoldi M, Diehlmann A, Bork S, Saffrich R, Benes V, Blake J, Pfister S, Eckstein V, Ho AD (2008) Replicative senescence of mesenchymal stem cells: a continuous and organized process. PLoS One 3:e2213
10. Miura M, Miura Y, Padilla-Nash HM, Molinolo AA, Fu B, Patel V, Seo BM, Sonoyama W, Zheng JJ, Baker CC, Chen W, Ried T, Shi S (2006) Accumulated chromosomal instability in murine bone marrow mesenchymal stem cells leads to malignant transformation. Stem Cells 24:1095–1103
11. Bartsch I, Willbold E, Witte F (2010) Monitoring the foreign body reaction using GFP macrophages and a multispectral acqusition and analysis system. Annual Meeting of the European Society for Biomaterials, Tampere
12. Moscona A, Moscona H (1952) The dissociation and aggregation of cells from organ rudiments of the early chick embryo. J Anat 86:287–301
13. Layer PG, Rothermel A, Willbold E (2001) From stem cells towards neural layers: a lesson from re-aggregated embryonic retinal cells. Neuroreport 12:A39–A46
14. Layer PG, Robitzki A, Rothermel A, Willbold E (2002) Of layers and spheres: the reaggregate approach in tissue engineering. Trends Neurosci 25:131–134
15. Schroeder M, Niebruegge S, Werner A, Willbold E, Burg M, Ruediger M, Field LJ, Lehmann J, Zweigerdt R (2005) Differentiation and lineage selection of mouse embryonic stem cells in a stirred bench scale bioreactor with automated process control. Biotechnol Bioeng 92:920–933
16. Kurosawa H (2007) Methods for inducing embryoid body formation: in vitro differentiation system of embryonic stem cells. J Biosci Bioeng 103:389–398
17. Reininger-Mack A, Thielecke H, Robitzki AA (2002) 3D-biohybrid systems: applications in drug screening. Trends Biotechnol 20:56–61
18. Bartholomä P, Gorjup E, Monz D, Reininger-Mack A, Thielecke H, Robitzki A (2005) Three-dimensional in vitro reaggregates of embryonic cardiomyocytes: a potential model system for monitoring effects of bioactive agents. J Biomol Screen 10:814–822
19. Guaccio A, Guarino V, Perez MAA, Cirillo V, Netti PA, Ambrosio L (2011) Influence of electrospun fiber mesh size on hMSC oxygen metabolism in 3D collagen matrices: experimental and theoretical evidences. Biotechnol Bioeng 108:1965–1976
20. Faucheux N, Zahm JM, Bonnet N, Legeay G, Nagel MD (2004) Gap junction communication between cells aggregated on a cellulose-coated polystyrene: influence of connexin 43 phosphorylation. Biomaterials 25:2501–2506
21. Sommerhage F, Baumann A, Wrobel G, Ingebrandt S, Offenhausser A (2010) Extracellular recording of glycine receptor chloride channel activity as a prototype for biohybrid sensors. Biosens Bioelectron 26:155–161
22. Wu C, Chen P, Yu H, Liu Q, Zong X, Cai H, Wang P (2009) A novel biomimetic olfactory-based biosensor for single olfactory sensory neuron monitoring. Biosens Bioelectron 24:1498–1502
23. Venkatesh S, Byrne ME, Peppas NA, Hilt JZ (2005) Applications of biomimetic systems in drug delivery. Expert Opin Drug Deliv 2:1085–1096

24. Sahoo S, Toh SL, Goh JCH (2010) A bFGF-releasing silk/PLGA-based biohybrid scaffold for ligament/tendon tissue engineering using mesenchymal progenitor cells. Biomaterials 31:2990–2998
25. Venkatesh S, Sizemore SP, Byrne ME (2007) Biomimetic hydrogels for enhanced loading and extended release of ocular therapeutics. Biomaterials 28:717–724
26. Petka WA, Harden JL, McGrath KP, Wirtz D, Tirrell DA (1998) Reversible hydrogels from self-assembling artificial proteins. Science 281:389–392
27. Luo G, Zhang Q, Del Castillo AR, Urban V, O'Neill H (2009) Characterization of sol-gel-encapsulated proteins using small-angle neutron scattering. ACS Appl Mater Interfaces 1:2262–2268
28. Lysaght MJ, Reyes J (2001) The growth of tissue engineering. Tissue Eng 7:485–493
29. Huebner EA, Strittmatter SM (2009) Axon regeneration in the peripheral and central nervous systems. Results Probl Cell Differ 48:339–351
30. Galvan V, Jin K (2007) Neurogenesis in the aging brain. Clin Interv Aging 2:605–610
31. Jin K, Galvan V (2007) Endogenous neural stem cells in the adult brain. J Neuroimmune Pharmacol 2:236–242
32. Stieglitz T (2009) Development of a micromachined epiretinal vision prosthesis. J Neural Eng 6:065005
33. Freudenberg U, Hermann A, Welzel PB, Stirl K, Schwarz SC, Grimmer M, Zieris A, Panyanuwat W, Zschoche S, Meinhold D, Storch A, Werner C (2009) A star-PEG-heparin hydrogel platform to aid cell replacement therapies for neurodegenerative diseases. Biomaterials 30:5049–5060
34. Kettenmann H, Ransom BR (1995) Neuroglia. Oxford University Press, New York
35. Saito A, Sawada K, Fujimura S (2011) Present status and future perspectives on the development of bioartificial kidneys for the treatment of acute and chronic renal failure patients. Hemodial Int 15:183–192
36. Ye SH, Watanabe J, Takai M, Iwasaki Y, Ishihara K (2005) Design of functional hollow fiber membranes modified with phospholipid polymers for application in total hemopurification system. Biomaterials 26:5032–5041
37. Ding F, Humes HD (2008) The bioartificial kidney and bioengineered membranes in acute kidney injury. Nephron Exp Nephrol 109:e118–e122
38. Rozga J, Malkowski P (2010) Artificial liver support: quo vadis? Ann Transplant 15:92–101
39. Gonez LJ, Knight KR (2010) Cell therapy for diabetes: stem cells, progenitors or beta-cell replication? Mol Cell Endocrinol 323:55–61
40. Sumi S (2010) Regenerative medicine for insulin deficiency: creation of pancreatic islets and bioartificial pancreas. J Hepatobiliary Pancreat Sci 18:6–12
41. Niebruegge S, Nehring A, Bar H, Schroeder M, Zweigerdt R, Lehmann J (2008) Cardiomyocyte production in mass suspension culture: embryonic stem cells as a source for great amounts of functional cardiomyocytes. Tissue Eng Part A 14:1591–1601
42. Laflamme MA, Murry CE (2011) Heart regeneration. Nature 473:326–335
43. Rodriguez-Granillo A, Rubilar B, Rodriguez-Granillo G, Rodriguez AE (2011) Advantages and disadvantages of biodegradable platforms in drug eluting stents. World J Cardiol 3:84–92
44. Lei L, Guo SR, Chen WL, Rong HJ, Lu F (2011) Stents as a platform for drug delivery. Expert Opin Drug Deliv 8:813–831
45. Hanjaya-Putra D, Gerecht S (2009) Vascular engineering using human embryonic stem cells. Biotechnol Prog 25:2–9
46. Ravi S, Chaikof EL (2010) Biomaterials for vascular tissue engineering. Regen Med 5:107–120
47. Avila G, Misch K, Galindo-Moreno P, Wang HL (2009) Implant surface treatment using biomimetic agents. Implant Dent 18:17–26
48. Zilberman M, Kraitzer A, Grinberg O, Elsner JJ (2010) Drug-eluting medical implants. Handb Exp Pharmacol 199:299–341
49. Shekaran A, Garcia AJ (2011) Extracellular matrix-mimetic adhesive biomaterials for bone repair. J Biomed Mater Res A 96:261–272

50. Thorey F, Weinert K, Weizbauer A, Witte F, Willbold E, Bartsch I, Hoffmann A, Gross G, Lorenz C, Menzel H, Windhagen H (2011) Coating of titanium implants with copolymer supports bone regeneration: a comparative in vivo study in rabbits. J Appl Biomater Biomech 9:26–33
51. Macdonald ML, Samuel RE, Shah NJ, Padera RF, Beben YM, Hammond PT (2011) Tissue integration of growth factor-eluting layer-by-layer polyelectrolyte multilayer coated implants. Biomaterials 32:1446–1453
52. Petzold C, Rubert M, Lyngstadaas SP, Ellingsen JE, Monjo M (2011) In vivo performance of titanium implants functionalized with eicosapentaenoic acid and UV irradiation. J Biomed Mater Res A 96:83–92
53. Simchi A, Tamjid E, Pishbin F, Boccaccini AR (2011) Recent progress in inorganic and composite coatings with bactericidal capability for orthopaedic applications. Nanomedicine 7:22–39
54. Bruellhoff K, Fiedler J, Moller M, Groll J, Brenner RE (2010) Surface coating strategies to prevent biofilm formation on implant surfaces. Int J Artif Organs 33:646–653
55. Messenger MP, Tomlins PE (2011) Regenerative medicine: a snapshot of the current regulatory environment and standards. Adv Mater 23:H10–H17

Interaction of Cartilage and Ceramic Matrix

K. Wiegandt, C. Goepfert, R. Pörtner and R. Janssen

Abstract As subchondral bone is often affected during cartilage injuries, the aim of research is to generate osteochondral implants in vitro using tissue engineering techniques. These constructs consist of a cartilage layer grown on top of a bone phase. In clinical applications, phosphate ceramics have gained acceptance as bone substitute materials because of their great affinity to natural bone. Furthermore, the interaction between cartilage and the underlying bone equivalent is essential for the development and success of osteochondral implants. Here, the influence of a carrier containing hydroxyapatite on the quality of cartilage constructs generated in vitro is investigated. Attempts are made to explain the effects described, by considering chemical and physical properties of the biomaterial.

Keywords Bioceramics · Cartilage · Hydroxyapatite · Osteochondral implants · Surface structure

Contents

1	Introduction	298
2	Relevant Substrate Properties for Tissue Engineering	299
3	Impact of a Hydroxyapatite Carrier on Cartilage Formation	300
4	Time Course of Cartilage Formation	305
5	Impact of Surface Structure	307
6	Conclusion	311
References		312

K. Wiegandt · R. Janssen (✉)
Institute of Advanced Ceramics, Hamburg University of Technology,
Hamburg, Germany
e-mail: janssen@tuhh.de

C. Goepfert · R. Pörtner
Institute of Bioprocess and Biosystems Engineering,
Hamburg University of Technology, Hamburg, Germany

1 Introduction

Articular cartilage defects have only a limited potential to heal spontaneously, which results in joint pain and restricted functioning [1, 2]. Small defects of the articular cartilage surface lead to progressive loss of proteoglycans and disruption of the collagen network. Often cartilage damage proceeds to a full-thickness defect, which also affects the subchondral bone [3]. Consequently, regeneration of the underlying bone has to be incorporated into cartilage repair. Additionally, as the fixation of in vitro generated cartilage often causes problems during and after implantation because of high shear stresses, e.g. of up to 1.7 times body weight in the knee joint, a simultaneous replacement of cartilage and bone is advisable even if the bone is not affected [4–6].

Tissue engineering approaches therefore focus on the generation of osteochondral implants [7]. Instead of transplanting a cartilage–bone cylinder from a non-load-bearing area of the joint into the defective site, as done during autologous osteochondral transplantation (AOT), cells are cultivated in combination with biomaterials in vitro. These biphasic constructs are designed to reconstruct cartilage tissue as well as the underlying bone after implantation [7, 8]. Integration of the carrier into the bone provides an anchorage for the cartilage constructs in the joint. According to Martin et al. [5], approaches can be divided into four strategies: (a) The bone phase is grown with a scaffold, but the cartilage is cultivated without a scaffold on top of the bone. (b) Different scaffolds are used for the bone and the cartilage phase and cultivated separately. Scaffolds are connected during implantation. (c) One scaffold is used for both phases. This scaffold has different structures or compositions for the bone and the cartilage tissue. (d) One homogenous scaffold is used for both phases.

Bioceramics such as bioactive glasses, hydroxyapatite and other calcium phosphates are often used for these approaches as they offer bioactive, osteoconductive and partly osteoinductive properties [9]. Calcium phosphate ceramics in particular are favoured as bone substitute or coating materials because of their natural occurrence in the human body [10, 11]. Additionally, bioceramics contribute good mechanical (compression strength, stiffness, wear resistance) and chemical stability. Against this, their disadvantages are their brittleness and low biodegradability.

During generation of osteochondral implants, biomaterials provide a scaffold both for bone ingrowth as well as for cartilage formation. Thus, the applicability of the bone equivalent to cartilage tissue has to be proven. As cultivation principles often include a proliferation step of chondrocytes which is connected with a dedifferentiation, as additional requirements biomaterials should exhibit not only a carrier for cell growth but also have to support chondrocyte differentiation and cartilage development.

2 Relevant Substrate Properties for Tissue Engineering

Many requirements for biomaterials have been defined for the attachment of cells, proliferation, differentiation or tissue development depending on the host tissue. At an elementary level, the biomaterial has to be non-toxic, biocompatible, biodegradable if needed and must not elicit an inflammatory response from the body [9]. Besides these basic demands, the structural properties of scaffolds have recently become relevant during development of the material. The role of biomaterials has changed from bio-inert cell carriers to bio-functional materials whose properties are used to interfere in cell behaviour, e.g. by mimicking the natural microenvironment [12]. Physical parameters such as the overall architecture, grain size, surface structure, porosity, pore size and mechanical properties, as well as the chemical composition, surface chemistry and charge of biomaterials, are known to influence the process and quality of in vitro generated tissue and its adhesion to the material [11, 13–15].

Physical characteristics of biomaterials have attracted attention for more than three decades, with clinical studies showing a more effective integration of implants into bone for rough than for smooth surfaces [16, 17]. As a consequence, several studies have investigated surface properties of biomaterials, not only of in vivo implants but also in the field of tissue engineering for a number of host tissues [18]. In cell culture experiments, it was found that several cell types react to modified substrate surface topographies, e.g. with an increase in adhesion, acceleration of cell movement, orientation, morphological changes of the cells, cytoskeletal changes, changes in contact inhibition, activation of phagocytosis or changes in gene expression [19–22]. Furthermore, porosity and pore size are relevant on the macroscopic scale to reduce mass transfer limitations of nutrients and waste products or to allow tissue ingrowth [15, 23].

However, it is difficult and often not possible to separate chemical and physical influences of biomaterials on cells. Hence, there is still an ongoing discussion of the importance of chemical and physical aspects [24, 25]. One study differentiating between physical and chemical effects showed that surface chemistry plays a role during short-term attachment of cells, while the topography can be correlated to the long-term adhesion [25]. Parameters which are influenced by substrate chemistry and physical characteristics such as wetting properties, surface energy and surface groups play a role in cell behavior [11, 13, 26–29]. It was further observed that differentiation of certain cell types was supported by substrate surface characteristics, even in the absence of differentiating factors added to the culture medium [25, 30]. These results confirm that the choice of an adequate scaffold during tissue engineering is an essential prerequisite for success [23, 31]. One reason for the reaction of cells to modified surfaces is the selective adsorption and arrangement of proteins which may induce cell attachment, proliferation or differentiation of cells [11]. Adsorption of molecules from culture liquid of body fluids, including proteins for cell attachment, growth factors, lipids, sugars and ions, is well studied. The adsorption of these molecules was shown to be specific for the physical and chemical properties of the biomaterial's surface [13, 32]. Biological aspects of cell reactions including signal mechanisms inside the cells will not be discussed here.

Thus, research aims to control cell reactions and tissue development with the help of an adequate material. Because of the various interactive factors influencing tissue formation—e.g. geometry, composition, porosity and surface structure including macroscopic topography, roughness and grain size—the appraisal of results is difficult and the effects of modified substrate structure are therefore not yet fully understood [18]. Furthermore, cell phenotypes and cells at different maturation states react differently to material characteristics, and culture conditions and different methods of analysis also influence the results [13, 18, 22, 25]. Literature review shows that little is known of how chondrocytes and cartilage tissue respond to different ceramic substrate properties [19]. Therefore, the following sections highlight the impact of a carrier on cartilage quality compared to unsupported tissue over a long cultivation period, important surface characteristics of biomaterials for cartilage tissue engineering, and the influence of carrier surface structure modifications.

3 Impact of a Hydroxyapatite Carrier on Cartilage Formation

The results presented in the following are based on a cultivation principle developed by Nagel-Heyer et al. [33]. Here, the osteochondral implant consists of a ceramic carrier which acts as bone equivalent. On top of this carrier, in vitro generated cartilage is cultivated without any scaffold. The principle combines the advantages of AOT and ACT (autologous chondrocyte transplantation), as autologous chondrocytes and synthetic scaffolds are used for the formation of the three-dimensional cartilage–carrier constructs. This concept allows controllable proliferation, differentiation and matrix production phases [33–36]. As shown in Fig. 1, tissue is explanted from the defective articular cartilage. After an enzymatic digestion, chondrocytes are expanded within three subcultivation steps in monolayer culture, as the initial cell number is limited by the small size of the biopsy (step a). However, proliferation is accompanied by dedifferentiation of cells. A portion of the dedifferentiated cells is cultivated on top of a ceramic carrier to form a cell layer (step b). In previous studies, it was found that this initial cell layer is necessary to improve bonding between in-vitro cartilage and the ceramic carrier [37]. The other cells are re-differentiated in an alginate gel supported by the addition of specific growth factors (step c). In step d, cells together with their cell-associated matrix are recovered from the gel, seeded on top of the cell-coated carrier and cultivated in a high-density cell culture. Afterwards, cartilage–carrier constructs can be implanted into the defective cartilage–bone site.

Chondrocytes from 4-to 6-month-old pigs were used in experiments. The concept was successfully applied in mini-pigs. After 1 year, the cartilage defect was completely closed and the recently formed cartilage showed no difference

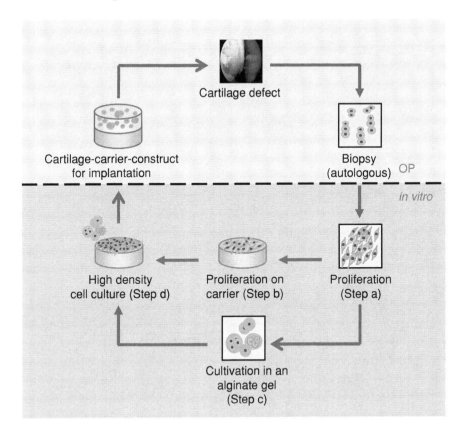

Fig. 1 Cultivation principle for the generation of cartilage–carrier constructs according to Meenen, Adamietz, Goepfert et al

from the surrounding tissue. However, no in vivo resorption or remodelling of the calcium phosphate carrier (Calcibon®, Biomet, Germany) by bone could be observed. As a result, the carrier was not integrated into the bone to an adequate level [34]. For comparison of this cultivation principle with other types of bilayered osteochondral scaffolds, we refer to the detailed review published by O'Shea et al. [9].

The commercially available carrier Sponceram HA® (Zellwerk, Germany) with diameter of 4.55 mm and thickness of 2 mm was used for the generation of osteochondral implants. The carrier consisted of hydroxyapatite, derived by sintering ground porcine bone. Hydroxyapatite was chosen as the bone substitute material as it offers biocompatible, bioactive, osteoconductive and in some cases even osteoinductive properties [38, 39]. The investigation of this carrier for cartilage tissue engineering—the second phase of osteochondral implants—is described in the following.

Important measurable parameters amongst others are porosity, surface structure, hardness, Young's modulus and the determination of surface groups.

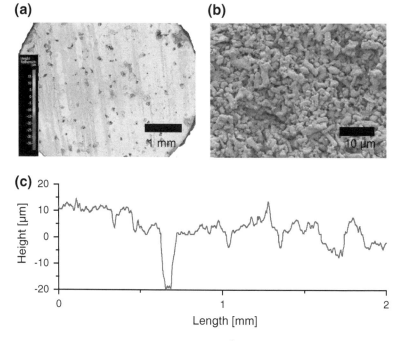

Fig. 2 Surface structure of the carrier Sponceram HA® determined by **a** focus variation with an InfiniteFocus microscope, **b** by scanning electron microscopy and **c** by profile measurements

From a macroscopic point of view, the surface of the Sponceram HA® received exhibits a wavy structure with large local defects (Fig. 2). Profile measurements determined defects with a diameter of up to 150 μm and a depth of up to 100 μm. The scanning electron microscopy (SEM) picture shows a disordered structure of coarse grains exhibiting a diameter of about 1–2 μm and a length of 5–10 μm.

Density determination revealed a porous structure with relative density of 43%. Relatively small macropores allow no ingrowth of cells; for bone tissue, for example, pore diameters of more than 300 μm are recommended [9]. As cartilage is an avascular tissue, a scaffold which supports angiogenesis by an interconnected porosity with adequate pore diameters is not required for nutrient supply [40].

Using the cultivation principle described above, cartilage–carrier constructs were generated on top of Sponceram HA® carrier and compared to cartilage cultivated without a carrier. Significant differences could be observed in the mass and the thickness of tissue-engineered cartilage. The wet weight of cartilage grown without a carrier compared to cartilage grown on top of a carrier was 33.4 ± 1.7 mg and 18.6 ± 4.3 mg, respectively. The thickness was determined to be 2.4 ± 0.1 mm for unsupported cartilage and 0.7 ± 0.2 mm for cartilage–carrier constructs. With a factor of 3.5 for the Young's modulus and 2.3 for the glycosaminoglycan (GAG) content, values were significantly lower for the cartilage grown on top of the carrier compared to cultivation without a carrier (Fig. 3).

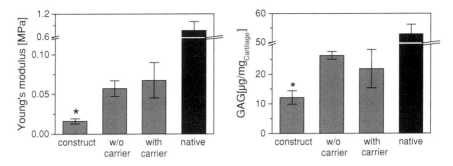

Fig. 3 Young's modulus and glycosaminoglycan (GAG) content of cartilage–carrier constructs (construct), cartilage without a carrier (w/o carrier) and cartilage (with carrier) cultivated in media containing a carrier ($n = 4$–5, $\alpha < 0.05$)

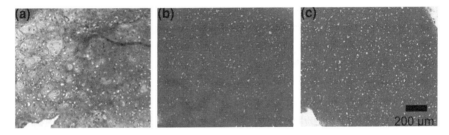

Fig. 4 Histological sections stained with Safranin O of **a** cartilage–carrier constructs, **b** cartilage without a carrier and **c** cartilage cultivated in media containing a carrier

To measure the absolute values, native porcine cartilage from which chondrocytes were isolated was investigated with the same analytical methods and is additionally illustrated in Fig. 3. While the Young's modulus for unsupported cartilage was only about 13% of native cartilage, the GAG content reached 43%.

To estimate tissue organization, histological sections of the in vitro generated cartilage tissue were prepared and glycosaminoglycans as cartilage extracellular matrix components were stained red with Safranin O as shown in Fig. 4. While the tissue of cartilage–carrier constructs showed areas with low glycosaminoglycan content, the histological sections of cartilage without using a carrier during cultivation showed a homogenous tissue in which glycosaminoglycans and chondrocytes were evenly distributed. Additionally, a more prominent staining for collagen type II than type I could be observed under all conditions (not shown here). The distribution of collagens is comparable to that seen with Safranin O staining. Histological slide in Fig. 4(a) shows that the different measured values of the GAG content (and with it possibly also of the increased Young's modulus) of cartilage–carrier constructs compared to unsupported cartilage arose not only from the tissue formed at the interface between biomaterial and cartilage, but from the entire engineered tissue.

Ongoing investigations aim to explain the negative effect of the hydroxyapatite carrier on cartilage formation. For calcium phosphate ceramics in cell culture media, several processes with effects on cell behavior can occur, like dissolution from the ceramic, precipitation from the solution onto the ceramic and diffusion into the ceramic or ion exchange at the ceramic–tissue interface [41]. To investigate if the material itself or some dissolution products of hydroxyapatite into cell culture media cause this negative influence on tissue formation, cartilage without a carrier was cultivated next to a carrier (Figs. 3, 4, notation:with carrier). No significant effects could be observed between cultivation of cartilage cultivated without and next to a carrier in the cell culture liquid, which is expected as hydroxyapatite is known to have a low dissolution rate [42]. However, the possibility cannot be ruled out that high local concentrations of dissolution products or a change in the pH near the tissue–biomaterial interface affect cartilage formation when cultivated on top of the carrier.

Although no influence of dissolution products of hydroxyapatite in the culture media could be observed in this study, surface properties of the carrier may still affect cell reactions. Curran et al. [43] found that $-NH_2$ and $-SH$ modified surfaces promote osteogenesis of mesenchymal stem cells, that $-OH$ and $-COOH$ modified surfaces support chondrogenesis and that mesenchymal stem cells grown on $-CH_3$ surfaces maintained their phenotype. These surface groups were found to modulate focal adhesion assembly, particularly selective integrin binding, and thereby regulate cellular response [44]. As expected for hydroxyapatite, determination of surface groups of the Sponceram HA® carrier by Fourier transform infrared spectroscopy (FTIR spectroscopy) showed, besides PO_4, predominantly OH functional groups. In addition, hydroxyl groups are hydrophilic and this facilitates wetting of the biomaterial with cell culture liquid and initial adherence of cells. Altogether, regarding chemical properties, it is assumed that the hydroxyapatite material of the carriers Sponceram HA® has no harmful effect during the formation of in vitro cartilage and that no disturbing impurities affect cartilage development.

When examining the influence of substrate properties on cartilage formation, further parameters should be kept in mind, such as the surface topography. This is discussed in the next section. In parallel to the chemical aspect of surface groups, an enhanced surface energy can be measured for surfaces modified with hydroxyl groups compared to the other functional groups mentioned [43]. Again, high surface energy indicates advanced adhesion strength of cells [28]. Furthermore, surface charge may be an additional factor which induces cell attachment and differentiation [13, 45]. It is assumed that negative charges which occur due to COOH groups interact strongly with serum proteins, but that OH groups offer a rather neutral surface charge. Surface charges may alter the membrane potential of cells and with it cause signaling inside the cell [13]. In addition, certain mechanical parameters of the biomaterial such as the elastic modulus support cell differentiation via focal adhesion complexes which act as mechano-transducers at the cellular level [15, 46–48]. With nanoindentation measurements, a Young's modulus of 1.7 ± 0.4 GPa for the Sponceram HA® carrier was measured, which on comparison to the Young's modulus for bovine native cartilage of 0.08–2.1 MPa

[49] does not provide a natural microenvironment for chondrocytes, at least for the cell layer near to the carrier. Teixeira et al. [39] could demonstrate that endochondral ossification of a growth plate cartilage to bone took place on biphasic calcium phosphate scaffolds because of their mechanical and chemical analogy to the mineral phase of hard tissue. Bone formation was also observed after implantation of calcium phosphate ceramics in ectopic tissues such as muscles [50], leading to the hypothesis that calcium phosphates as hydroxyapatite support hard tissue synthesis.

During discussion of the reduced cartilage quality when cultured on top of a carrier compared to unsupported tissue, mass transfer limitations due to the carrier have to be kept in mind. Tissue generation on top of a carrier is affected by a insufficient nutrient supply from the bottom. Despite porosity of more than 50% of the Sponceram HA® carrier, diffusion of nutrients and gases is hindered by the carrier itself and additionally by the impermeable clamping device used to fix the carrier in the liquid. Therefore, waste products such as lactate and CO_2 can only be slowly removed from the matrix, which may lead to a decrease in pH. Low pH may result in a reduction of the osmotic pressure which induces cell reactions [51]. Mass transport limitations of the cartilage tissue can be improved by using bioreactor systems such as a flow chamber or loading reactors [35, 52, 53].

4 Time Course of Cartilage Formation

Although cartilage–carrier constructs and cartilage without a carrier are cultivated under the same conditions, cartilage grown without a carrier attained a significantly higher quality of the matrix. Knowledge of the progress of cartilage may help to identify differences in cartilage development and further optimize cultivation protocols by defining time windows for biochemical and biomechanical stimulations. Additionally, in consideration of the clinical applications of in vitro generated tissue, it is necessary to reduce the cultivation time before implantation. The development of cartilage was thus studied over a long time period. Constructs cultivated with and without carriers were harvested after 1–5 weeks of cartilage high-density cell culture (step d) as shown in Figs. 5 and 6.

Only slight changes in cartilage quality could be observed during cultivation of cartilage–carrier constructs. In contrast, biochemical and biomechanical properties varied over time during cultivation without carriers. By increasing cultivation time, the masses of cartilage increased significantly ($\alpha < 0.05$). Young's modulus and GAG content showed a maximum after 3 weeks of cultivation. Significant differences in the Young's modulus were detected between the first to the third week and second to fifth week ($p < 0.001$), and there was also a significant difference in the GAG content between the third to the first and fifth weeks ($p = 0.002$). Changes in Young's modulus may be correlated with the GAG content in tissue engineered cartilage, as zonal organization and orientation of the collagen network could not be observed in unsupported cartilage using the

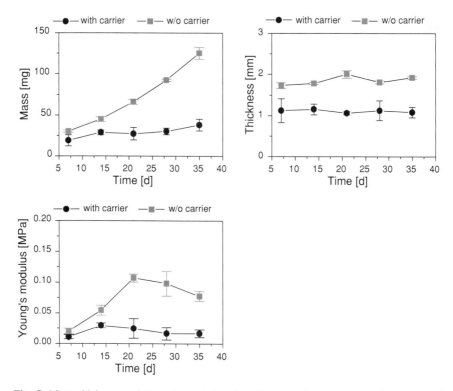

Fig. 5 Mass, thickness and Young's modulus of cartilage–carrier constructs and unsupported cartilage cultivated for 5 weeks. Significant differences ($\alpha < 0.05$) are not shown ($n = 3$ for the Young's modulus, $n = 5$ for the mass and thickness)

cultivation procedure presented above [54]. The DNA content relating to cartilage mass significantly decreased with cultivation time ($p < 0.001$), probably because distances between cells increased due to the enhanced amount of extracellular matrix. Consequently, GAG-to-DNA ratio increased significantly ($p < 0.001$). Ofek et al. [55] in a long-term study also observed no further improvement of cartilage properties (GAG per wet weight, Young's modulus and collagen per wet weight) after 4 weeks of cultivation. It is assumed that after this period exogenous stimulation mechanisms such as mechanical loading or additional growth factors are necessary to further enhance cartilage characteristics.

Although a decrease in DNA content related to cartilage mass could be observed, the total DNA and GAG contents increased during the cultivation time (Table 1), so that a stable growth of cartilage is achieved even over 5 weeks of in vitro cultivation. With a specific DNA content of 6.05 pg per cell for porcine chondrocytes, the cell number was 1.9×10^6 cells for a construct cultivated on top of a carrier and 2.4×10^6 cells for a construct cultivated without a carrier after 5 weeks. With a starting cell number of 1.8×10^6 cells per construct, cells hardly proliferated during this period, which is desired as cartilage contains only about 2–10 vol.% of chondrocytes [56].

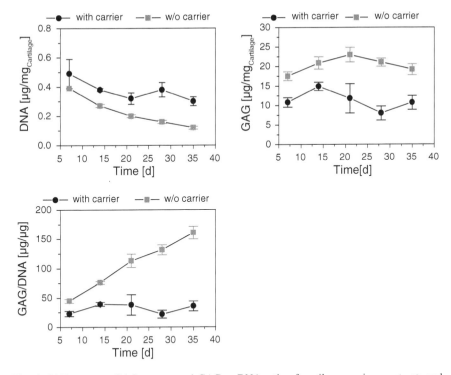

Fig. 6 DNA content, GAG content and GAG-to-DNA ratio of cartilage–carrier constructs and cartilage grown without a carrier cultivated for 5 weeks. Significant differences ($\alpha < 0.05$) are not shown ($n = 4$–5)

Differences in the development of cartilage–carrier constructs and unsupported cartilage can be already observed after 7 days of cultivation. In particularly, mass, Young's modulus, GAG content and GAG-to-DNA ratio developed differently depending on culture conditions. While a decrease in Young's modulus and GAG content of cartilage cultivated without a carrier appeared after 3 weeks, the same but less pronounced effect could be observed for cartilage–carrier constructs after only 2 weeks. While the total GAG content of unsupported cartilage was increased by a factor of 4.5, the GAG content of cartilage–carrier constructs only doubled within 5 weeks (Table 1). Thus, exogenous stimulation of cells must be applied already after 2 weeks during cultivation of cartilage without a carrier, while cultivation of cartilage–carrier constructs must be optimized after 3 weeks.

5 Impact of Surface Structure

As many studies have already shown, the surface structure of biomaterials considerably influences cell behaviour and with it the success of tissue engineering methods. For example, Hallab et al. [28] measured enhanced adhesion shear

Table 1 Total DNA and GAG contents of cartilage–carrier constructs and cartilage without carriers according to cultivation time ($n = 4$–5)

Time [d]	With carrier		Without carrier	
	DNA [µg]	GAG [µg]	DNA [µg]	GAG [µg]
7	9.05 ± 1.39	207.7 ± 55.4	11.70 ± 0.44	523.2 ± 17.7
14	10.61 ± 0.44	399.9 ± 22.9	13.02 ± 0.45	994.6 ± 61.7
21	9.52 ± 1.53	362.3 ± 182.4	13.50 ± 0.47	1568.5 ± 84.5
28	10.80 ± 2.32	236.8 ± 91.1	15.62 ± 0.26	2005.4 ± 124.4
35	11.58 ± 1.62	415.9 ± 107.1	14.50 ± 0.54	2339.5 ± 134.8

strength of fibroblasts predominantly with increasing surface energy and, in the second instance, with increasing roughness on metallic and polymeric substrates. Dos Santos et al. [24] observed that, by culturing osteoblasts on hydroxyapatite, surface topography influenced cell differentiation more than changing surface chemistry. In particular, it could be shown that roughness on the micro- and nanometer scale influences cell morphology, cytoskeletal organisation, proliferation or differentiation of various cell types [24]. Again, the reason for cell behaviour on different surfaces was unclear. Ponche et al. [18] described that, considering the relevant scale, the arithmetic roughness parameter R_a correlated with the wettability of titanium surfaces, which may influence selective adsorption and arrangement of proteins. However, other effects on cell behavior on modified surfaces structures can also be found in the literature. For example, Papenburg et al. [29] stated that surface topography seems to have a predominant effect versus wettability on the morphology of pre-myoplasts cultured on various materials.

To eliminate the effects of biomaterial composition, surface structures were modified by various methods such as photolithography and chemical treatment [57, 58], glancing angle deposition [22], laser blasting or deep reactive ion etching [59]. With these techniques, highly organized surface structures of different geometrically forms such as pillars, cubes or grooves in the nano- and micrometer scale range can be produced [18, 25]. However, these methods are sophisticated and consequently the carriers are very different from carriers which can be used as implants in the clinical routine. To work according to clinical applications and to ensure exactly the same chemistry of the biomaterial, in the following study the above-mentioned commercially available hydroxyapatite carrier was used for the investigation of different surface topographies. Because of the low mechanical stability of the carriers—a low hardness of 14.0 ± 5.5 MPa and Young's modulus of 1.7 ± 0.4 GPa were determined by nanoindentation measurements—conventional grinding/polishing failed to create smooth and homogeneous surfaces. Soft materials like plain paper and ink jet polymer sheet foils were therefore used as grinding tools to modify the surface topography in the direction of smooth (using plain paper) or structured (using rough polymer sheets), when compared to the rough and inhomogeneous untreated carrier (Fig. 7). These methods did not influence the nano-scale roughness, grain size and porous structure, but the macroscopic surface topographies of the hydroxyapatite.

Interaction of Cartilage and Ceramic Matrix 309

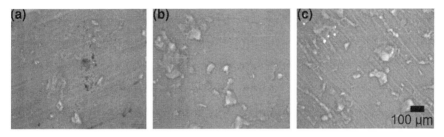

Fig. 7 InfiniteFocus images of commercial grade Sponceram HA® **a** as received, **b** after surface polishing using plain paper and **c** using polymer sheet as grinding tools (adapted from Wiegandt et al. [60])

Fig. 8 SEM pictures of a cell layer grown on modified carrier surfaces: **a** cell layer on the rough and inhomogeneous carrier as received, **b** cell layer on the smooth carrier polished using plain paper, **c** cell layer on the structured carrier polished using polymer sheet (adapted from Wiegandt et al. [60])

Expanded dedifferentiated chondrocytes were seeded onto carriers with modified surface structures for proliferation. The formation of a cell layer was visualized by scanning electron microscopy after 2 weeks of cultivation. While the untreated carrier showed a closed cell layer, only a low number of cells is visible on top of the modified surfaces (Fig. 8). In particular, 200,000 cells were seeded in all three samples. After two weeks, 290,000 ± 40,000 cells were detected on the substrate as received, whereas on the polished samples the cell number was below the detection limit. Thus, no great expansion was observed and, as a consequence, on the polished surface cell deaths even occurred.

Next, cartilage–carrier constructs as described above were prepared using carriers with the different surface structures. The adhesive strength between carrier and tissue was estimated by applying subjective values, detaching cartilage from the biomaterial with the help of tweezers. The results indicate that the connection between tissue and biomaterial was weaker for constructs on top of the polished surfaces than for untreated surfaces. Afterwards, the remaining cells of the carrier were also stained using DAPI as shown in Table 2. Five times more cells were found on top of the untreated carrier than on the polished surfaces. It was assumed that, in contrast to the modified surfaces, the tissue was disrupted during adhesion strength estimation because of a strong bond between cartilage and biomaterial.

Table 2 Remaining cells per carrier with different surface structures after removing constructs ($n = 4$)

Surface modification	As received	Polished using plain paper	Polished using polymer sheet
Surface	Rough and inhomogeneous	Smooth	Structured
Cells per carrier	$31.8 \times 10^3 \pm 6.1 \times 10^3$	$6.4 \times 10^3 \pm 4.3 \times 10^3$	$5.2 \times 10^3 \pm 2.4 \times 10^3$

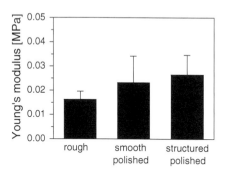
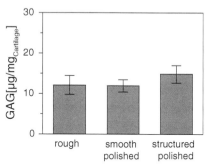

Fig. 9 Young's modulus and GAG content of cartilage–carrier constructs cultured on rough and inhomogeneous surfaces of the carriers as received, on smooth surfaces of the carriers polished with plain paper and on structured surfaces of the carriers polished with polymer sheet ($n = 4$–6)

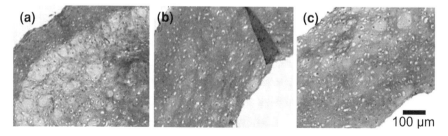

Fig. 10 Histological sections stained with Safranin O of cartilage cultured on **a** rough and inhomogeneous surfaces of the carriers as received, **b** smooth surfaces of the carriers polished with plain paper and **c** structured surfaces of the carriers polished with polymer sheet

Results are comparable with the formation of the cell layer which is necessary to trigger connection between carrier and tissue.

Young's modulus and GAG content were also measured as quality parameters. Constructs using carriers with modified surface structures showed slightly higher Young's moduli than those using untreated carriers. Differences in the biochemical parameters could hardly be observed when comparing cartilage–carrier constructs with modified surface topographies among each other (Fig. 9). Comparing histological sections, glycosaminoglycans are unequally distributed in constructs grown on the top of untreated carriers (Fig. 10) [60].

In summary, inhomogeneous and rough surfaces of hydroxyapatite support the proliferation and adhesion of chondrocytes and cartilage tissue, but lead to inferior quality of the engineered matrix. The negative influence of the carrier shown above was perhaps increased by a stronger adhesive strength to the tissue, compared to the modified smooth and structured surface topographies.

6 Conclusion

For the repair of cartilage defects, osteochondral implants offer the possibility of restoring cartilage tissue and additionally also the subchondral bone, and of anchoring the cartilage effectively in the joint. In the example presented here, a hydroxyapatite carrier was used as bone equivalent substrate and autologous cartilage tissue, which besides bone is the second tissue in osteochondral implants, was cultivated without a scaffold on top.

It was shown that cultivation on top of this ceramic carrier has a significant negative influence on the quality of the in vitro generated cartilage compared to unsupported tissue and, surprisingly, this effect was observed not only at the interface but also in the entire cartilage formed. Results indicated that the composition and surface chemistry did not negatively influence cartilage quality, whereas already moderate modification of surface structure causes slight changes in the biomechanical quality, distribution of glycosaminoglycans, adhesive strength between cartilage and biomaterial, and in attachment and proliferation of a chondrocyte monolayer. These studies produced interesting results, but the negative effect of cultivation on top of the carrier still remains inexplicable. For example, the nutrition supply of the cartilage tissue grown on top of a carrier was not investigated and may be limited by the impermeable carrier and holding device. Improvements in holding device, culture conditions, porosity and pore size of the carrier are needed to increase mass transfer.

In conclusion, the application of bioceramics—especially hydroxyapatite—seems to be suitable for the generation of the osteochondral implants described, but a further adaptation of the carrier to cartilage cultivation is an essential prerequisite for success. Modulation of physical and chemical properties may support differentiation of chondrocytes and tissue formation. In addition, integration in the subchondral bone has to be proven and the pore diameter for bone ingrowth has to be increased. With regard to the progress of the cultivation protocol, unsupported cartilage tissue with a glycosaminoglycan content of nearly 45% compared to native porcine cartilage could be generated in vitro, but the desired biomechanical properties of the tissue cannot be achieved at present.

Acknowledgments We would like to thank Kerstin Michael, Ditte Siemesgelüss, Teresa Richter, Daniel Fritsch and Nadja Holstein for their excellent technical support. The project was kindly supported by DFG (PO 413/7-1) and by BWF-FHH (Tissue Engineering).

References

1. Petersen JP, Ruecker A, von Stechow D, Adamietz P, Poertner R, Rueger JM, Meenen MN (2003) Present and future therapies of articular cartilage defects. Eur J Trauma (29):1–10
2. Hangody L, Füles P (2003) Autologous osteochondral mosaicplasty for the treatment of full-thickness defects of weight-bearing joints: ten years of experimental and clinical experience. J Bone Joint Surg Am 85-A(Suppl 2):25–32
3. Hunziker EB (2002) Articular cartilage repair: basic science and clinical progress. a review of the current status and prospects. Osteoarthr Cartil 10(6):432–463. doi:10.1053/joca.2002.0801
4. Uvehammer J (2001) Knee joint kinematics, fixation and function related to joint area design in total knee arthroplasty. Taylor & Francis, Stockholm
5. Martin I, Miot S, Barbero A, Jakob M, Wendt D (2007) Osteochondral tissue engineering. J Biomech 40(4):750–765. doi:10.1016/j.jbiomech.2006.03.008
6. Waldman SD, Grynpas MD, Pilliar RM, Kandel RA (2002) Characterization of cartilagenous tissue formed on calcium polyphosphate substrates in vitro. J Biomed Mater Res 62(3): 323–330. doi:10.1002/jbm.10235
7. Risbud MV, Sittinger M (2002) Tissue engineering: advances in in vitro cartilage generation. Trends Biotechnol 20(8):351–356
8. Mano JF, Reis RL (2007) Osteochondral defects: present situation and tissue engineering approaches. J Tissue Eng Regen Med 1(4):261–273
9. O'Shea TM, Miao X (2008) Bilayered scaffolds for osteochondral tissue engineering. Tissue Eng Part B 14(4):447–464
10. Schwartz Z, Braun G, Kohavi D, Brooks B, Amir D, Sela J, Boyan B (1993) Effects of hydroxyapatite implants on primary mineralization during rat tibial healing: Biochemical and morphometric analyses. J Biomed Mater Res 27(8):1029–1038
11. Deligianni DD, Katsala ND, Koutsoukos PG, Missirlis YF (2001) Effect of surface roughness of hydroxyapatite on human bone marrow cell adhesion, proliferation, differentiation and detachment strength. Biomaterials 22(1):87–96
12. Pennesi G, Scaglione S, Giannoni P, Quarto R (2011) Regulatory influence of scaffolds on cell behavior: how cells decode biomaterials. Curr Pharm Biotechnol 12(2):151–159
13. Boyan BD, Hummert TW, Dean DD, Schwartz Z (1996) Role of material surfaces in regulating bone and cartilage cell response. Biomaterials 17(2):137–146
14. Schwartz Z, Martin JY, Dean DD, Simpson J, Cochran DL, Boyan BD (1996) Effect of titanium surface roughness on chondrocyte proliferation, matrix production, and differentiation depends on the state of cell maturation. J Biomed Mater Res 30(2): 145–155. doi:10.1002/(SICI)1097-4636(199602)30:2<145:AID-JBM3>3.0.CO;2-R
15. Raghunath J, Rollo J, Sales KM, Butler PE, Seifalian AM (2007) Biomaterials and scaffold design: key to tissue-engineering cartilage. Biotechnol Appl Biochem 46(2):73–84
16. Boyan BD, Lincks J, Lohmann CH, Sylvia VL, Cochran DL, Blanchard CR, Dean DD, Schwartz Z (1999) Effect of surface roughness and composition on costochondral chondrocytes is dependent on cell maturation state. J Orthop Res 17(3):446–457
17. Jarcho M (1981) Calcium phosphate ceramics as hard tissue prosthetics. Clin Orthop Relat Res 157:259–278
18. Ponche A, Bigerelle M, Anselme K (2010) Relative influence of surface topography and surface chemistry on cell response to bone implant materials. Part 1: physico-chemical effects. Proc Inst Mech Eng H224(12):1471–1486
19. Hamilton DW, Riehle MO, Rappuoli R, Monaghan W, Barbucci R, Curtis ASG (2005) The response of primary articular chondrocytes to micrometric surface topography and sulphated hyaluronic acid-based matrices. Cell Biol Int 29(8):605–615. doi:10.1016/j.cellbi.2005.03.013
20. Curtis ASG, Wilkinson CDW (1998) Reactions of cells to topography. J Biomater Sci 9(12):1313–1329

21. Costa Martínez E, Escobar Ivirico JL, Muñoz Criado I, Gómez Ribelles JL, Monleón Pradas M, Salmerón Sánchez M (2007) Effect of poly(L-lactide) surface topography on the morphology of in vitro cultured human articular chondrocytes. J Mater Sci Mater Med 18(8):1627–1632. doi:10.1007/s10856-007-3038-1
22. Pennisi CP, Dolatshahi-Pirouz A, Foss M, Chevallier J, Fink T, Zachar V, Besenbacher F, Yoshida K (2011) Nanoscale topography reduces fibroblast growth, focal adhesion size and migration-related gene expression on platinum surfaces. Colloids Surf B 85(2):189–197
23. Spiteri CG, Pilliar RM, Kandel RA (2006) Substrate porosity enhances chondrocyte attachment, spreading, and cartilage tissue formation in vitro. J Biomed Mater Res A 78(4):676–683. doi:10.1002/jbm.a.30746
24. dos Santos EA, Farina M, Soares GA, Anselme K (2009) Chemical and topographical influence of hydroxyapatite and β-tricalcium phosphate surfaces on human osteoblastic cell behavior. J Biomed Mater Res 89A(2):510–520
25. Anselme K, Ponche A, Bigerelle M (2010) Relative influence of surface topography and surface chemistry on cell response to bone implant materials. Part 2: biological aspects. Proc Inst Mech Eng H224(12):1487–1507
26. Lim JY, Liu X, Vogler EA, Donahue HJ (2004) Systematic variation in osteoblast adhesion and phenotype with substratum surface characteristics. J Biomed Mater Res 68A(3):504–512
27. dos Santos EA, Farina M, Soares GA, Anselme K (2008) Surface energy of hydroxyapatite and beta-tricalcium phosphate ceramics driving serum protein adsorption and osteoblast adhesion. J Mater Sci Mater Med 19(6):2307–2316. doi:10.1007/s10856-007-3347-4
28. Hallab NJ, Bundy KJ, O'Connor K, Moses RJ, Jacobs JJ (2001) Evaluation of metallic and polymeric biomaterial surface energy and surface roughness characteristics for directed cell adhesion. Tissue Eng 7:55–71
29. Papenburg BJ (2010) Insights into the role of material surface topography and wettability on cell-material interactions. Soft Matter 18(6):4377–4388
30. Moroni L, Habibovic B, Monney DJ, van Blitterswijk CA (2010) Functional tissue engineering through biofunctional macromolecules and surface design. MRS Bull 35: 584–590
31. Kieswetter K, Schwartz Z, Hummert TW, Cochran DL, Simpson J, Dean DD, Boyan BD (1996) Surface roughness modulates the local production of growth factors and cytokines by osteoblast-like MG-63 cells. J Biomed Mater Res 32(1):55–63
32. Zeng H, Chittur KK, Lacefield WR (1999) Analysis of bovine serum albumin adsorption on calcium phosphate and titanium surfaces. Biomaterials 20(4):377–384
33. Nagel-Heyer S, Goepfert C, Morlock MM, Pörtner R (2005) Relationship between physical, biochemical and biomechanical properties of tissue-engineered cartilage-carrier-constructs. Biotechnol Lett 27(3):187–192. doi:10.1007/s10529-004-7859-4
34. Petersen JP, Ueblacker P, Goepfert C, Adamietz P, Baumbach K, Stork A, Rueger JM, Poertner R, Amling M, Meenen NM (2008) Long term results after implantation of tissue engineered cartilage for the treatment of osteochondral lesions in a minipig model. J Mater Sci Mater Med 19(5):2029–2038. doi:10.1007/s10856-007-3291-3
35. Nagel-Heyer S, Goepfert C, Feyerabend F, Petersen J, Adamietz P, Meenen N, Pörtner R (2005) Bioreactor cultivation of three-dimensional cartilage-carrier-constructs. Bioprocess Biosyst Eng 27(4):273–280
36. Wiegandt K, Goepfert C, Pörtner R (2007) Improving in vitro generated cartilage-carrier-constructs by optimizing growth factor combination. Open Biomed Eng J 1:85–90. doi:10.2174/1874120700701010085
37. Goepfert C, Böer R, Nagel-Heyer S, Toykan D, Adamietz P, Janssen R, Poertner R (2004) Formation of tissue-engineered cartilage on different types of calcium phosphate ceramics. Cytotherapy: 270–271
38. Rodrigues CVM, Serricella P, Linhares ABR, Guerdes RM, Borojevic R, Rossi MA, Duarte MEL, Farina M (2003) Characterization of a bovine collagen-hydroxyapatite composite scaffold for bone tissue engineering. Biomaterials 24(27):4987–4997

39. Teixeira CC, Nemelivsky Y, Karkia C, Legeros RZ (2006) Biphasic calcium phosphate: a scaffold for growth plate chondrocyte maturation. Tissue Eng 12(8):2283–2289. doi:10.1089/ten.2006.12.2283
40. Chung C, Burdick JA (2008) Engineering cartilage tissue: emerging trends in cell-based therapies. Adv Drug Deliv Rev 60(2):243–262
41. Anselme K, Bigerelle M, Noel B, Dufresne E, Judas D, Iost A, Hardouin P (2000) Qualitative and quantitative study of human osteoblast adhesion on materials with various surface roughnesses. J Biomed Mater Res 49(2):155–166
42. LeGeros RZ, Parsons JR, Daculsi G, Driessens F, Lee D, Liu ST, Metsger S, Peterson D, Walker M (1988) Significance of the porosity and physical chemistry of calcium phosphate ceramics. Biodegradation-bioresorption. Ann NY Acad Sci 523:268–271
43. Curran JM, Chen R, Hunt JA (2006) The guidance of human mesenchymal stem cell differentiation in vitro by controlled modifications to the cell substrate. Biomaterials 27(27):4783–4793. doi:10.1016/j.biomaterials.2006.05.001
44. Keselowsky BG, Collard DM, García AJ (2005) Integrin binding specificity regulates biomaterial surface chemistry effects on cell differentiation. Proc Natl Acad Sci USA 102(17):5953–5957. doi:10.1073/pnas.0407356102
45. Lee MH, Ducheyne P, Lynch L, Boettiger D, Composto RJ (2006) Effect of biomaterial surface properties on fibronectin-[alpha]5[beta]1 integrin interaction and cellular attachment. Biomaterials 27(9):1907–1916
46. Toh WS, Spector M, Lee EH, Cao T (2011) Biomaterial-mediated delivery of microenvironmental cues for repair and regeneration of articular cartilage. Mol Pharm
47. Liu SQ, Tian Q, Hedrick JL, Po Hui JH, Rachel Ee PL, Yang YY (2010) Biomimetic hydrogels for chondrogenic differentiation of human mesenchymal stem cells to neocartilage. Biomaterials 31(28):7298–7307
48. Engler AJ, Sen S, Sweeney HL, Discher DE (2006) Matrix elasticity directs stem cell lineage specification. Cell 126(4):677–689
49. Wu JZ, Herzog W (2002) Elastic anisotropy of articular cartilage is associated with the microstructures of collagen fibers and chondrocytes. J Biomech 35(7):931–942
50. Moroni L, Habibovic B, Monney DJ, van Blitterswijk CA (2010) Functional tissue engineering through biofunctional macromolecules and surface design. MRS Bull 35:584–590
51. Luppa D (2000) Biochemie und Pathochemia des hyalinen Gelenkknorpels. KCS 1(12):29–39
52. Heyland J, Wiegandt K, Goepfert C, Nagel-Heyer S, Ilinich E, Schumacher U, Pörtner R (2006) Redifferentiation of chondrocytes and cartilage formation under intermittent hydrostatic pressure. Biotechnol Lett 28(20):1641–1648
53. Hoenig E, Winkler T, Mielke G, Paetzold H, Schuettler D, Goepfert C, Machens HG, Morlock MM, Schill AF (2011) High amplitude direct compressive strain enhances mechanical properties of scaffold-free tissue-engineered cartilage. Tissue Eng Part A 17(9–10):1401–1411
54. Wiegandt K, Pörtner R, Müller R (2009) Einfluss hydrostatischer Druckbelastung während der in vitro Herstellung von dreidimensionalen Knorpelimplantaten. urn:nbn:de:gbv:830-tubdok-8126
55. Ofek G, Revell CM, Hu JC, Allison DD, Grande-Allen KJ, Athanasiou KA (2008) Matrix development in self-assembly of articular cartilage. PLoS One 3(7):2795. doi:10.1371/journal.pone.0002795
56. Kuettner KE, Aydelotte MB, Thonar EJ (1991) Articular cartilage matrix and structure: a minireview. J Rheumatol Suppl 27:46–48
57. Clark P, Connolly P, Curtis AS, Dow JA, Wilkinson CD (1990) Topographical control of cell behaviour: II. Multiple grooved substrata. Development 108(4):635–644
58. Brown A, Burke G, Meenan BJ (2011) Modeling of shear stress experienced by endothelial cells cultured on microstructured polymer substrates in a parallel plate flow chamber. Biotechnol Bioeng 108(5):1148–1158

59. Elter P, Weihe T, Lange R, Gimsa J, Beck U (2011) The influence of topographic microstructures on the initial adhesion of L929 fibroblasts studied by single-cell force spectroscopy. Eur Biophys J 40(3):317–327
60. Wiegandt K, Goepfert C, Richter T, Fritsch D, Janßen R, Pörtner R (2008) In vitro generation of cartilage-carrier-constructs on hydroxylapatite ceramics with different surface structures. Open Biomed Eng J 2:64–70. doi:10.2174/1874120700802010064

Bioresorption and Degradation of Biomaterials

Debarun Das, Ziyang Zhang, Thomas Winkler, Meenakshi Mour, Christina I. Günter, Michael M. Morlock, Hans-Günther Machens and Arndt F. Schilling

Abstract The human body is a composite structure, completely constructed of biodegradable materials. This allows the cells of the body to remove and replace old or defective tissue with new material. Consequently, artificial resorbable biomaterials have been developed for application in regenerative medicine. We discuss here advantages and disadvantages of these bioresorbable materials for medical applications and give an overview of typically used metals, ceramics and polymers. Methods for the quantification of bioresorption in vitro and in vivo are described. The next challenge will be to better understand the interface between cell and material and to use this knowledge for the design of "intelligent" materials that can instruct the cells to build specific tissue geometries and degrade in the process.

Keywords Biodegradation · Bioresorption · Implants · Osteoclasts · Tissue engineering

Contents

1 Introduction	318
1.1 Stable and Degradable Biomaterials	318
1.2 Medical Applications for Resorbable Biomaterials	318
2 Biodegradable Materials	319
2.1 Metals	319
2.2 Ceramics and Glasses	321
2.3 Polymers	322

D. Das · Z. Zhang · M. Mour · C. I. Günter · H.-G. Machens · A. F. Schilling (✉)
Department of Plastic Surgery and Hand Surgery, Klinikum Rechts der Isar,
Technische Universität München (TUM), Ismaninger Str. 22,
81675 Munich, Germany
e-mail: a.schilling@tum.de

T. Winkler · M. M. Morlock
Biomechanics Section, Hamburg University of Technology (TUHH),
Hamburg, Germany

3 Measurements of Bioresorption 322
   3.1 ISO Medical Biomaterials Degradation Assays 322
   3.2 Cell-Based Resorption Assays 325
   3.3 In Vivo Tests of Biomaterial Degradation 327
4 Conclusions and Outlook 327
References 328

1 Introduction

1.1 Stable and Degradable Biomaterials

By definition, biomaterials come into close contact with tissues of the body [1]. As soon as this happens, an interaction starts between the body and the material. Body fluids represent an extremely corrosive mixture of fluid, ions and proteins [1–4]. Movement of the body leads to mechanical stresses in the form of bending and shearing forces, which further promote the degradation of the material [5]. Additionally to this "passive" degradation, there is an inflammatory cellular response to foreign materials, which includes immune cells like monocytes, neutrophils, lymphocytes, macrophages or osteoclasts, and these cells can actively participate in the bioresorption (For review see [1]). The degree and nature of this foreign body response also depends on the properties of the device, such as material composition, three-dimensional morphology including porosity, surface structure and surface chemistry [6–8]. There are two main strategies for dealing with this tendency of the body to attack external materials: (1) try to find and use materials that are as inert as possible and do not induce an adverse inflammatory reaction; (2) integrate the body's innate repair mechanisms into the design of temporary devices, which can be slowly degraded or replaced by healing tissue when their work is done. The advantages and disadvantages of these two approaches are directly linked to the intended application, the site of implantation and the specific patient. An implant that is designed to last inertly for 20 years inside the body is a good option for a 70-year-old patient. In a 30-year-old patient, or even more in growing children, it is obvious that a strategy which involves regeneration may be a better solution. A necessary precondition, limiting the use of temporary supporting or augmenting biomaterials, however, is sufficient healing or regeneration capacity of the tissue, ensuring that after healing the implant is no longer needed [9].

1.2 Medical Applications for Resorbable Biomaterials

Consequently, bioresorbable implants are of great use for the temporary management of tissue defects in tissues with a usually good healing ability like skin, bone or tendon. Here fixation of the tissue improves the healing process and after the tissue is regenerated it can support its own structure without the need of the biomaterial [10]. For the same purpose, non-resorbable materials can also be used, but these have to be actively removed after the treatment, resulting in additional operations with all the complications for the patient, including pain and costs [10].

Widely used applications of biodegradable materials are therefore sutures and bands for the closing of soft tissue wounds or in the repair of tendons and ligaments [11]. Gauzes, felts and velour dressings made of biodegradable fibers are used in the treatment of burns or traumatic injuries [12]. Degradable polymeric meshes have been studied for their use in arterial grafting [13–16] and for splenorrhaphy and pelvic peritoneum reconstruction [17].

Bioresorbable implants have also been used in orthopedics and traumatology [18]. The use of bioabsorbable osteosynthetic screws, pins and nails in patients with peri- and intra-articular fractures has been shown to lead to bony union without any abnormal blood levels, infections or foreign body reactions [10, 19, 20].

In bone, controlled degradation can have an additional benefit. Bone is constantly remodeled and adapted to the load it has to bear [21]. A rigid fixation with a stiff implant will lead to decreases in stress levels inside the bone (stress shielding) which results in active degradation of the bone tissue by osteoclasts with deleterious effects on the mechanical properties of the bone [1, 22–25]. Ideally, controlled degradation of a temporary material leads to consecutive loss of mechanical strength of the device which in turn leads to slowly rising forces in the healing tissue, thereby enhancing the healing process and avoiding the unwanted consequences of stress shielding.

For medical devices in the cardiovascular system like intravascular stents, the mechanical forces are also important. An implant made of such a material should be able to function fully under constant blood flow until the diseased vessel completely recovers. After that, the stent should be gradually dissolved, consumed or absorbed [26].

Apart from these long-used applications, progress in the development of novel biomaterials has enabled the invention of a variety of novel medical technologies, like drug and gene delivery systems, tissue engineering and scaffold-based cell therapies, organ printing, nanotechnology-based diagnostic systems and microelectronic devices [27].

2 Biodegradable Materials

The materials utilized for temporary biomedical devices include metals, ceramics, glasses, polymers and composites of the former [1,28].

2.1 Metals

After decades of designing strategies to minimize the corrosion of metallic biomaterials, there is a growing interest in using corrodible alloys in a number of medical applications [29]. When metals are used in implants, corrosion is always a concern and has therefore been studied extensively in metallic biomaterials [30, 31]. This corrosion is facilitated by anions, cations, biological macromolecules and mechanical load, which are available in abundance in the physiological environment into which biomaterials are implanted [32]. On the side of the implant, geometric, metallurgical and surface properties play a role [30]. Degradation may result from electrochemical dissolution phenomena, wear, or a synergistic combination of the two [30]. The electrochemical degradation can be generalized, affecting the whole surface, localized at areas that are shielded from tissue fluids (crevice corrosion), or random (pitting corrosion) [30]. Examples of interactions between electrochemical and mechanical influences include stress corrosion cracking, corrosion fatigue and fretting corrosion, which may cause premature structural failure and accelerated release of degradation products [30]. These degradation products include wear debris, colloidal organometallic complexes (specifically or non-specifically bound), free metallic ions, inorganic metal salts or oxides, and precipitated organometallic storage forms [33]. These products can form complexes that sometimes elicit hypersensitivity responses [33]. There are also reports indicating possible side effects due to release and accumulation of metal ions into the surrounding tissues [34]. Thus the current aim of researchers and engineers is to control the rate of degradation/corrosion by developing novel alloys while minimizing the effects of the products of degradation.

2.1.1 Iron

Cardiovascular diseases account for the majority of the deaths worldwide, and the numbers are only expected to grow in the coming years [35]. This has led to sizeable resources being allocated to research in developing better arterial stents, which help combat coronary heart diseases. It is well know that iron is an essential element in the human body. For this reason iron is thought to offer acceptable biocompatibility when used in degradable materials investigated for stent applications [36]. Iron has good mechanical properties and a relatively low degradation rate. Its biodegradation leads to oxidation of iron into ferrous and ferric ions, which then dissolve in biological fluids [37]. Iron stents can be fabricated by casting or using electroformed iron (E-Fe), which has a faster degradation speed with a uniform degradation mechanism. E-Fe also inhibits cell proliferation of smooth muscle cells without decreases in metabolic activity [36]. This may be due to the increased degradation speed, probably leading to higher concentrations of ferrous ions around the cells, which has been shown to reduce the proliferation of smooth muscle cells under in vitro conditions [37]. For application as stent material, this may be an additional benefit, as it may inhibit neointimal hyperplasia and in-stent restenosis [37].

2.1.2 Magnesium

The use of magnesium in biological systems dates back to 1906 when it found application as fracture fixation plates [38]. Thus when novel applications for biodegradable implants were developed recently, magnesium was an obvious choice for evaluation [39]. The modulus of elasticity of magnesium is closer to bone than the elastic modulus of commonly used metal implants like titanium or cobalt–chrome–molybdenum–alloys. This property of magnesium makes it interesting for orthopedic applications. The corrosion rate of pure magnesium is however too high for application as an implant material [40]. Studies have shown that corrosion as well as mechanical properties can be positively influenced when magnesium is alloyed with rare earth metals [40–44], The properties of these magnesium alloys may help develop implants that can act as a scaffold on which new bone can grow, as well as fixtures to hold together bone long enough to allow natural healing to take place [41]. Magnesium has also recently been investigated as a possible material for intravascular stent applications [41, 45]. It has been shown that hemolysis and platelet adhesion can be positively influences by certain alloys [45].

2.1.3 Tungsten

Degradable coils made of tungsten were used in a series of pediatric patients [46] for the occlusion of pathological vessels. On follow-up, fluoroscopic analysis showed a decreased radio-opacity indicating degradation of the coils. Correspondingly, there was a marked increase in serum levels of tungsten and the previously occluded vessels were recanalized. Although in vitro analysis of the tungsten coils, which interestingly was performed after the clinical application, showed a slow degradation and low toxicity of tungsten [47], its further application for this indication is not recommended [46] by the authors.

2.2 Ceramics and Glasses

When considering ceramics, glasses and glass–ceramics as biomaterials, a wide range of characteristics are observed especially in terms of stability in physiological environments [1]. They can be categorized into three groups based on their surface reactivity: essentially inert materials, soluble materials, and intermediate materials with limited or controlled surface reactivity [48]. Ceramics, typically alumina [49] and certain hydroxyapatites (HA), are biologically inert, especially dense calcium hydroxyapatite [48], the naturally occurring mineral phase of bones and teeth [1]. Close relatives of these HA such as tri-calcium phosphate (TCP) or calcium-alkali-orthophosphates are biodegradable [50–53]. Different factors have been determined that influence the degradation behavior of ceramic biomaterials. These include: physical forms (degrees of micro- and macro-porosity; density); composition (TCP vs. HA; glass composites vs. HA ceramic) and crystallinity (e.g. coralline HA vs. HA ceramic) [54]. In materials made from powders, disintegration is mainly governed by the solubility product of the necks connecting the powder particles after crystallization [55]. Consequently, when the necks are dissolved, the materials break down into the original particles. It has been shown that those particles can induce inflammatory responses depending on their size [56, 57]. On the other hand, controlled surface reactivity can be used to provide the surrounding cells with the building blocks, such as calcium and phosphate, for the new tissue or stimulate the cells with leaching bioactive components [1]. Osteoclasts and macrophages are involved in phagocytosis and in resorption of bioactive ceramics and glasses [54]. The non-degraded materials tend to accumulate to form extracellular deposits [58].

2.3 Polymers

The development of artificial polymers has changed almost all fields of modern science and engineering. Biodegradable implants are no exception to this. Biodegradable polymers typically contain linkages susceptible to hydrolysis. If such a material is hydrophilic and capable of absorbing water, the physiological environment will lead to its degradation [59]. Tissue enzymes can greatly influence this process. Enzymes are catalysts for specific biochemical reactions. Certain enzymes are able to induce or accelerate hydrolytic degradation of polymers, which are normally not degraded at body temperature [1, 60–62]. In addition, certain cells of the immune system are involved in the degradation and resorption process, either by their attachment to polymer surfaces and the release of destructive enzymes on to the surface [63] or by the ingestion of fragments of the polymer. Degradable polymers include: polyglycolide acid (PGA) [64, 65], poly(L-lactide) (PLL), PLGA [66, 67], PLLA [68, 69], poly(D-lactide) (PDL), poly(c-caprolactone) (PCL) [70], polyphosphazenes [71, 72], poly(orthoester) (POE) [73], poly(beta-hydroxybutyrate) (PHB) [64], polyhydroxyalkanoates (PHA) [74, 75], polyesters based on fumaric acid (PPF) [76–79], tyrosine-based polycarbonates [80, 81], rosin-based polymers [82–84], and the naturally occurring polymer collagen [85].

3 Measurements of Bioresorption

To be able to develop new biomaterials with tailored degradation characteristics, it is necessary to have methods available to quantify the extent of the degradation under different conditions (Table 1).

3.1 ISO Medical Biomaterials Degradation Assays

The International Organization for Standardization (ISO) is a worldwide federation which mainly focuses on establishing standards worldwide (www.iso.org). The main advantage of this approach for testing of biodegradation is that even if

Bioresorption and Degradation of Biomaterials

Table 1 Biomaterial resorption assays

	Analyzing principle	Pros	Cons
In vitro			
ISO 10993- (13–15)	Weight changes after direct contact with simulated in vivo fluids	Standard assay, easy to conduct and analyze	Many of the in vivo biomaterial resorption factors (e.g., growth factors, body temperature, cells, hormones) are not considered
Cell-based resorption assays	Analysis of material surface change after cell mediated bioresorption	Cells, which are a major resorption factor are considered; individualized assessment possible	Not all the in vivo factors can be included; it is difficult to standardize the procedure; resorption analysis is sometimes difficult
In vivo			
Implantation in animals	To mimic the in vivo human environments as closely as possible, evaluation by histology or imaging	Complete physiological environmental conditions	Ethical issues, expensive
Clinical studies	To detect human-specific effects with all the clinical level tests	Valid results	Ethical issues, safety issue for clinical trial candidates

materials are developed and tested in different countries by different groups, the results can still be compared to each other. ISO 10993 is a series of such standards designed to evaluate the biocompatibility of medical devices before clinical evaluation. It is now the generally accepted standard procedure for the evaluation of medical devices. There are three parts to the ISO 10993 used for analyzing biomaterial degradation: (1) ISO 10993-13: Identification and quantification of degradation products from polymeric medical devices; (2) ISO 10993-14: Identification and quantification of degradation products from ceramics; (3) ISO 10993-15: Identification and quantification of degradation products from metals and alloys. These three parts cover all the important current degradable materials for medical purposes.

3.1.1 ISO Degradation Assays for Polymeric Materials

For polymeric biomaterials, such as collagen, PLA, PCL or PGA, ISO 10993-13 is the suitable test. ISO 10993-13 has two major parts: (1) the accelerated degradation test and (2) the real-time degradation test. Accelerated degradation testing is mainly used for a first assessment of general degradability of such biomaterials in vivo. The principle of the accelerated test is to immerse the material in a simulated body fluid, at a temperature higher than 37 °C and lower than the melting temperature for a period of between 2 days and 60 days. This test will lead to sample degradation, which can be analyzed by sample mass balance and molecular balance. If the information gathered from the accelerated test is not sufficient to justify the use of the material, the real-time test will additionally be applied. This test is designed to mimic the real-time in vivo environment. Therefore, it is conducted at normal body temperature (37 °C) and four different time periods are required. After these periods, the same mass balance analysis and molecular mass analysis will be applied and the information on degradation characteristics will be derived from these data. Both accelerated degradation and normal degradation tests are mainly designed to stimulate hydrolytic and oxidative degradation.

ISO Degradation Assays for Ceramic Materials

Ceramic materials, which are commonly used in dental and orthopedic applications, are analyzed differently from polymers. ISO 10993-14 is mainly designed for testing in vitro degradation of ceramic materials in solutions of different pH. The test consists of two parts: (1) degradation in an extreme solution, typically citric acid buffer solution (pH 3) which represents the worst possible environment in vivo for the material; (2) degradation in a solution that represents the normal in vivo environment, typically Tris-HCl buffer with a pH of 7.4 ± 0.1. This test is very easy to conduct and thus is adopted frequently in pre-clinical testing of ceramic materials.

ISO Degradation Assays for Metals

Since metals or alloys possess high electrical conductivity, the degradation test for metals or alloys measures not only the possible chemical degradation by immersion testing but also the potentiodynamic and potentiostatic ability of the materials, trying to predict the in vivo electrochemical behavior. The electrolyte used for analysis can be an isotonic aqueous solution of 0.9% sodium chloride, artificial saliva or artificial plasma. The surface of the material, mass balance and electrochemical behavior are measured as parameters to estimate the possible in vivo degradation.

3.1.2 Degradation Assays for Composite Materials

In recent years, more and more medical devices have been constructed from more than one class of biomaterials. In this case, therefore, a combination of the above tests is usually necessary and an assessment is needed of whether each part of the medical device needs to be analyzed separately or if a combined analysis is reasonable.

3.2 Cell-Based Resorption Assays

For reasons of standardization, ISO testing of material degradation disregards many of the influences observed in vivo, such as active cellular influences. This may be an explanation why some materials which were shown to be fully degradable under ISO conditions have still been detected after years of in vivo implantation [86, 87]. Therefore, to improve the prediction of in vivo behavior, cell-based degradation testing has been developed (Fig. 1).

Several kinds of cells have the ability to resorb foreign materials such as osteoclasts and macrophages [88, 89]. The most commonly used cells for analyzing resorption in vitro are osteoclasts. This is probably due to the widespread clinical application of such materials in both dental and orthopedic patients. Primary osteoclasts are found in abundance in the bones of neonatal animals [90]. In 1984, Chambers and colleagues isolated osteoclasts from neonatal rabbit long bones by fragmenting the bone structure in HEPES-buffered medium [91]. Similar isolation procedures were then transferred to neonatal rat and neonatal chicken [92]. In addition, Jones et al. [93] found that osteoclasts isolated by the same approach showed no differences in species (rat, chicken or rabbit) with regard to biomaterial resorption. This method was considered by some researchers to be the best for the isolation of large numbers of primary osteoclasts [94]. Besides primary osteoclasts, differentiated osteoclasts, especially differentiated human osteoclasts, have recently been studied for this purpose, because they promise to permit individualized testing of the material with the cells of specific patients [52, 53]. Two factors are essential for osteoclast differentiation: receptor activator of nuclear factor kappa-B ligand (RANKL) and macrophage colony-stimulating factor (M-CSF). Both RANKL and M-CSF are secreted by osteoblasts for activation of osteoclast differentiation. M-CSF is a hematopoietic growth factor involved in the proliferation and differentiation of monocytes [95]. Although indispensable for osteoclast differentiation, high concentrations of M-CSF actually inhibit osteoclast formation in vitro. The identification of RANKL as a differentiation factor for osteoclasts and its recombinant production has had a major impact on cell-based resorption assays [96]. Isolation of mononuclear precursors from different sources and species and differentiation into osteoclasts has since become more and more popular in cell-based resorption assays.

Since cellular resorption typically leads to only subtle changes in the surface structure of the material, the evaluation of the material mass is usually not suitable for cell-based resorption assays and specialized methodology is necessary.

The analysis of surface resorption of biomaterials is commonly based on several classical and also advanced optical methods such as light microscopy (LM), scanning electron microscopy (SEM), confocal laser scanning microscopy (CLSM) and infinite focus microscopy (IFM). For two-dimensional (2D) analysis, LM is still the first choice. This is partly due to the wide availability of such equipment and also the ease of the procedures. However, the LM resorption assay

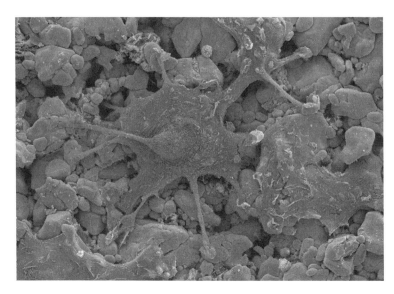

Fig. 1 Cell-based resorption assay with an osteoclast (marked in *pink*) on the surface of a calcium-orthophosphate (scanning electron microscopy ×2300)

is critically dependent on the detection of resorption pits, which typically have to be stained to be visible. The same is true for CLSM. If resorption is not detectable in LM and CLSM due to lack of staining and low contrast of the material, scanning electron microscopy can be used [97]. For this, the specimen has to be specially prepared to be suitable for the necessary vacuum environment during the analysis. Typically, cells are first removed from the material surface; the specimen is dried in an increasing ethanol series and then sputtered with gold to enhance the surface contrast. Several parameters such as pit number and pit area can be analyzed using 2D imaging in combination with image analysis software [88]. For 3D parameters like pit depth and pit volume, other methodologies are necessary. For this purpose, 3D-SEM based on stereophotogrammetry methods has been tested [97]. This technology uses images obtained from different angles of the pit to calculate the volume of the pit [98], but is relatively labor-intensive. Our group has recently introduced infinite focus microscopy as a possible tool for measuring bone substitute resorption [52, 53]. The IFM system makes use of the relatively small focal plane of light microscopes and is based on algorithms that can automatically detect focused areas in a series of images. The algorithm then combines only the focused parts of each image with the position information of the image in the image stack. From these data the 3D surface can be reconstructed with a high resolution (<1 μm). No sample pre-processing step is necessary and the size of the detection area is limited only by the computing power.

3.3 In Vivo Tests of Biomaterial Degradation

Even cell-based in vitro tests cannot completely model the in vivo situation and therefore in vivo biomaterial degradation is also monitored. Different methods are used depending on the type of material. The most widely applied methodology so far has been histology combined with different analysis technologies. This, however, requires explantation of the tissue in which the material was embedded and can therefore only be used pre-clinically. Assessment of biodegradation and bioresorption has been reported in terms of changes in macro-porosity, density, changes in pore and grain diameters and average crystal size with time from the surface and the core of implant materials as observed by scanning (SEM) and transmission (TEM) electron microscopy [54, 99]. For more accurate and specific analysis, energy-dispersive X-ray analysis (EDX) and Fourier transform infrared (FT-IR) spectrophotometry are employed. Elemental analysis can be carried out by EDX, which is connected to SEM. Chemical analysis of composites can be carried out by FT-IR spectrophotometry [100]. With the application of synchrotron-radiation-based microtomography (SRµCT) in attenuation mode, not only can the amount of biomaterial degradation be monitored in 3D, but also the spatial distribution of alloying elements during in vivo corrosion [101, 102].

For nondestructive in vivo analysis of degradation of radiodense biomaterials, X-ray µCT systems can be used to determine biodegradation, bioresorption, and the extent of reaction of the surrounding tissue [103]. X-ray µCT has a spatial resolution in the micrometer range depending on the area of interest and is commonly used in small animal testing. X-ray µCT or magnetic resonance imaging (MRI) systems combined with positron emission tomography (PET) or single photon emission computed tomography (SPECT) scanners can be used to monitor the functional processes of biologically active molecules in the body by injecting a short-lived radioactive isotope tracer [104]. Non-irradiation techniques such as micro MRI (µMRI) and functional MRI (fMRI) also have resolutions in the micrometer scale and can therefore be used for analysis of the degradation of non-radiodense biomaterials [105].

4 Conclusions and Outlook

The human body is a composite structure, completely constructed of biodegradable materials, and can last—with good maintenance—for more than 100 years. It achieves this goal by constantly actively repairing skin, bones, gut, liver and many other tissues. This repair process includes bioresorption of the old structure and replacement of it with new tissue. As this is the result of millions of years of evolution, it seems only sensible to adapt this method for the use of biomaterials too. In recent years there has been huge progress in this area. Apart from the materials mentioned above, bionic nano-methods now allow the production of collagen-like structures [106] as well as composite structures between ceramics and polymers that mimic bone or teeth [107]. It is probably only a matter of time before the properties of artificial materials surpass those of their natural prototype. At present resorbable biomaterials are mostly used for the repair of diseased or injured tissues, but other applications are already under discussion, such as drug delivery nano-systems, biodegradable nano-robots, multi-functional particles, biogalvanic batteries, and electronic components like sensors that dissolve and disappear [108–114].

One of the next challenges will probably be to better understand the interface between cell and material and to use this knowledge for the design of materials tailored to specific cellular environments. So far, most of the research has concentrated on understanding how the physiological environment changes the material. It seems at least as important to understand how the surrounding cells react to the degrading material. It has long been known that cells react to the surface properties of the materials they are attached to by changing their shape or movement [115]. Recently it has also been shown that the differentiation of stem cells is dependent on the elastic modulus of the material on which they are cultivated [116]. At present, when a material is introduced into the body, an unwanted foreign body reaction is principally expected. It can be envisioned, however, that the right combination of materials combined with advanced 3D design may one day lead to "intelligent" materials that can instruct stem cells to build specific tissue geometries and degrade in the process, leaving only materials that the cells can use to regenerate. To reach this distant goal a close collaboration of cellular and molecular biologists with engineers, doctors and entrepreneurs will be necessary.

Acknowledgments T.W. was supported by a grant from the German Federal Ministry of Education and Science (BMBF, 16SV5057). M.M.M. is consultant to DePuy and Zimmer. Institutional support was received from Ceramtec, Aesculap, DePuy, and Zimmer. No royalties. A.F.S. has in the past 5 years provided consulting services to Biomet, Curasan, Eucro, Heraeus, IPB, and Johnson & Johnson. No royalties.

References

1. Williams DF (1987) Review: tissue-biomaterial interactions. J Mater Sci 22:3421–3445
2. Brown SA, Merritt K (1981) Fretting corrosion in saline and serum. J Biomed Mater Res 15:479–488
3. Williams DF, Clark GCF (1982) The corrosion of pure cobalt in physiological media. J Mater Sci 17:1675–1682
4. Williams DF (ed) (1985) Critical review of biocompatibility. CRC, Boco Raton
5. Konttinen YT, Zhao D, Beklen A, Ma G, Takagi M, Kivelä-Rajamäki M, Ashammakhi N, Santavirta S (2005) The microenvironment around total hip replacement prostheses. Clin Orthop Relat Res 430:28–38
6. Arshady R (2003) Polymeric biomaterials: chemistry, concepts, criteria. In: Arshady R (ed) Introduction to polymeric biomaterials: the polymeric biomaterials series. Citus Books, London

7. Ratner BA, Horbett TA (2004) Some background concepts. In: Ratner BD, Schoen FJ, Lemons JE (eds) Biomaterials science: an introduction to materials in medicine. Elsevier, San Diego
8. Schoen FJ, Anderson JM (2004) Host response to biomaterials and their evaluation. In: Ratner BD, Schoen FJ, Lemons JE (eds) Biomaterials science: an introduction to materials in medicine. Elsevier, San Diego
9. Törmälä P, Laiho J, Helevirta P, Rokkanen P, Vainionpää S, Böstman O, Kilpikari J (1986) Resorbable surgical devices. In: Proceedings of the fifth international conference on polymers in medicine and surgery, Leeuwenhost Congress Centre, The Netherlands, pp 16/1–16/6
10. Törmälä P, Pohjonen T, Rokkanen P (1998) Bioabsorbable polymers: materials technology and surgical applications. Proc Instn Mech Engrs 212:101–111
11. Schedl R, Fasol P (1979) Achilles tendon repair with the plantaris tendon compared with repair using polyglycol threads. J Trauma 19:189–194
12. Schmitt EE, Polistina RA (1975) Surgical dressing of absorbable polymers. US Pat. 3875937
13. Bowald S, Busch C, Erikson I (1978) Arterial grafting with polyglactin mesh in pigs. Lancet 311:153
14. Bowald S, Busch C, Erikson I (1979) Arterial regeneration following polyglactin 910 suture mesh grafting. Surgery 86:722–729
15. Audell L, Bowald S, Busch C, Erikson I (1980) Polyglactin mesh grafting of the pig aorta. Acta Chir Scand 146:97–99
16. Greisler HP, Kim DU, Price JB, Voorhes AB (1985) Arterial regenerative activity after prosthetic implantation. Arch Surg 120:315–323
17. Delany HM, Solanki B, Driscoll WB (1985) Use of absorbable mesh for splenorrhaphy and pelvic peritoneum reconstruction. Contemp Surg 27:11–15
18. Rokkanen P, Böstman O, Vianionpää S, Vihtonen K, Törmälä P, Laiho J, Kilpikari J, Tamminmäki M (1985) Biodegradable implants in fracture fixation: early results of treatment of fractures of the ankle. Lancet 1:1422–1424
19. Yamamuro T, Matsusue Y, Uchida A, Shimada K, Shimozaki E, Kitaoka K (1994) Bioabsorbable osteosynthetic implants of ultra-high strength poly-l-lactide: a clinical study. Int Orthop 18:332–340
20. Pelto-Vasenius K, Hirvensalo E, Vasenius J, Partio EK, Böstman O, Rokkanen P (1998) Redisplacement after ankle osteosynthesis with absorbable implants. Arch Orthop Trauma Surg 117:159–162
21. Wolff J (2010) The classic: on the theory of fracture healing. Clin Orthop Relat Res 468:1052–1055
22. Landry M, Fleisch H (1964) The influence of immobilization on bone formation as evaluated by osseous incorporation of tetracycline. J Bone Joint Surg 46:764–771
23. Williams DF, Gore LF, Clark GCF (1983) Quantitative microradiography of cortical bone in disuse osteoporosis following fracture fixation. Biomaterials 4:285–288
24. Uthoff HK, Dubuc FL (1971) Bone structure changes in the dog under rigid internal fixation. Clin Orthop 81:165–170
25. Woo SLY, Akeson WH, Courts RD, Rutherford L, Doty D, Jemmott GF, Amiel D (1976) A comparison of cortical bone atrophy secondary to fixation with plates with large differences in bending stiffness. J Bone Joint Surg Am 58:190–195
26. Song G, Song S (2007) A possible biodegradable magnesium implant material. Adv Mater Eng 9:298–302
27. Williams DF (2009) Leading opinion: on the nature of biomaterials. Biomaterials 30:5897–5909
28. Williams DF (1981) Biocompatibility of clinical implant materials. CRC, Boca Raton
29. Williams D (2006) New interests in magnesium. Med Device Technol 17:9–10
30. Jacobs JJ, Gilbert JL, Urban RM (1998) Current concepts review—corrosion of metal orthopaedic implants. J Bone Joint Surg Am 80:268–282

31. Williams DF (ed) (1983) Biocompatibility of orthopaedic implant materials. CRC, Boca Raton
32. Brown SA, Merritt K (1981) Fretting corrosion in saline and serum. J Biomed Mater Res 15:479–488
33. Hallab N, Merritt K, Jacobs JJ (2001) Metal sensitivity in patients with orthopaedic implants. J Bone Joint Surg Am 83:428–436
34. Jonas L, Fulda G, Radeck R (2001) Biodegradation of titanium implants after long time insertion used for the treatment of fractured upper and lower jaws through osteosynthesis: elemental analysis by electron microscopy and EDX or EELS. Ultrastruct Pathol 25: 375–383
35. Murray CJL, Lopez AD (1996) The global burden of disease. World Health Organization, Geneva
36. Moravej M, Purnama A, Fiset M, Couet J, Mantovani D (2010) Acta Biomater 6:1843–1851
37. Mueller PP, May T, Perz A, Hauser H, Peuster M (2006) Control of smooth muscle cell proliferation by ferrous iron. Biomaterials 27:2193–2200
38. Lambotte MA (1932) L'utilisation du magnesium comme materiel perdu dans l'osteosynthèse. Societé nationale de chirurgie 1325–1334
39. Staiger M, Pietak A, Huadmai J, Dias G (2006) Magnesium and its alloys as orthopedic biomaterials: a review. Biomaterials 27:1728–1734
40. Cardarelli F (2000) Less common non-ferrous metals. Materials handbook. Springer, London, pp 99–107
41. Heublein B, Rohde R, Kaese V, Niemeyer M, HartungW, Haveric A (2003) Biocorrosion of magnesium alloys: a new principle in cardiovascular implant technology? Heart 89: 651–656
42. Mani G, Feldman MD, Patel D, Agrawal CM (2007) Coronary stents: a materials perspective. Biomaterials 28:1689–1710
43. Witte F, Kaese V, Haferkamp H, Switzer E, Meyer-Lindenberg A, Wirth CJ, Windhagen H (2005) In vivo corrosion of four magnesium alloys and the associated bone response. Biomaterials 26:3557–3563
44. Erne P, Schier M, Resink TJ (2006) The road to bioabsorbable stents: reaching clinical reality. Cardiovasc Intervent Radiol 29:11–16
45. Gu X, Zheng Y, Cheng Y, Zhong S, Xi T (2009) In vitro corrosion and biocompatibility of binary magnesium alloys. Biomaterials 30:484–498
46. Peuster M, Fink C, von Schnakenburg C, Hausdorf G (2002) Dissolution of tungsten coils does not produce systemic toxicity, but leads to elevated levels of tungsten in the serum and recanalization of the previously occluded vessel. Cardiol Young 12:229–235
47. Peuster M, Fink C, von Schnakenburg C (2003) Biocompatibility of corroding tungsten coils: in vitro assessment of degradation kinetics and cytotoxicity on human cells. Biomaterials 24:4057–4061
48. Hench LL, Etheridge EC (1982) Biomaterials: an interfacial approach, Academic, New York
49. Davidge RW (1984) Structural degradation of ceramics. Biomaterials 5:37–41
50. Klein CP, Driessen AA, de Groot K, van den Hoof A (1983) Biodegradation behavior of various calcium phosphate materials in bone tissue. J Biomed Mater Res 17:769–784
51. Klein CP, de Groot K, Driessen AA, van der Lubbe HB (1985) Interaction of biodegradable beta-whitlockite ceramics with bone tissue: an in vivo study. Biomaterials 6:189–192
52. Winkler T, Hoenig E, Huber G, Janssen R, Fritsch D, Gildenhaar R, Berger G, Morlock MM, Schilling AF (2010) Osteoclastic bioresorption of biomaterials: two- and three-dimensional imaging and quantification. Int J Artif Organs 33:198–203
53. Winkler T, Hoenig E, Gildenhaar R, Berger G, Fritsch D, Janssen R, Morlock MM, Schilling AF (2010) Volumetric analysis of osteoclastic bioresorption of calcium phosphate ceramics with different solubilities. Acta Biomater 6:4127–4135
54. LeGeros RZ (1993) Biodegradation and bioresorption of calcium phosphate ceramics. Clin Mater 14:65–88

55. Lu J, Descamps M, Dejou J, Koubi G, Hardouin P, Lemaitre J, Proust JP (2002) The biodegradation mechanism of calcium phosphate biomaterials in bone. J Biomed Mater Res 63:408–412
56. Lange T, Schilling AF, Peters F, Mujas J, Wicklein D, Amling M (2011) Size dependent induction of proinflammatory cytokines and cytotoxicity of particulate beta-tricalciumphosphate in vitro. Biomaterials 32:4067–4075
57. Lange T, Schilling AF, Peters F, Haag F, Morlock MM, Rueger JM, Amling M (2009) Proinflammatory and osteoclastogenic effects of beta-tricalciumphosphate and hydroxyapatite particles on human mononuclear cells in vitro. Biomaterials 30:5312–5318
58. Klein CP, de Groot K, Driessen AA, Ven der Lubbe HBM (1985) Interaction of biodegradable β-whitlockite with bone tissue: an in vivo study. Biomaterials 6:189–192
59. Lucas N, Bienaime C, Belloy C, Queneudec M, Silvestre F, Nava-Saucedo JE (2008) Polymer biodegradation: mechanisms and estimation techniques. Chemosphere 73:429–442
60. Smith R, Williams DF (1985) The degradation of a synthetic polyester by a lysomal enzyme. J Mater Sci Lett 4:547–549
61. Smith R, Oliver C, Williams DF (1987) The enzymatic degradation of polymers in vitro. J Biomed Mater Res 21:991–1003
62. Marchant RE, Miller KM, Anderson JM (1984) In vivo biocompatibility studies. V. In vivo leukocyte interactions with biomer. J Biomed Mater Res 18:1169–1190
63. Williams DF, Smith R, Oliver C (1986) The degradation of 14C-labelled polymers by enzymes. In: Christel P, Meunier A, Lee AJC (eds) Biological and biomechanical performance of biomaterials. Elsevier, Amsterdam
64. Pişkin E (1994) Review: biodegradable polymers as biomaterials. J Biomater Sci Polym Ed 6:775–795
65. Herrmann JB, Kelly RJ, Higgins GA (1970) Polyglycolic acid sutures. Arch Surg 100: 486–490
66. Göpferich A (1996) Mechanisms of polymer degradation and erosion. Biomaterials 17:103–114
67. Wu L, Ding J (2004) In vitro degradation of three-dimensional porous poly(D,L-lactide-co-glycolide) scaffolds for tissue engineering. Biomaterials 25:5821–5830
68. Rokkanen PU, Böstman O, Hirvensalo E, Mäkelä EA, Partio EK, Pätiälä H, Vainionpää SI, Vihtonen K, Törmälä P (2000) Bioabsorbable fixation in orthopaedic surgery and traumatology. Biomaterials 21:2607–2613
69. Pihlajamäki H, Böstman O, Hirvensalo E, Törmälä P, Rokkanen P (1992) Absorbable pins of self-reinforced poly-L-lactic acid for fixation of fractures and osteotomies. J Bone Joint Surg Br 74:853–857
70. Pitt CG (1990) In: Chasin M, Langer R (eds) Biodegradable polymers as drug delivery systems. Marcel Dekker, New York
71. Allock HR (1990) In: Chasin M, Langer R (eds). Biodegradable polymers as drug delivery systems. Marcel Dekker, New York
72. Qui LY, Zhu KJ (2000). Novel biodegradable polyphosphazenes containing glycine ethyl ester and benzyl ester of amino acethydroxamic acid as cosubstituents: synthesis, characterization and degradation properties. J Appl Polym Sci 77:2955–2987
73. Ng SY, Vandamme T, Tayler MS, Heller J (1997) Synthesis and erosion studies of self-catalysed poly(orthoester)s. Macromolecules 30:770–772
74. Doi Y, Kitamura S, Abe H (1995) Microbial synthesis and characterization of poly(3-hydroxybutyrate-co-3-hydroxyhexanoate). Macromolecules 28:4822–4828
75. Li HY, Du RL, Chang J (2005) Fabrication, characterization, and in vitro degradation of composite scaffolds based on PHBV and bioactive glass. J Biomater Appl 20:137–155
76. Gunatillake PA, Adhikari R (2003) Biodegradable synthetic polymers for tissue engineering. Eur Cell Mater 5:1–16
77. Rezwan K, Chen QZ, Blaker JJ, Boccaccini AR (2006) Biodegradable and bioactive porous polymer/inorganic composite scaffolds for bone tissue engineering. Biomaterials 27:3413–3431

78. Temenoff JS, Mikos AG (2000) Injectable biodegradable materials for orthopaedic tissue engineering. Biomaterials 21:2405–2412
79. Ishaug-Riley SL, Crane-Kruger GM, Yaszemski MJ, Mikos AG (1998) Three-dimensional culture of rat calvarial osteoblasts in porous biodegradable polymers. Biomaterials 19: 1405–1412
80. Tangpasuthadol V, Pendharkar SM, Kohn J (2000) Hydrolytic degradation of tyrosine-derived polycarbonates, a class of new biomaterials: part I study of model compounds. Biomaterials 21:2371–2378
81. Tangpasuthadol V, Pendharkar SM, Peterson RC, Kohn J (2000) Hydrolytic degradation of tyrosine-derived polycarbonates, a class of new biomaterials. part II: study of model compounds. Biomaterials 21:2379–2387
82. Mandaogade PM, Satturwar PM, Fulzele SV, Gogte BB, Dorle AK (2002) Rosin derivatives: novel film forming materials for controlled drug delivery. React Funct Polym 50:233–242
83. Satturwar PM, Mandaogade PM, Fulzele SV, Darwhekar GN, Joshi SB, Dorle AK (2002) Synthesis and evaluation of rosin based polymers as film coating materials. Drug Dev Ind Pharm 28:383–389
84. Sahu NH, Mandaogade PM, Deshmukh AM, Meghre VS, Dorle AK (1999) Biodegradation studies of rosin-glycerol ester derivative. J Bioact Compat Polym 14:344–360
85. Friess W (1998) Collagen-biomaterial for drug delivery. Eur J Pharm Biopharm 45:113–136
86. Klammert U, Ignatius A, Wolfram U, Reuther T, Gbureck U (2011) In vivo degradation of low temperature calcium and magnesium phosphate ceramics in a heterotopic model. Acta Biomater 7:3469–3475
87. Walton M, Cotton NJ (2007) Long-term in vivo degradation of poly-L-lactide (PLLA) in bone. J Biomater Appl 21:395–411
88. Schilling AF, Linhart W, Filke S, Gebauer M, Schinke T, Rueger JM, Amling M (2004) Resorbability of bone substitute biomaterials by human osteoclasts. Biomaterials 25: 3963–3972
89. Xia Z, Triffitt JT (2006) A review on macrophage responses to biomaterials. Biomed Mater 1:1–9
90. Hoebertz A, Arnett TR (2003) Isolated osteoclast cultures. Methods Mol Med 80:53–64
91. Chambers TJ, Revell PA, Fuller K, Athanasou NA (1984) Resorption of bone by isolated rabbit osteoclasts. J Cell Sci 66:383–399
92. Boyde A, Jones SJ (1987) Early scanning electron microscopic studies of hard tissue resorption: their relation to current concepts reviewed. Scanning Microsc 1:369–381
93. Jones SJ, Boyde A, Ali NN (1984) The resorption of biological and non-biological substrates by cultured avian and mammalian osteoclasts. Anat Embryol (Berl) 170:247–256
94. Oursler MJ, Collin-Osdoby P, Anderson F, Li L, Webber D, Osdoby P (1991) Isolation of avian osteoclasts: improved techniques to preferentially purify viable cells. J Bone Miner Res 6:375–385
95. Cao JJ, Wronski TJ, Iwaniec U, Phleger L, Kurimoto P, Boudignon B, Halloran BP (2005) Aging increases stromal/osteoblastic cell-induced osteoclastogenesis and alters the osteoclast precursor pool in the mouse. J Bone Miner Res 20:1659–1668
96. Yasuda H, Shima N, Nakagawa N, Yamaguchi K, Kinosaki M, Mochizuki S, Tomoyasu A, Yano K, Goto M, Murakami A, Tsuda E, Morinaga T, Higashio K, Udagawa N, Takahashi N, Suda T (1998) Osteoclast differentiation factor is a ligand for osteoprotegerin/osteoclastogenesis-inhibitory factor and is identical to TRANCE/RANKL. Proc Natl Acad Sci U S A 95:3597–3602
97. Fuller K, Ross JL, Szewczyk KA, Moss R, Chambers TJ (2010) Bone is not essential for osteoclast activation. PLoS One 5:e12837
98. Boyde A, Ali NN, Jones SJ (1985) Optical and scanning electron microscopy in the single osteoclast resorption assay. Scan Electron Microsc 3:1259–1271
99. Salgado AJ, Coutinho OP, Reis RL (2004) Bone tissue engineering: state of the art and future trends. Macromol Biosci 4:743–765

100. Murugan R, Ramakrishna S (2004) Bioresorbable composite bone paste using polysaccharide based nano hydroxyapatite. Biomaterials 25:3829–3835
101. Witte F, Eliezer A, Cohen S (2010) The history, challenges and the future of biodegradable metal implants. Adv Mater Res 95:3–7
102. Witte F, Fischer J, Beckmann F, Störmer M, Hort N (2008) Three-dimensional microstructural analysis of Mg–Al–Zn alloys by synchrotron-radiation-based microtomography. Scripta Materialia 58:453–456
103. Shikinami Y, Matsusue Y, Nakamura T (2005) The complete process of bioresorption and bone replacement using devices made of forged composites of raw hydroxyapatite particles/poly l-lactide (F-u-HA/PLLA). Biomaterials 26:5542–5551
104. Ritman EL (2004) Micro-computed tomography-current status and developments. Annu Rev Biomed Eng 6:185–208
105. Pihlajamäki H, Kinnunen J, Böstman O (1997) In vivo monitoring of the degradation process of bioresorbable polymeric implants using magnetic resonance imaging. Biomaterials 18:1311–1315
106. Shoulders MD, Raines RT (2009) Collagen structure and stability. Annu Rev Biochem 78:929–958
107. Kalfakakou V, Simons TJ (1990) Anionic mechanisms of zinc uptake across the human red cell membrane. J Physiol 421:485–497
108. Yun Y, Dong Z, Shanov V, Schulz M, Heineman W, Kumta P, Sfeir C, Yarmolenko S (2008) Mg nanowires for biology and nanomedicine. Invention disclosure UC 108-072
109. Shanov V, Witte F, Schulz M, Yun Y, Kumta P, Sfeir, Yarmolenko (2008) Composition and method for producing magnesium based biodegradable composite implants. Invention disclosure UC 108-91
110. Mast D, Shanov V, Jayasinghe C, Schulz M (2008) Use of carbon nanotube thread, ribbon, and arrays for the transmission and reception of electromagnetic signals and radiation. Invention disclosure UC 109-35
111. Schulz M, Shanov V, Sundaramurthy S, Yun Y, Wagner W, Nagy P, Fox C, Witte F, Xu Z, Yarmolenko s (2009) Corrosion measurement for biodegradable metal implants. Invention disclosure UC 109-85
112. Shanov V, Schulz M, Yun Y, Rai D, Xue D (2009) Composition and method for magnesium biodegradable material for medical implant applications. Invention disclosure UC 109-89
113. Schulz MJ, Shanov VN, Sankar J, Witte F, Wagner W, Borovetz H, Kumta P, Sfeir C (2009) Permanent and biodegradable responsive implants that expand and adapt to the human body. Invention disclosure UC 109-111
114. Chen L, Xie J, Srivatsan M, Varadan VK (2006) Magnetic nanotubes and their potential use in neuroscience applications. Proc SPIE Int Soc Opt Eng 6172:61720
115. Conserva E, Lanuti A, Menini M (2010) Cell behavior related to implant surfaces with different microstructure and chemical composition: an in vitro analysis. Int J Oral Maxillofac Implants 25:1099–1107
116. Engler AJ, Sen S, Sweeney HL, Discher DE (2006) Matrix elasticity directs stem cell lineage specification. Cell 126:677–689

Index

A
Adipose-derived stem cells (ADSCs), 76
Agar, 164
 diffusion test, 128
Agarose derivatives, 164
Albumin, 3
Alginate–sulfate, 246
Alginates, 164, 235, 242, 273
Amyotrophic lateral sclerosis (ALS), 184
Angiogenesis, 195
 bioactive glass, 213
Annexin V/anchorin CII, 73
Apocynin, 132
Artificial organs, 285
ASTM F748, 117
Autologization, spontaneous in situ, 110
Autologous osteochondral transplantation (AOT), 298

B
Biglycan, 71
Bioactive factors, 153
Bioactive glasses, 195
 osteoconduction/osteoinduction, 198
Bioactivity, 117, 133
Bioartificial organs, 105
Bioceramics, 297
Biocompatibility, 117, 285
Biodegradation, 317
Biohybrids, 285
 biosensors, 290
 cardiovascular, 292
 classification, 289
 kidney/liver, 291
 musculoskeletal, 292
 neuronal, 291
 pancreas, 292
Biomaterials, 67, 153, 285
 3D, 263
 resorbable, 318
 stiffness, 270
Biomimetic surfaces, 169
Bioreactors, dynamic stem cell microenvironment, 251
Bioresorption, 317
 measurements, 322
Blood, 3
Body fluid mimicry, 140
Bone, 153, 195
Bone-marrow-derived stem cells, 75, 228, 265
Bone morphogenetic proteins (BMPs), 4, 84
 BMP-2, 170, 232
Bone regeneration, 157
 scaffolds, 163
Bone sialoprotein, 23, 166

C
Calcein AM (CaAM), 55
Calcium-alkaliorthophosphates, 321
Calcium phosphates (TP), 164, 196
Carbon nanotubes, 167
Carboxymethylchitosan/poly(amido amine), 167
Cardiovascular tissue engineering, 105
Carnitine, 175
Cartilage, 67, 297, 300
 articular, 67
 carrier constructs, 302

C (cont.)
  formation, time course, 305
Cell adherence, 136
Cell adhesion, in vitro surfaces, 35
Cell differentiation, 138
Cell homing, 80
Cell migration, 53
Cell proliferation, 137
Cell–ECM interactions, 177
Cells, 67
  adhesion, 33
  migration, 33
  spreading, 33
Cell–surface interactions, 1, 33
Cell–surface junction, 33, 39
Cell-type specificity, 133
Cellulose, 164
Centrifugation assay, 47
Ceramic matrix, 297
Ceramic-based composite biomaterials, 274
Ceramics, 164, 321
Chitosan, 164
Chlorpromazine, 132
Chondrocytes, 70, 74, 300
  ECM interactions, 73
Chondrogenesis, 70, 80
Chondroitin sulfate (CS), 71, 92
Cisplatin, 123
Collagen, 23, 71, 164, 303
Compatibility, bioactive compounds, 288
  structural compounds, 287
Composites, 153, 273
Controlled release delivery, 242
Cortical bone, 269
Critical-size defects (CSD), 181
Cultures, 117
Cyclic adenosine monophosphate (cAMP), 179
Cyclophosphamide, 123
Cytokines, 12, 232
  secretion, 179
Cytotoxicity, 117, 121

D
DCF test, 130
Decellularized biological materials, 105
Decorin, 71
Degradable coils, tungsten, 321
Degradation, 317
  assays, 322

Dental, 153
Dental follicle cells (DFCs), 163
Dental periodontal ligament stem cells (DPLSCs), 163
Dental pulp stem cells (DPSCs), 163
Dental tissue, 182
Dentin, 183
Dermatan sulfate, 71
Detachment assays, 46
Dexamethasone, 172
Diabetes, 292
Dichlorodihydrofluorescein (H2DCF), 130
Dichlorofluorescein (DCF), 130
Differentiation, 153
Direct contact test, 127
DNA delivery, 87
Drug delivery, 285
  biohybrids, 290

E
ECM, 227
Electric cell–substrate, 33
  impedance sensing (ECIS), 49
Electropolished stainless steel (EPSS), 7
Embryoid bodies (EBs), 77, 242
Embryonic stem cells (ESCs), 77, 158, 231
Enamel, 183
Encapsulating scaffold biohybrids, 290
Epitope presentation, 241
ERK, 277
ERK/MAPK pathway, 20
Exposure period, 132
Extracellular matrix (ECM), 4, 71, 105, 265
  synthetic, 234
Extracellular-signal-regulated kinase (ERK)/mitogen-activated protein kinase (MAPK) pathway, 20

F
Fibrin, 164, 235, 273
Fibroblast growth factor (FGF), 78, 244
Fibroblasts, 79
Fibronectin, 5, 164, 239
Filter diffusion test, 128
Fluorescence interference contrast microscopy (FLIC), 41
Focal adhesion, 263
Focal adhesion kinase (FAK), 21
  pathway, 276

Index

Foreign body reaction (FBR), 287
Formazan, 129
Fused deposition modelling (FDM), 168

G

Gelatin, 92, 164, 235, 273
    microparticles (GMPs), 84
Gene delivery, 86
Geometry, 263
Glass-based scaffolds, 195
Glass–ceramics, 202, 321
Glasses, 321
    bioactive, 195
Glutathione S-transferase, 129
Glycosaminoglycans (GAGs), 71, 244, 303
Grafts, decellularized, 108
Grafts, native, 107
Graft-versus-host-reaction (GvHR), 288
Growth factors, 12
GTPase RhoA, 181

H

Haematoma, 4
Hard tissue, 153
Heart valve substitutes, 112
Heat-inactivated serum, 136
Hematopoietic progenitor cells (HPCs), 251
Heparin binding peptide (HBP), 238
Heparin sulfate, 71, 245
Hepatocyte growth factor (HGF), 246
HIV-1 Tat, 48
Human dermal fibroblasts (hDFs), 79
Human embryoid bodies (hEBs), 232
Human leukocyte antigen (HLA), 112
Hyaluronates, 164
Hyaluronic acid (HA), 71, 232, 235
Hydrodynamic flow experiments, 48
Hydrogels, 81, 166, 236
    bilayered, 91
Hydrophilicity/hydrophobicity, 169
Hydroxyapatite (HA), 164, 196, 266, 297, 300, 321

I

Immunogenicity, 108
    extracellular matrices, 112
Impedance sensing, 33
Implantation, initial interactions, 3
Implants, 317
    bioresorbable, 318

In vitro methods, 117
Indirect contact test, 128
Indolactam V, 233
Induced pluripotent stem cells (iPS), 158, 231
Insulin-like growth factor (IGF), 79
Integrated biomimetic systems, 171
Integrins, 4, 37, 239
Interface structure/composition, 22
Interleukins, 4
Iron, electroformed, 320
ISO degradation assays, medical biomaterials, 322
ISO10993, 117
Isophenphos, 123

J

JAK-STAT pathway, 233

K

Keratin sulfate, 71

L

Laminin, 54
Layer-by-layer (LbL) method, 173
LB monolayers, 174
Limitations, 117

M

Magnesium, 320
Major histocompatibility complex (MHC), 288
Mass transport limitations, 251
Matrices, chemistry, 247
    complex hierarchical, 250
Matrigel, 235
Matrix metalloproteinase (MMP), 240
    derived peptide, 82
Mechanotransduction, 1, 19, 275
Mesenchymal stem cells (MSCs), 70, 153, 161, 229, 263, 289
    artificial surfaces, 176
    dental tissues, 182
Metals, 319
Microenvironment, 227
    sensing, 275
Micromotion, 33
Microparticles, 242
Microroughness, 1
MTS, 130
MTT test, 129
Muscle-derived stem cells (MDSCs), 78

N

NAD(P)H-oxidoreductases, 129
Nanodendrimers, 167
Nanofiber mesh scaffold, 89
Nanofibers, 167
Nanoimprinting, 176
Nanomaterials, 167
Nanoscaled drug release, 171
Nanosphere drug release, 170
Nervous system, 291
Neural stem cells (NSC), 241
Neuronal prostheses, biocompatibility, 291
Neutral red uptake test, 129
Notch signaling, 179

O

Oligo(poly(ethylene glycol) fumarate) (OPF), 85, 242, 250
Osseointegration, 12
Osteochondral implants, 297
Osteochondral tissue regeneration, 67, 90
Osteoclasts, 317
Osteopontin, 23
Oxide layer, 7
Oxygen tension, 140

P

P receptors, 179
PCL/TCP scaffolds, 171
Periodontal defects, 182
Periosteum-derived stem cells (PDSCs), 79
Platelet-derived growth factor (PDGF), 245
Polyanionic co-polysaccharides, 164
Polyelectrolytes, LbL, 174
Polyesters, 173
Polymers, 153, 164, 322
    foams, 167
    surfaces, biocompatibility testing, 60
Polyphosphazenes, 322
Polysaccharides, 164, 235
    sulfation, 244
Polystyrene (PS), surfaces, 60
Poly(beta-hydroxybutyrate) (PHB), 322
Poly(-caprolactone) (PCL), 80, 165, 322
    resveratrol-conjugated, 166
Poly(carnitine allylester), 175
Poly(carnitine), 175
Poly(croton betain), 175
Poly(dimethylsiloxane) gratings, 173
Poly(ethylene glycol) (PEG)
    hydrogels, 76, 236
Poly(ethylene oxide) (PEO), 166
Poly(glycolic acid) (PGA), 165, 236, 322
Poly(hydroxyalkanoate) (PHA), 82, 322
Poly(hydroxymethylglycolide-co--caprolactone) (PHMGCL), 169
Poly(N-isopropyl-acrylamide-co-acrylic acid), 238
Poly(L,D-lactic-co-glycolic acid) (PLGA), 86, 167, 236, 273
Poly(lactic acid) (PLA), 165, 322
Poly(D,L-lactide) (PDLLA), 211, 236
Poly(lactides)/apatite, 170
Poly(orthoester) (POE), 322
Poly(poly(ethylene oxide)terephthalate-co-(butylene) terephthalate) (PEOT/PBT), 166
Pore size, 268
Porosity, 268
Proliferation, 153
Prostaglandins, 12
Protein adsorption, 8
Protein denaturation, 135
Protein-coated resonators, 56
Proteoglycans, 71, 244

Q

Quartz crystal microbalance (QCM), 33, 56
    actuator mode, 61

R

Reactive oxygen species (ROS), 130
Reflection interference contrast microscopy (RICM), 40
Regenerative medicine, 153
Release kinetics, 132
Reseeding, 109
Resorption assays, cell-based, 325
Retinoic acid (RA), 178
RGD peptide, 4, 71, 82, 238
ROCKII, 181
Roughness, 15, 88, 263

S

Scaffold–cell interaction, 169
    hard tissue, 181
Scaffolds, 153, 157, 195
    peptide-modified, 237
Selective laser ablation (SLA), 168
Selective laser sintering (SLS), 168
Self-assembled composites, 173
Shear stress, 61

Index

Short amino acid sequences, 4, 71, 82, 238, 275
Signaling molecules, affinity binding, 243
  release, 241
Single wall carbon nanotubes, 123
Sodium hyaluronate, 92
Soluble factors, 232
Spatio-temporal presentation, 241
Stauprimide, 233
Stem cells, 156
  adult, 161
  apical papilla (SCAPs), 163
  differentiation, 267
  ectomesenchymal, 161
  embryonic (ESCs), 77, 158, 231
  fate determination, 227
  hematopoietic (HSCs), 162, 229
  human exfoliated deciduous teeth (SHEDs), 163
  microenvironment, 177, 227
  pluripotent, 158
Stereolithography, 168
Stiffness, 263
Stromal cell-derived factor-1, 170
Surface plasmon resonance (SPR), 33, 43, 245
  microscopy (SPRM), 45
Surface-dependent response, 6
Surfaces, biological modifications, 9
  biomimetic, 169
  chemistry, 6
  conditioning, 3
  hydrophilic/hydrophobic, 36
  patterning, 173
  structure, 297
  topography, 1, 11
  modification, 172

T

Tension-induced proteins (TIPs), 178
Test battery, 117
Testing strategy, 117
Tetraethylorthosilicate (TEOS), 201
Tissue culture treated polystryrene (TCPS), 138
Tissue engineered medical products (TEMPs), 118
Tissue engineering, 1ff
Tissue regeneration, osteochondral, 67
Tissue rejection, 107, 157
Tissue–implant interface, 1, 22
Titanium surfaces, 10
Tooth decay, 182
Tooth replacement, 156
Total internal reflection aqueous fluorescence microscopy (TIRAF), 42
Total internal reflection fluorescence microscopy (TIRF), 42
Toxicity, 121
Trabecular bone (spongy bone), 269
Transcription factors, 158
Transforming growth factor β (TGFβ), 4, 75, 85, 170, 238, 242
Transient receptor potential (TRP), 178
Tricalcium phosphate, 164, 171, 266, 321
Tumour necrosis factor, 4
Tungsten, 321

V

Vascular prosthesis, 110
Viability, MTT test, 129
Vitronectin, 5

W

Water layer, implant, 3
Wnt/β-catenin, 179
Wound healing scratch assay, 53

Z

Zonal cartilage engineering, 90